European Consortium for
Mathematics in Industry

U. Helmke, D. Prätzel-Wolters
and E. Zerz (Eds.)

Operators, Systems,
and Linear Algebra

# European Consortium for Mathematics in Industry

Edited by

Leif Arkeryd, Göteborg
Heinz Engl, Linz
Antonio Fasano, Firenze
Robert M. M. Mattheij, Eindhoven
Pekka Neittaanmäki, Jyväskylä
Helmut Neunzert, Kaiserslautern

Within Europe a number of academic groups have accepted their responsibility towards European industry and have proposed to found a European Consortium for Mathematics in Industry (ECMI) as an expression of this responsibility.

One of the activities of ECMI is the publication of books, which reflect its general philosophy; the texts of the series will help in promoting the use of mathematics in industry and in educating mathematicians for industry. They will consider different fields of applications, present casestudies, introduce new mathematical concepts in their relation to practical applications. They shall also represent the variety of the European mathematical traditions, for example practical asymptotics and differential equations in Britian, sophisticated numerical analysis from France, powerful computation in Germany, novel discrete mathematics in Holland, elegant real analysis from Italy. They will demonstrate that all these branches of mathematics are applicable to real problems, and industry and universities in any country can clearly benefit from the skills of the complete range of European applied mathematics.

# Operators, Systems, and Linear Algebra

Three Decades of
Algebraic Systems Theory

Edited by

Uwe Helmke, Universität Würzburg

Dieter Prätzel-Wolters and Eva Zerz,
Universität Kaiserslautern

Springer Fachmedien Wiesbaden GmbH 1997

Die Deutsche Bibliothek – CIP-Einheitsaufnahme

**Operators, systems and linear algebra** : three decades
of algebraic systems theory / ed. by Uwe Helmke ...
(European Consortium for Mathematics in Industry).
ISBN 978-3-663-09824-9    ISBN 978-3-663-09823-2 (eBook)
DOI 10.1007/978-3-663-09823-2



Cover design by Peter Pfitz, Stuttgart.

*Dedicated to Paul A. Fuhrmann*
*on the occasion of his 60th birthday*

# Preface

This book contains the plenary lectures presented at the Workshop 'Operators, Systems and Linear Algebra: Three Decades of Algebraic Systems Theory,' held in Kaiserslautern, Germany, September 24–26, 1997. It is a Festschrift honoring the impact of the work of Paul Fuhrmann in operator and control theory. The book includes essays written by prominent scientists to present their views on some of the most recent developments in the area of mathematical systems theory and its applications. The papers cover a wide range of theoretical and applied topics, with emphasis on computational aspects and relations to engineering problems. The impact of Paul Fuhrmann's work can be traced through many parts of the volume. Polynomial models and shift realizations, the algebraic structure of linear state feedback, partial realization theory and the recursive inversion of Hankel and Toeplitz operators, spectral factorization, and parametrizations of classes of rational functions are some of the topics in the book, where explicit references are made to his work.

It is impossible to give in this short preface any substantial review of Paul Fuhrmann's numerous and major contributions to finite and infinite dimensional linear systems theory. Instead we like to emphasize that his work demonstrates in a clear way the underlying attempt to unreveal the basic unity of mathematics. His work is guided by the paradigm 'forest before trees.' As a striking example for this we mention the unifying concept of polynomial models, which led to a synthesis of Kalman's module-theoretic approach to the realization problem, state space models and the polynomial matrix method of Rosenbrock for linear systems in differential operator representation. Polynomial models have their roots in operator theory. Paul Fuhrmann's elegant book *Linear Systems and Operators in a Hilbert Space* established the basic relations between operator and systems theory in a novel and general way. The recent monograph *A Polynomial Approach to Linear Algebra* presents the areas of linear algebra and linear systems theory in an excitingly new coherent way, stressing the fundamental role played by shift operators and polynomial models in these contexts.

It is a pleasure to thank all the authors for the enthusiasm with which they accepted the invitation to contribute to this volume. Their work covers diverse topics in systems and control theory, such as Toda Lattices, Noncommutative Power Series in Nonlinear Control, Computer Vision, the Partial Realization Problem, Lyapunov Theory, Discontinuous Dynamics, 2D Systems, Spectral Factorizations, Supervisory Control, Modeling and Interpolation. The papers in this book present an excellent perspective on prominent trends in systems and control theory, and their relations to a variety of mathematical areas. It also underlines the point of view Paul Fuhrmann has expressed so clearly in 1981:

'Mathematics has always benefited from the transplanting of ideas and motivations from other fields. It seems to me that system theory besides being intellectually exciting is today one of the richest sources of ideas for the mathematician as well as a major area of application of mathematical knowledge.'

It has been a great honor and privilege to have the opportunity of celebrating the 60th birthday of Paul Fuhrmann by holding this workshop in Kaiserslautern. With admiration and affection we honor a colleague and friend of enormous creativity, energy, courage and mathematical strength. Paul Fuhrmann has been most influential on the development of the German systems and control community. We are convinced that he will continue to work vigorously for many years to come and we extend to him on this occasion our best wishes for future health and happiness.

The editors thank the 'Graduiertenkolleg Technomathematik' and the 'Institut für Techno- und Wirtschaftsmathematik' at the University of Kaiserslautern for generous support of the workshop. We are grateful to the 'European Consortium for Mathematics in Industry' and Dr. Peter Spuhler at Teubner Verlag for their fast and encouraging response to our planning of this volume. Cordial thanks are especially due to Prof. Joachim Rosenthal who helped us in writing parts of the introduction. Finally, we wish to express once more our gratitude towards all authors, for their willingness to participate in this meeting and for providing their manuscripts in a timely way.

Kaiserslautern, September 1997

Uwe Helmke, Dieter Prätzel-Wolters, Eva Zerz

# Contents

# Introduction

The purpose of this section is to introduce readers not familiar with mathematical systems theory to some of its basic areas, methods and applications. The goal of mathematical systems and control theory is to develop the mathematical foundations for the interdisciplinary area of systems and circuit theory, signal processing and control, and their applications. Starting from the early work of Wiener in cybernetics as well as the pioneering work of Bellman, Kalman and Pontryagin is has quickly developed into a rich and fast growing research area for mathematicians and engineers, giving impetus to almost all areas of science. Important application areas of systems and control theory include electrical engineering, process control, automation and mechanical engineering, but also the natural sciences.

A basic object of mathematical control theory are dynamical systems described by linear or nonlinear systems of differential equations in state space form

$$\begin{aligned} \dot{x}(t) &= f(x(t), u(t)) \\ y(t) &= h(x(t)). \end{aligned}$$

In contrast to the classical theory of nonlinear dynamical systems, such as developed by Poincaré and Birkhoff, the presence of external input and output variables in such differential equations leads to further questions and phenomena that necessitate the introduction of new concepts and ideas. In fact, a natural question that arises for such control systems is that of determining the fundamental possibilities and limitations of changing the system dynamics by appropriate choices of the input variables $u(t)$. Similarly one would like to know in how far one can determine the evolution of the state vector $x(t)$ from measurements of the output variables $y(t)$. An important further issue is that of identification and modeling. Thus, given a set of input output trajectories $(u(t), y(t))$, one is asked to find a dynamical control system that generates this set via certain initial conditions.

In order to formulate and answer such questions in precise mathematical terms, Kalman has developed in the sixties an axiomatic framework for systems theory with the state space, controllability, and observability as the central concepts of the theory. Of course, the precise form in which such questions are stated depends very much on the particular type of dynamical systems under consideration. Dynamical systems can be described in continuous or discrete time and the state and control variables can assume continuous or discrete values (such as in e.g. automata, discrete event systems). In mechanics and circuit theory the systems that arise are often best described by implicit differential equations, involving algebraic constraints on the state variables. Other types of models include stochastic systems, systems with delays or those described by partial differential equations.

Mathematical systems theory nowadays has broadened into a vast research area, with its own concepts and methods, research journals, textbooks and conferences. The mathematical tools for analyzing control systems stem from practically all areas of pure and applied mathematics: real and complex (functional) analysis, ordinary and partial differential equations, discrete mathematics, stochastics, numerical analysis, optimization, operator theory, commutative algebra, linear algebra, algebraic and differential geometry, graph theory, and even number theory.

From the very beginnings of state space theory tools from linear algebra and operator theory have played an important role. It has been one of the main achievements of P. A. Fuhrmann's work to combine concepts from operator theory with the module theory of polynomial rings into a particularly powerful and computationally effective tool for analyzing linear control systems. Based on this polynomial model approach there is a still ongoing stream of research into the foundations of linear systems theory. The analysis of control systems with algebraic methods is called algebraic systems theory. Although the main focus has been so far on linear systems, more recently nonlinear systems have been analyzed as well using tools from commutative and differential algebra.

In order to illustrate some of the basic ideas of control theory we focus on linear time-invariant systems. Most design methods in control work particularly well for linear systems, thus restriction to this class of systems is of considerable practical interest. A linear time-invariant system in state space form is

$$\begin{aligned} \dot{x}(t) &= Ax(t) + Bu(t) \\ y(t) &= Cx(t). \end{aligned}$$

Here $x(t)$ is the $n$-dimensional state vector, $u(t)$ is the $m$-dimensional input vector and $y(t)$ denotes the $p$-dimensional output vector. Taking the Laplace-transform of the input-output operator yields the transfer matrix

$$G(s) = C(sI - A)^{-1}B,$$

which is a $p \times m$-matrix of strictly proper rational functions. Conversely, by Kalman's realization theorem, any matrix $G(s)$ of strictly proper rational functions is the transfer function of a controllable and observable system $(A, B, C)$.

In practice however such systems often occur as coupled systems of higher order differential equations, such as e.g.

$$T(D)z(t) = U(D)u(t), \quad y(t) = V(D)z(t),$$

where $D = \mathrm{d}/\mathrm{d}t$ and $U(D), V(D), T(D)$ are higher order differential operators. System representations of this form have been studied by Rosenbrock and Wolovich. A more abstract and comprehensive theory of such polynomial system representations has been subsequently achieved by Fuhrmann. In his theory of polynomial models a state space

$$X_T = \mathbb{F}^m[s] \cap s^{-1}T(s)\mathbb{F}^m[[s^{-1}]]$$

is associated to any nonsingular polynomial matrix $T(s)$.

This explicit construction of the state space allows an extremely elegant solution of the state space realization problem. Thus, for any polynomial matrix fraction decomposition $V(s)T(s)^{-1}U(s)$ of a strictly proper rational matrix function $G(s)$, a canonical state space realization of $G(s)$ is defined as

$$A : X_T \to X_T, \quad f \mapsto \pi_T(sf)$$

$$B : \mathbb{F}^m \to X_T, \quad x \mapsto \pi_T(Ux)$$

$$C : X_T \to \mathbb{F}^p, \quad C_T f = (VT^{-1}f)_{-1}.$$

Here $\pi_T$ denotes a canonical projection operator onto $X_T$ and $(\cdot)_{-1}$ denotes the residue operation. In particular, reachability and observability of $(A, B, C)$ turns out to be equivalent to left coprimeness of $(T, U)$ and right coprimeness of $(V, T)$, respectively.

Proceeding in a different direction, Rosenbrock's work has been extended by Willems and his school to the so-called behavioral approach to systems theory. For reasons of conceptual and technical clarity, it is advantageous to identify a system not with the representing equations, but with the corresponding solution spaces. Thus, a dynamical system is viewed as a set of trajectories, and the generating equations or operators are considered as representations of these behaviors. Moreover, no a priori distinction is made between inputs and outputs, instead it is deduced from the model which system variables are free and which are determined by others.

Starting from each of the above definitions of a linear system a comprehensive structure theory of linear systems has been developed. In each point of view, modules of polynomials play a central role. From a historical perspective it may be interesting to observe that several tools from algebraic system theory have been developed in pure mathematics already a century ago. For example, in the classical work of Dedekind and Weber on the theory of algebraic functions in one complex variable so-called normal basis for modules over a polynomial ring are introduced. These appear to be completely equivalent to the so-called minimal basis occurring in systems theory. Moreover, Dedekind and Weber proved the existence of the Wiener-Hopf factorization for invertible rational transfer matrices – all well more than 100 years ago!

It is beyond the scope of this short introduction to consider in more detail the enormous progress which has been made in mathematical systems and control theory during the last two decades: robust and adaptive control, $H_\infty$-control, identification, nonlinear control, and so on. Instead we like to highlight some of the most relevant application areas for algebraic systems theory.

**Convolutional Codes and Systems Theory**: In the present computer age it is crucial that data are stored and transmitted such that errors can be easily detected and corrected. It is the purpose of coding theory to accomplish this goal. Convolutional codes belong to the most widely used codes in data transmission and have been successfully implemented in conventional phone systems, in the area of wireless communication, and in the transmission of images from deep space. Since the late sixties it has been widely known that convolutional codes and linear systems which are defined over a finite field are essentially the same objects. Despite this natural connection there has been little interaction between coding theory and systems theory during the seventies and eighties. On the coding theory side, Forney has developed an algebraic theory for convolutional codes. In this theory convolutional codes are identified with vector spaces over the field of rational functions. Recent work has stressed various deeper connections between coding theory, systems theory, and symbolic dynamics. A point of view that has emerged from this interaction is that of interpreting convolutional codes are viewed as modules over a polynomial ring $R$, with coefficients in a finite field $\mathbb{F}$. Of particular importance are submodules of $R^n$. By duality those submodules define linear behaviors in the sense of Willems. The construction, the encoding and the decoding of convolutional codes can be described by 'polynomial methods.' It thus appears that the polynomial approach in systems theory put forward by P. A. Fuhrmann is of significance for coding theory as well.

**Hybrid Systems and Robotics**: The term hybrid system refers to a system incorporating discrete and continuous state variables. A simple physical example of such a mechanical hybrid control system is a computer controlled system, e.g., a plotter. This device transforms logical variables obtained by the computer into mechanical commands to move the pen to a desired position. Other examples include mechanical systems with inequality constraints, algebraic differential systems, and even traffic control problems. The analysis of such systems often requires a mixture of analytical and algebraic methods.

One important control problem in robotics is the inverse kinematics problem. Here the actuators in the joints of a robot arm are to be controlled in such a way that the hand follows a specified trajectory. Such tasks lead to difficult nonlinear control problems (target tracking, path following, stabilization) on Lie groups. Another important problem is that of motion planning, i.e., to control the motion of a rigid body if obstacles are present in the configuration space (piano movers problem). Problems of this kind can be tackled by differential geometry methods, as well as using tools from differential algebra.

**Analog Circuit Design**: A major problem in electronic design automation is the lack of flexible strategies and systematic techniques for computer-aided analysis and design of analog circuits. Computer support is becoming a key factor for these applications, with the time required for designing an analog circuit being 4–10 times greater than the time to design a digital circuit. Thus the design time of mixed-signal integrated circuits is dominated by the analog parts.

Modeling analog circuits leads to a description in terms of systems of algebro-differential equations, i.e., to state space models in descriptor form. These models contain the basic network parameters such as resistances and capacitances, in a specific parametrized form. Any 'pointwise' simulation of the network for given parameter values falls short in exhibiting the behavior of the system under changes of the parameters. Thus for the simulation and design of such circuits it has become important to carry out the necessary calculations in a symbolic way. A systematic analysis und synthesis of such parametrized descriptor systems is a considerable challenge, as even for moderately small networks, the symbolic computation of the network can easily fail within Mathematica or Maple. Further progress in this area crucially depends on an efficient combination of advanced methods from the theory of systems over rings, computer algebra and numerical linear algebra.

The following contributions are written by outstanding researchers in systems and control theory, and linear algebra. The papers present new advances in linear and nonlinear systems, linear algebra, stochastic systems, operator theoretic methods and applications in computer vision. The proceedings therefore offer a notable cross section of recent work in the vital field of mathematical systems theory.

# On the Approximation of Hankel Matrices[1]

A. C. Antoulas
Rice University, Houston, USA

*Dedicated to Professor P. A. Fuhrmann on the occasion of his 60th birthday*

**Abstract**: In this paper we examine the problem of optimal approximation in the 2-norm, of a finite square Hankel matrix by a Hankel matrix of rank one, and provide a necessary and sufficient condition for its solvability.

## 1   Introduction

It is a great pleasure to contribute to the Festschrift honoring Paul Fuhrmann on the occasion of his 60th birthday. I have known Paul through his papers from the very beginning of my studies as a graduate student. A few years later I met him personally during one of his visits to the ETH. My dissertation was actually based on what are known as *polynomial models*, which were introduced in [4], and I had called the *Fuhrmann realization* (see [6]). Through Paul's work I have gained deep insight into numerous facets of the algebraic structure of linear systems. The papers which I have studied most and were most influential for my research are: his 1976 paper on an analyst's view of algebraic system theory [4], and his 1991 paper on a polynomial approach of Hankel-norm approximation [5].

The present note is related to this latter paper. Through the work of Adamjan, Arov, and Krein [1], [2], it is known that an infinite Hankel matrix which has finite rank and whose associated impulse response sequence is square summable, can be approximated optimally by an infinite Hankel matrix of lower rank in the induced 2-norm. This remarkable result is known *not* to hold in general for finite Hankel matrices (see [3]). In the sequel we will try to elucidate a small part of this problem. We will address namely, the optimal approximation of Hankel matrices in the induced 2-norm, for the special case of rank-one approximants.

[1]This research was initiated while the author was visiting the Australian National University, in February 1996. Numerous discussions with Brian Anderson on the topic of low-rank approximations of Hankel matrices are gratefully acknowledged.

## 2  The problem

Consider the real Hankel matrix $H \in \mathbb{R}^{n \times n}$. Let the eigenvalue decomposition of $H$ be:

$$
\begin{aligned}
H &= U\Lambda U^* \quad \text{where} \quad U = [u_1 \ \cdots \ u_n], \ UU^* = I_n & (2.1) \\
\Lambda &= \operatorname{diag}(\lambda_1, \cdots, \lambda_n), \quad \lambda_1 > |\lambda_2| > |\lambda_3| \geq \cdots \geq |\lambda_n| \geq 0 & (2.2)
\end{aligned}
$$

that is: the eigenvalue of largest modulus is assumed (without loss of generality) positive; furthermore $\lambda_1$ and $\lambda_2$ are assumed for simplicity of the exposition to have multiplicity one. In the sequel we will use the notation

$$
\sigma_2 := |\lambda_2|
$$

We will investigate the existence and construction of approximants $\tilde{H}$ of $H$ which satisfy:

$$
\begin{aligned}
&\tilde{H} \in \mathbb{R}^{n \times n} \text{ has Hankel structure} & (2.3) \\
&\operatorname{rank} \tilde{H} = 1 & (2.4) \\
&\| H - \tilde{H} \| = \sigma_2 & (2.5)
\end{aligned}
$$

The norm used above is the induced 2-norm, i.e. the largest singular value of $H - \tilde{H}$. By the Schmidt-Mirsky theorem, the last condition thus implies that we are looking for approximants which are *optimal* in this norm.

A useful parametrization of all Hankel matrices of rank one is given in terms of two real parameters as follows:

$$
\tilde{H} = kv(\gamma)v(\gamma)^*, \quad v(\gamma) := [1 \ \ \gamma \ \cdots \ \gamma^{n-1}]^*, \quad k, \gamma \in \mathbb{R} \tag{2.6}
$$

The problem thus reduces to: given $H$, find conditions on $k$, $\gamma$, such that (2.5) is satisfied where the approximant $\tilde{H}$ is given by (2.6).

## 3  The solution

In order to state the solution, we need to introduce the following notation. To every eigenvector $u_i = [u_{i1} \ \cdots \ u_{in}]^*$ of $H$ we associate a polynomial of degree $n - 1$:

$$
\pi_i(\gamma) := v_i(\gamma)^* u_i = u_{i1} + u_{i2}\gamma + \cdots + u_{in}\gamma^{n-1} \tag{3.1}
$$

We are now ready to state

**Theorem 3.1 Optimal rank-one approximation of Hankel matrices**
*Consider the real $n \times n$ Hankel matrix $H$ with eigenvalue decomposition (2.1),*

*(2.2). $\tilde{H}$ given by (2.6) is an optimal rank-one Hankel approximant of $H$ if, and only if, the following conditions are satisfied:*

$$\exists \, \gamma \in \mathbb{R} \; \ni \; \pi_2(\gamma) = 0 \tag{3.2}$$

$$\frac{\pi_1^2(\gamma)}{\lambda_1^2 - \lambda_2^2} \geq \frac{\pi_3^2(\gamma)}{\lambda_2^2 - \lambda_3^2} + \cdots + \frac{\pi_n^2(\gamma)}{\lambda_2^2 - \lambda_n^2} \tag{3.3}$$

*If the above conditions are satisfied, all optimal approximants $\tilde{H}$ are given by (2.6), where $\gamma$ satisfies (3.2) and $k > 0$ satisfies:*

$$\frac{1}{k} \geq \frac{\pi_1^2(\gamma)}{\lambda_1 + \sigma_2} + \frac{\pi_3^2(\gamma)}{\lambda_3 + \sigma_2} + \cdots + \frac{\pi_n^2(\gamma)}{\lambda_n + \sigma_2} \tag{3.4}$$

$$\frac{1}{k} \leq \frac{\pi_1^2(\gamma)}{\lambda_1 - \sigma_2} + \frac{\pi_3^2(\gamma)}{\lambda_3 - \sigma_2} + \cdots + \frac{\pi_n^2(\gamma)}{\lambda_n - \sigma_2} \tag{3.5}$$

**Example.** This result will now be illustrated by means of a simple example. Let

$$H = \begin{pmatrix} 1 & 0 & a \\ 0 & a & 0 \\ a & 0 & 1 \end{pmatrix}$$

The eigenvalue, eigenvector pairs $(\mu_i, x_i)$ of $H$ are:

$$\mu_1 = 1 + a, \; x_1 = \frac{1}{\sqrt{2}} \begin{pmatrix} 1 \\ 0 \\ 1 \end{pmatrix}; \quad \mu_2 = a, \; x_2 = \begin{pmatrix} 0 \\ 1 \\ 0 \end{pmatrix}; \quad \mu_3 = 1 - a, \; x_3 = \frac{1}{\sqrt{2}} \begin{pmatrix} 1 \\ 0 \\ -1 \end{pmatrix};$$

For $a \geq 0$, the ordered eigendecomposition is: $\lambda_1 = \mu_1$, $\lambda_2 = \mu_2$, $\lambda_3 = \mu_3$, and correspondingly $u_1 = x_1$, $u_2 = x_2$, $u_3 = x_3$; for $a \leq 0$, we have: $\lambda_1 = \mu_3$, $\lambda_2 = \mu_2$, $\lambda_3 = \mu_1$, and $u_1 = x_3$, $u_2 = x_2$, $u_3 = x_1$.

The roots of the polynomial $\pi_2 = \gamma$, considered as a polynomial of degree two, are zero and infinity, the corresponding vectors $v(\gamma)$ being (1 0 0) and (0 0 1). We will treat the first case; the second one can be treated similarly.

From definition (3.1), it follows that $\pi_1(0) = \pi_3(0) = \frac{1}{\sqrt{2}}$. Hence, for $a \geq 0$, condition (3.3) reduces to $a \leq \frac{1}{2}$. Similarly, for $a \leq 0$, the same condition reduces to $a \geq -\frac{1}{2}$. Thus $H$ can be optimally approximated by a rank-one Hankel matrix in the induced 2-norm, iff

$$-\frac{1}{2} \leq a \leq \frac{1}{2}$$

It follows from (3.4) and (3.5) that the range of $k > 0$, corresponding to the above values of $a$, is:

$$\frac{1 - 2|a|}{1 - |a|} \leq k \leq \frac{1 + 2|a|}{1 + |a|}$$

■

# 4 The proof

The proof of the theorem above is based on the following three results.

**Proposition 4.1** *Let $A = \mathrm{diag}\,(\alpha_1, \cdots, \alpha_k)$, $b = (\beta_1 \cdots \beta_k)^*$, and $\hat{A} := A + kbb^*$. It follows that*

$$\det \hat{A} \;=\; \prod_i \alpha_i + k \left( \beta_1^2 \prod_{i \neq 1} \alpha_i + \cdots \beta_k^2 \prod_{i \neq k} \alpha_i \right) \tag{4.1}$$

$$=\; \det A \left[ 1 + k \left( \frac{\beta_1^2}{\alpha_1} + \cdots + \frac{\beta_k^2}{\alpha_k} \right) \right] \tag{4.2}$$

*where the second equality holds for nonsingular $A$.*

**Proof.** There holds

$$\det \hat{A} = \det \begin{pmatrix} A & -b \\ kb^* & 1 \end{pmatrix}$$

The first equality follows by direct expansion of the determinant of the matrix given above, for instance, with respect to the last column.  ∎

**Corollary 4.1** *Let $\lambda$, $u$ be an eigenvalue and the corresponding eigenvector of the real symmetric matrix $M$. $\lambda$ is an eigenvalue of the rank-one perturbation $\hat{M} = M + kvv^*$, where $v$ is a real column vector, if, and only if, $u$ and $v$ are orthogonal, i.e. $v^*u = 0$. In this case $u$ is also an eigenvector of $\hat{M}$ corresponding to the eigenvalue $\lambda$.*

**Proof.** Let $M = XDX^*$ be the eigenvalue decomposition of $M$, and let $\lambda = d_i$, $u = x_i$, where $x_i$ denotes the i-th column of $X$; By applying the above proposition, where $A := D - \lambda I$ and $b := X^*v$, it follows that $\det(\hat{A} - \lambda I) = 0$ iff $\beta_i = v^*x_i = v^*u = 0$; this proves the assertion.  ∎

It is clear that if $\tilde{H}$ is an optimal rank-one approximant of $H$, i.e. (2.5) is satisfied, the difference $H - \tilde{H}$ *must* have an eigenvalue equal *either* $\lambda_2$ or $-\lambda_2$. However, because of the interlacing property of the eigenvalues of a symmetric matrix and a rank-one update thereof, the latter case can be ruled out.

**Lemma 4.1** *Recall the eigenvalue decomposition of $H$ given by (2.1), (2.2). Let $\tilde{H}$ be a Hankel matrix satisfying (2.4) and (2.5). Then $H - \tilde{H}$ has an eigenvalue equal to $\lambda_2$.*

**Proof.** The interlacing property asserts that the ordered eigenvalues of the symmetric $n \times n$ matrices $A$ and $\hat{A} = A - kvv^*$, $k > 0$, satisfy:

$$\lambda_i(A) \geq \lambda_i(\hat{A}) \geq \lambda_{i+1}(A), \quad i = 1, \cdots, n$$

where $\lambda_{n+1}(A)$ is defined as minus infinity.

Thus if $\lambda_2(H) > 0$, the largest eigenvalue of the difference $H - \tilde{H}$ must lie between $\lambda_1(H)$ and $\lambda_2(H)$, which because of assumption (2.5), implies the statement made. Otherwise, if $\lambda_2(H) < 0$, $\lambda_2(H)$ is the smallest eigenvalue of $H$ and therefore by the interlacing property the smallest eigenvalue of $H - \tilde{H}$ is less than or equal to $\lambda_2(H)$. Again because of (2.5), the validity of the lemma is established. ∎

The above lemma together with the corollary prove the necessity of condition (3.2). If this condition is satisfied, there remains to determine $k > 0$, if possible, so that $\sigma_2 = |\lambda_2|$ is the largest eigenvalue of $H - \tilde{H}$, in absolute value. This is equivalent with:

$$\sigma_2 I_n \geq H - kv(\gamma)v(\gamma)^* \geq -\sigma_2 I_n$$

If we multiply this set of inequalities on the left by $U^*$, and on the right by $U$, we obtain

$$\sigma_2 I_n \geq \begin{pmatrix} \lambda_1 & & & \\ & \lambda_2 & & \\ & & \ddots & \\ & & & \lambda_n \end{pmatrix} - k \begin{pmatrix} \pi_1(\gamma) \\ \pi_2(\gamma) \\ \vdots \\ \pi_n(\gamma) \end{pmatrix} \begin{pmatrix} \pi_1(\gamma) & \pi_2(\gamma) & \cdots & \pi_n(\gamma) \end{pmatrix} \geq -\sigma_2 I_n$$

$$(4.3)$$

Keeping in mind that $\pi_2(\gamma) = 0$, these inequalities reduce respectively, to

$$\Gamma_-(k) \geq 0, \ \Gamma_+(k) \geq 0 \tag{4.4}$$

where $\Gamma_-(k) :=$

$$\begin{pmatrix} \sigma_2 - \lambda_1 & & & \\ & \sigma_2 - \lambda_3 & & \\ & & \ddots & \\ & & & \sigma_2 - \lambda_n \end{pmatrix} + k \begin{pmatrix} \pi_1(\gamma) \\ \pi_3(\gamma) \\ \vdots \\ \pi_n(\gamma) \end{pmatrix} \begin{pmatrix} \pi_1(\gamma) & \pi_3(\gamma) & \cdots & \pi_n(\gamma) \end{pmatrix}$$

$$(4.5)$$

and $\Gamma_+(k) :=$

$$\begin{pmatrix} \sigma_2 + \lambda_1 & & & \\ & \sigma_2 + \lambda_3 & & \\ & & \ddots & \\ & & & \sigma_2 + \lambda_n \end{pmatrix} - k \begin{pmatrix} \pi_1(\gamma) \\ \pi_3(\gamma) \\ \vdots \\ \pi_n(\gamma) \end{pmatrix} \begin{pmatrix} \pi_1(\gamma) & \pi_3(\gamma) & \cdots & \pi_n(\gamma) \end{pmatrix}$$

$$(4.6)$$

Finally a moment's reflection shows that $\Gamma_-(k)$ and $\Gamma_+(k)$ are positive semi-definite iff their determinants are non-negative; these determinants can in turn be computed using the proposition stated earlier:

$$\det \Gamma_-(k) = \prod_{i \neq 2}(\sigma_2 - \lambda_i) \cdot \left[ 1 + k \sum_{i \neq 2} \frac{\pi_i^2(\gamma)}{\sigma_2 - \lambda_i} \right] \geq 0 \ \Rightarrow \ 1 + k \sum_{i \neq 2} \frac{\pi_i^2(\gamma)}{\sigma_2 - \lambda_i} \leq 0$$

$$\det \Gamma_+(k) \;=\; \prod_{i\neq 2}(\sigma_2 + \lambda_i) \cdot \left[1 - k \sum_{i\neq 2} \frac{\pi_i^2(\gamma)}{\sigma_2 + \lambda_i}\right] \geq 0 \;\Rightarrow\; 1 - k \sum_{i\neq 2} \frac{\pi_i^2(\gamma)}{\sigma_2 + \lambda_i} \geq 0$$

This implies that $k > 0$ satisfies conditions (3.4) and (3.5). This in turn implies that the set of $k$'s satisfying these inequalities is non-empty iff condition (3.3) is fulfilled.

This completes the proof of the theorem. ∎

# References


[1] V.M. Adamjan, D.Z. Arov, and M.G. Krein, *Analytic properties of Schmidt pairs for a Hankel operator and the generalized Schur-Takagi problem*, Math. USSR Sbornik, **15**: 31-73 (1971).

[2] V.M. Adamjan, D.Z. Arov, and M.G. Krein, *Infinite block Hankel matrices and related extension problems*, American Math. Society Transactions, **111**: 133-156 (1978).

[3] A.C. Antoulas, *Approximation of linear operators in the 2-norm*, Linear Algebra and Appl., Special Issue on "Challenges in Linear Algebra", (to appear).

[4] P.A. Fuhrmann, *Algebraic system theory: An analyst's point of view*, J. Franklin Institute, **301**: 521-540 (1976).

[5] P.A. Fuhrmann, *A polynomial approach to Hankel norm and balanced approximations*, Linear Algebra and its Applications, **146**: 133-220 (1991).

[6] P.A. Fuhrmann, *A polynomial approach to linear algebra*, Springer Verlag (1996).



A. C. Antoulas
Department of Electrical and Computer Engineering
Rice University
Houston, Texas 77251-1892, USA
e-mail: aca@rice.edu
fax: +1-713-524-5237

# The Reduction of Multiplicity in Systems Modeling

Tomas Björk
Stockholm School of Economics, Stockholm, Sweden

Andrea Gombani
LADSEB-CNR, Padova, Italy

**Abstract**: We discuss here some open issues concerning the problem of modeling signals with low multiplicity within the context of systems theory. In particular we focus on the issue of approximating a stochastic process with a low multiplicity one, and we propose some applications to the theory of interest rates models.

## 1   Introduction

We discuss here some issue concerning the problem of modeling systems with low multiplicity and we suggest an application to interest rates models.

In many phenomena in engineering and economics, we observe time series or continuous signals with a lot of components; nevertheless we would like to explain, at least approximately, all these data with a small number of parameters: typical examples are time series analysis of economic data, and multichannel transmission (cellular phones). If we assume to choose a linear gaussian stochastic system as our candidate for the modeling, say, in discrete time, the problem could be seen as a dynamic counterpart to the well known Frisch scheme: given a time series $\{y(t)\}_{t\in \mathbb{N}}$, decompose it into the output of stochastic linear system *with a low number of inputs* and an unnormalized white noise. Clearly, the assumption that the model is a linear gaussian process leads to just one of the many approaches which are present in the literature. In an alternative approach this problem has been initially addressed by Willems [19] and a detailed study of the behavioral setup to the problem has been carried out by Roorda in his thesis [18].

If we assume that the model is linear gaussian and time invariant, then this means that its spectral density $\Phi$ of $y$ can be decomposed as

$$\Phi = \hat{\Phi} + \Delta \tag{1}$$

where $\hat{\Phi}$ has low rank and $\Delta$ is constant (we try to keep notation to a minimum and, for the time being, our observations hold both for discrete and continuous time). Again, even within this much more limited class of models, several path have been explored. For example, in the case $\Delta$ is diagonal, a pointwise analysis has been carried out in a series of papers by Anderson and Deistler (see e.g. [4]). But no claim is made about the rationality of $\Phi$.

Nevertheless, if we want to represent our model as a finite dimensional linear system, we have to impose rationality of the spectral density $\Phi$. In this case, the spectral density can be determined from the data by an extensive number of different methods. Most of them though will produce a rational, but full rank density. If we want a decomposition like (1) (provided it exists) we have to look for spectral factors of $\Phi$ of the form

$$W = \hat{W} + \tilde{D} \tag{2}$$

where the rows of $\hat{W}$ and those of $\tilde{D}$ are pointwise orthogonal (i.e. $W(\omega)\tilde{D}^* = 0$ for $\omega \in \mathbb{I}$ or $\omega \in \mathbb{T}$ depending on whether we are in continuous or discrete time). The problem then becomes to characterize those spectral factors for which the decomposition (2) has minimal rank in the dynamic part $\hat{W}$. We call dynamic rank of $W$ the minimal rank of $\hat{W}$ in (2). Let $W = \left( \begin{array}{c|c} A & B \\ \hline C & D \end{array} \right)$ be a minimal realization of $W$. Then it's easy to see that, if there exists a projection matrix $\pi$ such that $B\pi = B$, then we can write

$$W = (C(sI - A)^{-1}B + D)\pi + D(I - \pi)$$

In conclusion, the rank of $\hat{W}$ is equal to the rank of $B$, and therefore characterizing ·decompositions (2) with rank of $\hat{W}$ minimal is equivalent to finding the spectral factors $W = \left( \begin{array}{c|c} A & B \\ \hline C & D \end{array} \right)$ with $B$ of minimal rank. It should be noted that the first obvious guess, the minimum-phase spectral factor, does not work, as the following example shows. Let $W$ be as follows: $A$ is the companion matrix of the polynomial with roots $(-1/, 1/3, 1/(2+i), 1/(2-i))$;

$$B = \begin{bmatrix} 1 & 0 & 0 & 0 \\ 0 & 0 & 0 & 0 \\ 0 & 0 & 0 & 0 \end{bmatrix} \qquad C = \begin{bmatrix} 1/2 & 2 & 8 & 1 \\ 1 & 1 & -1 & 1/2 \end{bmatrix} \qquad D = \begin{bmatrix} 1 & 1 & 2 \\ 0 & 0 & -1 \end{bmatrix}$$

So, $B$ and thus $\hat{W}$ have rank 1. Then the computation of the minimum-phase factor $W_-$ yields

$$B_- = \begin{bmatrix} -0.0229 & -0.8363 & 0 \\ -0.2952 & -0.7922 & 0 \\ -0.0699 & -0.4477 & 0 \\ 0.3056 & 0.0903 & 0 \end{bmatrix} \qquad D_- = \begin{bmatrix} 3.6751 & -0.5095 & 0 \\ -1.2930 & -1.4482 & 0 \end{bmatrix}$$

and it can be easily seen that $B_-$ has rank two. In a similar way, it can be checked that none of the internal factors (those of dimension $2 \times 2$) will yield a dynamic rank one model.

This shows that our problem does not seem to have an easy answer. In fact, if we were trying to proceed backwards in the above example, there would be no reason to guess, from the realization of $W_-$, that there exists a spectral factor $W$ of dynamic rank one. The above case, nevertheless, can be solved by a very elegant method devised by Fijalkov and Loubaton [6], which works though, only when $\hat{W}$ has rank one. If again $\Delta$ is diagonal and the rank $m_0$ of the $p \times p$ spectral density $\hat{\Phi}$ is such that $2m_0 < p$, then the problem has been solved (together with some other cases) in Deistler and Scherrer [5].

## 2  Some ideas for a general solution

The general case remains thus unsolved. Therefore the approach we envisage is the following. We first define, given a $p \times p$ spectral density $\Phi$ of rank $m_0$, the set $\mathcal{W}^m$ of all spectral factors $W$ of $\Phi$ of dimension $p \times m$, endowed with the topology induced by any euclidian distance on the coefficients of $W$ as rational functions. Then the problem can be stated as follows:

1. *parametrize all the spectral factors $\mathcal{W}^m$ of given dimension $p \times m$*

2. *show that the set $\mathcal{W}^m$ is a manifold and produce an atlas*

3. *use gradient algorithms to minimize the rank of $B$.*

The approach we outline here is based on stochastic realization theory, for which a lot of the fundamental work was done by Lindquist and Picci (see e.g. [14] and [15]). In fact the case $m = m_0$ has been worked out by Lindquist and Picci, whereas the case $m = n + p$, where $2n$ is the degree of $\Phi$ goes back to Anderson [1].

The idea we want to exploit for the first and second point above stems from the following result:

**Theorem 2.1** *Let $W$ be a $p \times m$ spectral factor of the $p \times p$ spectral density $\Phi$ of rank $m_0$. Then there exists an $m \times m_0$ function $\hat{Q}$ which is rigid (i.e. $\hat{Q}^*\hat{Q} = I_p$) and such that:*

$$W\hat{Q} = W_+$$

*where $W_+$ denotes the maximum-phase factor.*

This result is due to Lindquist and Picci [14] for the case $p = m_0$ and has been extended in Fuhrmann and Gombani [11] to the case $p > m_0$. We recall that a stable function is *inner* if $Q^*Q = I$. It is well known now that if $\hat{Q}$ is rational rigid, then there exists an essentially unique $m \times m$ inner function $Q$ of the same degree such that the first $m_0$ columns coincide with $\hat{Q}$. Thus, from what we just said, a minimal $p \times m$ stable spectral factors $W$ uniquely determines an inner function $Q$ such that $W = [W_+, 0]Q^*$ (where obviously the 0 matrix is $p \times (m - m_0)$).

Conversely, each inner function $Q$ such that $[W_+, 0]Q^*$ is stable determines uniquely a spectral factor. Nevertheless, it might happen that there exist two such inner functions $Q_1, Q_2$ such that $[W_+, 0]Q_1^* = [W_+, 0]Q_2^*$. But it can be shown that if we impose the degree of $Q$ to be minimal, then the correspondence between spectral factors and inner functions is a bijective function : call this function $\Psi$. Then, we can endow $\mathcal{Q}^m$ with the topology induced by that of $\mathcal{W}^m$ by $\Psi$; since $\psi$ is differentiable with its inverse, this map is a diffeomorphism. So, we can make the following

**Definition 2.1** *We set $\mathcal{Q}^m$ to be the set of $m \times m$ inner functions $Q$ such that $[W_+, 0]Q^*$ is stable and which have minimal degree.*

The set $\mathcal{Q}^m$ has a very rich and complex structure, which is studied in detail in Fuhrmann and Gombani [11]. Nevertheless for our present purpose all we need is a result from Baratchart and Gombani [3]: denote by $r$ the maximal degree of the inner functions in $\mathcal{Q}^m$ (it can be shown that this number coincides with the number of transmission zeros of $W_+$: we assume that these zeros are simple).

**Theorem 2.2** *Let The set $\mathcal{Q}^m$ (and hence the set $\mathcal{W}^m$) is a smooth manifold diffeomorphic to a product of $r$ $(m - m_0)$-dimensional spheres.*

Since an atlas for a sphere is easy to construct, we can construct an atlas $\{\mathbf{V}_i^{\mathcal{Q}}, \phi_i^{\mathcal{Q}}\}_{i \in \mathcal{I}}$ for $\mathcal{Q}^m$ ($\mathcal{I}$ is a suitable set of indexes) and hence an atlas $\{\mathbf{V}_i^{\mathcal{W}}, \phi_i^{\mathcal{W}}\}_{i \in \mathcal{I}}$ for $\mathcal{W}^m$. In particular, it's well known that we can easily obtain realizations of the different spectral factors with the same $A$ and $C$. Thus, we can write $W = \left( \begin{array}{c|c} A & B_W \\ \hline C & D_W \end{array} \right)$, where $A$ and $C$ are always the same, and $B$ and $D$ are uniquely determined by $W$. Thus if $W \in \mathbf{V}_i^{\mathcal{W}}$ and $\xi = \phi_i^{\mathcal{W}}(W)$, we can define, for each $i \in \mathcal{I}$:

$$B_i(\xi) := B_{\phi_i^{\mathcal{W}}(\xi)}$$
$$D_i(\xi) := D_{\phi_i^{\mathcal{W}}(\xi)}$$

Thus a minimization procedure on the rank of $B(\xi)$ is theoretically feasible (for instance by minimizing recursively the singular values of $B(\xi)$. There are, though, several problems which remain open, like

- Does there exist a global criterion for minimization?

- How do the critical points look like?

- What is the best criterion to perform the minimization if there exists no low rank model (and thus we have to take an approximaxion)?

- Although the scheme we outline can provide a solution, the numerical procedure is not the most satisfactory from a conceptual point of view, at least for the exact case. Do there exist algebraic or geometric conditions which characterize the spectral factors of dynamic low rank?

To indicate that the last question might have a positive answer, we would like to give another formulation to the above problem. Let $\overline{W} = WK^*$ and $\overline{W}_+ = W_+K_+^*$ be the Douglas-Shapiro-Shields factorizations of $W$ and $W_+$ and let $\overline{W} = \overline{W}_+\overline{Q}^*$ be the inner-outer factorization of $\overline{W}$ in the space of antistable rational functions (whose $L^2$ closure is the Hardy space $H_-^2$). Then, since we can write

$$\overline{W} = [W_+, 0]QK^* = [W_+K_+^*, 0]\overline{Q}^*$$

we obtain, setting $K_+^e := \begin{bmatrix} K_+ & 0 \\ 0 & I \end{bmatrix}$, after some simple verifications ($W_+$ is only left invertible: we refer to Fuhrmann and Gombani [11] for details)

$$K_+^e Q = \overline{Q}K \tag{3}$$

with $K_+^e$, $\overline{Q}$ left coprime and $Q$ and $K$ right coprime. The functions $K$ and $Q$ are called *skew-prime* and factorization (3) is called *skew-prime factorization*; the properties of this construction for non rational inner functions have been studied in detail in Fuhrmann [8]. It is quite intuitive (although the proof is not straight-forward) that instead of parametrizing $Q \in Q^m$, we can parametrize $\overline{Q} \in \overline{Q}^m$. We recall that the multiplicity of an inner function is defined as the multiplicity of the space $QH_+^2$ (see Fuhrmann [7]). In the rational case, if $Q$ has realization $Q = \left(\begin{array}{c|c} A & B \\ \hline C & D \end{array}\right)$, then the multiplicity coincides with the rank of $B$. Our problem can thus be formulated as follows:

*Given $W_+ = \overline{W}_+K_+$, find $\overline{Q} \in \overline{Q}^m$ such that, in the skew-prime factorization (3), the multiplicity of $K$ is minimal.*

# 3  An application to interest rates models

To see a concrete application of this philosophy, we turn to a newer field of research. In recent years, a new theory for the modeling of interest rates has been developed in the context of a no arbitrage market in *continuous time*. The idea is that in the market there exist *locally riskless asset* $B(t)$ with dynamics:

$$dB(t) = r(t)B(t)dt$$

The value $r(t)$ is called spot rate at time $t$ and the dynamics of $r$ depends on the time $t$. Then the assumption of no arbitrage simply means that - under a suitable measure $Q$, called *martingale measure* - the price of any other asset $\Pi(t)$ discounted by $B(t)$ is a martingale, i.e.

$$\frac{\Pi(t)}{B(t)}$$

is a $Q$-martingale.

If the dynamics of $r$ is stochastic, i.e. if $r$ obeys a stochastic differential equation:

$$dr(t) = \mu(r, t)dt + \sigma(r, t)dW(t)$$

where $W$ is a Browniam Motion, then $r$ is called *interest rates process*. Then the price of any asset depending on the interest rates (like a Treasury Bond) becomes a random variable, and the question is to determine a stochastic model for the bond prices which is compatible with the no arbitrage assumption. We denote by $q(t, x)$ the price at time $t$ of a zero-coupon bond expiring at time $t + x$, and introduce the *forward rate process* $r(t, x) := -\frac{\partial}{\partial x} \ln q(t, x)$.

Then it can be shown (see [13], [16]) that the no arbitrage assumption in the bond market has the following very elegant formulation: suppose the forward rates process $r(t, x)$ has dynamics, (with respect to the martingale measure $Q$):

$$dr(t, x) = \alpha(t, x)dt + \sigma(t, x)dW^Q(t)$$

where $\sigma$ is an $m$ dimensional row vector and $W$ is an $m$ dimensional Brownian Motion. Then there is no arbitrage, i.e. the process $\frac{q(t,x)}{B(t)}$ is a $Q$-martingale if the following relation holds:

$$\alpha(t, x) = \frac{\partial}{\partial x}r(t, x) + \sigma(t, x) \int_0^x \sigma'(t, u)du \tag{4}$$

A system theoretic approach to the previous theory leads to new theoretical framework for the modeling of the term structure as well as new problems in identification theory. A first result in this direction is the following (see [3])

**Theorem 3.1**     *1. The spot forward rates dynamics is given by*

$$dr(t, x) = [\frac{\partial}{\partial x}r(t, x) + \sigma(x) \int_0^x \sigma(\tau)'d\tau]dt + \sigma(x)dW(t) \tag{5}$$

$$r(0, x) = r^*(0, x) \tag{6}$$

*2. The spot forward rates dynamics is given by*

$$\begin{cases} d\xi(t) = A\xi(t)dt + BdW(t) \\ r_i(t, x) = C_i e^{Ax}\xi(t) + D_i^r(t, x) + r_i^*(0, t + x) \end{cases} \tag{7}$$

*with initial condition $\xi(0) = 0$ and with*

$$D(t, x) = \frac{1}{2}\left\{ \left| \int_0^{t+x} \sigma(u)du \right|^2 - \left| \int_0^x \sigma(u)du \right|^2 \right\}$$

A first problem is obviously the one of identifying $A, B, C$ (or, in the terminology of mathematical finance, to *calibrate* the model). Two approaches are currently present in the literature: the first is to identify $\sigma$ from past data: although we do *not* observe the data under the measure $Q$, it can be shown that the measure under which we observe the data and the measure $Q$ are related by a Girsanov transformation. Thus $\sigma$ does not change and it can be estimated by standard estimation methods. Nevertheless, it might be conceivable that better results can be obtained if we exploit the linearity of state dynamics: formulation (7) would allow in this case for the use of linear identification methods. The second approach is to derive the *implied volatility*: each choice of model will determine (via a nonlinear function) the price of all interest rates options. Some of these options are very heavily traded and constitute therefore a good basis for a model matching procedure. Both of these approaches are currently being investigated by the authors. It should be noticed that, if $B$ is $n \times m$, with $m > 1$, then the parametrization of the pairs $A, B$ becomes nontrivial. A solution for this problem can be found in Fuhrmann and Helmke [12].

When we consider a multivariable system, we naturally come to another issue, connected with the multiplicity of a process. Suppose we have several interest rates processes $r_0, r_1, ..., r_k$ (corresponding to different currencies markets: the rate $r_0$ is the one of the currency by which we decide to measure our world and is thus called *domestic market*, while the others are called foreign markets) defined, according to our model, by the volatilities $\sigma_i(x) = C_i e^{A_i x} B_i$, for $i = 0, ..., k$, where $A_i$ is $n_i \times n_i$, $B_i$ is $n_i \times m$ and $C_i$ is $1 \times n_i$. Then we can write, for $i = 0, ..., k$

$$\begin{cases} d\xi_i(t) = A_i \xi_i(t)dt + B_i dW_i(t) \\ r_i(t, x) = C_i e^{A_i x} \xi_i(t) + D_i(t, x) + r_i^*(0, t + x) \end{cases}$$

where

$$D_i(t, x) = \frac{1}{2} \left\{ \left| \int_0^{t+x} \sigma_i(u)du \right|^2 - \left| \int_0^x \sigma_i(u)du \right|^2 \right\}$$

By construction, the interest rates dynamics are with respect to the respective martingale measures. It turns out that these measures are connected by the exchange rates processes $Q^i(t)$ which describe the exchange rate between the $i - th$ currency and the domestic one. So, it can be shown (see Musiela and Rutkowski [17]) that, if $\nu_i(t)$ is the volatility of the process $Q^i(t)$, then

$$dW_i = dW_0(t) - \nu_i(t)dt$$

Thus, the Heath-Jarrow-Morton condition for the foreign market becomes

$$\alpha_i(t, x) = \frac{\partial}{\partial x} r_i(t, x) + \sigma_i(t, x) \int_0^x \sigma_i'(t, u)du - \sigma_i(t, x)\nu_i(t) \tag{8}$$

In our setting, $\sigma_i$ does not depend on $t$; we can thus set

$$W := W_0 \qquad D_0^0(t, x) := D_0(t, x)$$

$$D_i^0(t, x) := D_i(t, x) - \sigma_i(x) \int_0^t \nu_i(u)du \qquad \text{for } i = 1, ..., k$$

and, in a fairly obvious manner,

$$A := \begin{bmatrix} A_0 & 0 & \ldots & 0 \\ 0 & A_1 & \ldots & 0 \\ \vdots & \vdots & \ddots & \vdots \\ 0 & 0 & \ldots & A_k \end{bmatrix} \qquad B := \begin{bmatrix} B_0 \\ B_1 \\ \vdots \\ B_k \end{bmatrix}$$

$$C := \begin{bmatrix} C_0, C_1, \ldots, C_k \end{bmatrix} \qquad D(t,x) := \begin{bmatrix} D_0^0(t,x) \\ D_1^0(t,x) \\ \vdots \\ D_k^0(t,x) \end{bmatrix} \qquad r(t,x) := \begin{bmatrix} r_0(t,x) \\ \vdots \\ r_l(t,x) \end{bmatrix}$$

Then the dynamics of the above interest rates $r_0, r_1, ..., r_k$ is also described by

$$\begin{cases} d\xi(t) = A\xi(t)dt + BdW(t) \\ r(t,x) = Ce^{Ax}\xi(t) + D(t,x) + r^*(0, t+x) \end{cases} \tag{9}$$

Our model has a block diagonal $A$ by construction; but there is no real reason to make this restriction. The model (9) still makes perfect sense for any choice of $A$. In particular, each row $r_i$ satisfies the Heath-Jarrow-Morton condition for the foreign market (8); but now it might happen that some state variables influence several interest rates (as is in reality).

The other relevant observation is that, while $m$ is determined by the number of rates we want to model, the choice of $k$ is basically free. It is quite clear that, for computational reasons, we would like to make this number as small as possible; on the other hand, if $k < m$ it means (roughly speaking) that we only have $m - k$ independent markets and, since this is not the case in reality, then our model is not arbitrage free. But the point we want to make is different: although every market can be modeled independently by means of its own martingale measure, there are two reasons to model more markets simultaneously. The first is that such model is needed to price currency options and this is why they are presently used. But there exists, in our view, a more subtle and yet not well exploited reason: if we choose to model all our markets by a *common* Markov process, as we have indicated above, we will not obtain a model containing more information, but some of this information might be shared (in the sense that a state contributes to more than one interest rate). In particular, it might seem quite sensible to believe that the, for example, the German rates influence the italian ones but not conversely. Therefore, in the model for the italian market we might expect to have as states the german rates *and* the italian rates (the market is dependent), but in the model of the german rates we would not expect to see the italian rates (the market is independent). The questions which arise then are the following:

- How do we recognize when a state is a forward rate for some currency? Is there a basis transformation which always achieves that (this is the case in the single currency market)?

- If the above question has a positive answer, how can we isolate independent and dependent markets? Or, in other words, is it possible to reduce the multiplicity of the input Wiener process so that the domestic rate dynamics remains the same?

It is quite obvious that these are quite important questions: if we could answer positively to them, this would not imply that we could model better the dependent markets, but we could discard, in the study of a market, the other currencies which do not contribute to the movements of that market.

# References


[1] B.D.O. Anderson, "Algebraic Properties of Minimal Degree Spectral Factors" *Automatica*, 1973, **9**, 491-500

[2] L. Baratchart and A. Gombani, "A parametrization of external spectral factors", *Preprints of the 10th IFAC Symposium on Identification and System Parameter Estimation*, Copenhagen, 1994.

[3] T. Biörk and A. Gombani, Minimal Realizations of Interest Rate Models, submitted for publication.

[4] M.Deistler and B.D.O. Anderson, Linear dynamic errors-in-variables models: some structural theory, *Journal of Econometrics*, 1991, **41**, pp. 39-63.

[5] M. Deistler and W.Scherrer, Identification of linear systems from noisy data, G.B.Di Masi, A. Gombani, A. Kurzhanski Eds., *Modeling, Estimation and Control of Systems with Uncertainty* , Birkäuser Boston, 1991.

[6] I.Fijalkov and Ph. Loubaton Identification of rank 1 rational spectral densities from noisy observations: a stochastic realization approach, to appear on *Systems and Control Letters*.

[7] P.A. Fuhrmann, *Linear Systems and Operators in Hilbert Space*, McGraw-Hill, New York, 1981.

[8] P.A. Fuhrmann, On skew primeness of inner functions, *Lin. Alg. Appl.*, 1994, **208-209**, 539-551.

[9] P.A. Fuhrmann, "On the characterization and parametrization of minimal spectral factors", *Journal of Mathematical Systems, Estimation and Control*, 1995, **5**, 383-444.

[10] P.A. Fuhrmann, "Geometric control revisited", in preparation.

[11] P.A. Fuhrmann and A. Gombani, "On a Hardy space approach to the analysis of spectral factors", submitted for publication.

[12] P.A. Fuhrmann and U. Helmke, "A homeomorphism between observable pairs and conditioned invariant subspaces", to appear in *Sys. & Contr. Letts.*.

[13] Heath, D., Jarrow, R. and Morton, A. Bond pricing and the term structure of interest rates. *Econometrica*, 1992, **60**, 77-106.

[14] A. Lindquist and G. Picci, "Realization theory for multivariable stationary Gaussian processes", *SIAM J. Contr. Optimiz.*, 1985, **23**, 809-857.

[15] A. Lindquist, G. Michaletzky and G. Picci, "Zeros of spectral factors, the geometry of splitting subspaces and the algebraic Riccati inequality", *SIAM J. Contr. Optimiz.*, 1995, **33**, pp. 365-401.

[16] M. Mu, Stochastic PDE:s and term structure models, *Journées Internationales de Finance*, IGR-AFFI. La Baule, June 1993.

[17] M. Musiela and M. Rutkowski, *Arbitrage Pricing of Derivatives Securities: Theory and Applications*, Springer-Verlag, 1997.

[18] B. Roorda, *Global Total Least Squares*, Ph.D. Thesis, Erasmus Univerity, Rotterdam, 1995.

[19] J.C.Willems, From time series from linear systems, part III: Approximate modeling, *Automatica*, 1987, **23**, 87-115.

Tomas Björk
Department of Finance
Stockholm School of Economics
Box 6501
113 83 Stockholm, Sweden
ph: +46-8-7369162
fax: +46-8-312327
e-mail: `fintb@hhs.se`

Andrea Gombani
LADSEB-CNR
Corso Stati Uniti 4
35127 Padova, Italy
ph: +39-49-8295756
fax: +39-49-8295763
e-mail: `gombani@ladseb.pd.cnr.it`

# A Rational Flow for the Toda Lattice Equations

Roger Brockett[1]
Harvard University, Cambridge, MA, USA

Dedicated to my long-time friend and colleague, Paul Fuhrmann,
on the occasion of his $60^{th}$ birthday

**Abstract**: We show how a certain class of Hamiltonian systems give rise to differential equations on spaces of matrices whose elements are rational functions. In particular, we reinterpret the results of Kac and van Moerbeke on the periodic Toda lattice in terms of such differential equations and relate the action-angle coordinates found by them to the evolution of a certain two by two symmetric matrix of rational functions flowing on a space of fixed McMillian degree and fixed Cauchy index. Realization theory is used to pass from a description of the flow in terms of rational matrices to a description in terms of the original coordinates.

## 1  Introduction

The theory of realizations of finite dimensional linear time invariant systems reached a certain maturity in the mid 1960's. The subsequent extensions to infinite dimensional systems and refinements involving the reactance signature and its relationship to topological ideas added to its appeal and gave additional confirmation of the idea that realization theory packages an interesting set of ideas in a natural way. When Paul Furhmann and I started to collaborate in 1973 he was quick to point out that the cornerstone of realization theory, the state space isomorphism theorem, could be viewed as being more or less equivalent to the spectral theorem, at least in the case of self-adjoint compact operators on a Hilbert space. Although I had been teaching and writing about the finite dimensional theory for some time, this point of view added significantly to my appreciation of the subject. In his elegant book *Linear Systems and Operators in Hilbert Space* one sees such relationships between system theory and operator theory developed in much more generality.

[1]This work was supported in part by the National Science Foundation under Engineering Research Center Program, NSF EEC 94-02384, the US Army Research Office under grants DAAL03-92-G-0115 and DAAG55-97-1-0114.

In this paper we use the following basic ideas from this theory, limiting the discussion to the situation in which all scalars are real. First of all, any matrix whose entries are rational functions of a complex variable $s$, proper in the sense that the rational functions go to zero at infinity, can be expressed as $G(s) = C^T(Is-A)^{-1}B$, with $A, B$ and $C$ all being constant matrices of the appropriate size. If, among all such representations of $G$, we pick one having the property that the dimension of the square matrix $A$ is minimal, then this dimension is the rank of the infinite Hankel matrix formed from the Laurent expansion of $G$,

$$H = \begin{bmatrix} \hat{G}_0 & \hat{G}_1 & \hat{G}_2 & \cdots & \cdots \\ \hat{G}_1 & \hat{G}_2 & \hat{G}_3 & \cdots & \cdots \\ \hat{G}_2 & \hat{G}_3 & \hat{G}_4 & \cdots & \cdots \\ \cdots & \cdots & \cdots & \cdots & \cdots \\ \cdots & \cdots & \cdots & \cdots & \cdots \end{bmatrix} \quad ;$$

with

$$G(s) = \hat{G}_0 s^{-1} + \hat{G}_1 s^{-2} + \hat{G}_2 s^{-3} + \dots$$

This integer is called the *McMillian degree* of $G(s)$. If it happens that the partial fraction expansion of $G(s)$ has no repeated poles then the McMillian degree equals the sum of the ranks of the residue matrices. If $G$ is real and symmetric then the Hankel matrix just displayed is real and symmetric as well. Its signature is called the *Cauchy index* of $G$. Of course the Cauchy index equals the McMillian degree if the Hankel matrix is positive semidefinite and equals the negative of the McMillian degree if it is negative semidefinite. In the former case there exists a representation of $G$ of the form $G(s) = C^T(Is - A)^{-1}C$ with $A = A^T$. Finally, we recall a basic fact about feedback. If $K$ is a constant matrix with suitable dimensions and $G(s) = C^T(Is - A)^{-1}B$ then $(I + GK)^{-1}G = C^T(Is - A + BKC^T)^{-1}B$ and thus the McMillian degree of $(I + GK)^{-1}G$ cannot exceed that of $G$.

The present paper investigates a new use of realization theory, now applied to the currently popular subject of completely integrable Hamiltonian systems. It brings together a number of system theoretic techniques with the goal of providing a new point of view on the relationships between completely integrable systems and flows on spaces of transfer functions. In particular, it is significant in that the techniques used serve to reduce a system of arbitrary order to a two by two problem involving rational functions. More specifically, we are interested in solution techniques for isospectral flows of the form

$$\dot{L} = [L, B(L)]$$

based on the introduction of matrix valued rational function formed from the resolvent of $L$. We will show that the introduction of $t$-dependent, matrix valued rational functions of the form

$$G(t, s) = C^T(Is - L(t))^{-1}C$$

leads to an efficient way of thinking about the important work of Kac and van Moerbeke [2] on the solution of the periodic Toda lattice. Notice that the particular form of the right-hand side of the differential equation implies that analytic functions of $L$ satisfy similar equations. If $\phi(L) = \sum \alpha_i L^i$ then

$$\dot{\phi}(L) = [\phi(L), B(L)]$$

Thus

$$\frac{d}{dt}(Is - L)^{-1} = [(Is - L)^{-1}, B(L)]$$

Pre and post multiplying $(Is - L)^{-1}$ by the constant matrices $C^T$ and $C$, respectively, we obtain a square matrix whose entries are rational functions of $s$,

$$G(s) = C^T(Is - L)^{-1}C$$

and whose derivative is

$$\frac{d}{dt}G(s) = C^T(Is - L)^{-1}B(L)C - C^T B(L)(Is - L)^{-1}C$$

If $L$ is symmetric and $B(L)$ is skew-symmetric then

$$\frac{d}{dt}G(s) = C^T(Is - L)^{-1}B(L)C + C^T B^T(L)(Is - L)^{-1}C$$

To pass from this expression for the derivative to a differential equation in $G$ we need to express $C^T(Is - L)^{-1}B(L)C$ in terms of $G$. The Laurent expansion

$$G(s) = G_0 s^{-1} + G_1 s^{-2} + \ldots = C^T C s^{-1} + C^T L C s^{-2} + \ldots$$

is useful for this purpose. In fact, whenever it is possible to express $B(L)C$ as

$$B(L)C = \sum \alpha_i(L)C\beta_i(G(s))$$

with $\alpha_i$ being polynomials we can construct a differential equation for $G$. The advantage of passing from $L$ to $G$ is most obvious when the dimension of $G$ is much smaller that the dimension of $L$. In the case of the periodic Toda lattice we have a situation in which $L$ is $n$ by $n$ with $2n$ free parameters whereas $G$ is just two by two.

## 2   The Aperiodic Toda Lattice

In reference [3] J. Moser provided an explicit solution for the aperiodic Toda lattice. In this paper he makes explicit use of representations in terms of rational functions, both in terms of partial fractions and in terms of continued fractions.

Later Krishnaprasad [4] and Faybusovich and the author [5] investigated the relationships between the aperiodic Toda lattice and flows on spaces of rational functions. In this section we revisit this problem, not because we have anything to add to the solutions already given, but because we want to present a point of view that makes the transition to the periodic Toda lattice easy. Let $C$ be the $n$ by two matrix $[e_1, e_n]$ with $e_i$ being the $i^{th}$ standard basis element of $R^n$. Define $n$ by $n$ matrices $L_0$ and $N$ as

$$L_0 = \begin{bmatrix} b_1 & a_1 & \dots & 0 & 0 \\ a_1 & b_2 & \dots & 0 & 0 \\ \dots & \dots & \dots & \dots & \dots \\ 0 & 0 & \dots & b_{n-1} & a_{n-1} \\ 0 & 0 & \dots & a_{n-1} & b_n \end{bmatrix} \; ; \; N = \begin{bmatrix} 1 & 0 & \dots & 0 & 0 \\ 0 & 2 & \dots & 0 & 0 \\ \dots & \dots & \dots & \dots & \dots \\ 0 & 0 & \dots & n-1 & 0 \\ 0 & 0 & \dots & 0 & n \end{bmatrix}$$

The equations of motion for the aperiodic Toda lattice in terms of the Flaschka-Henon coordinates can be written as a double Lie bracket equation $\dot{L} = [L, [L, N]]$. (See [6].) Observe that

$$L_0 C = \begin{bmatrix} b_1 & a_1 & \dots & 0 & 0 \\ a_1 & b_2 & \dots & 0 & 0 \\ \dots & \dots & \dots & \dots & \dots \\ 0 & 0 & \dots & b_{n-1} & a_{n-1} \\ 0 & 0 & \dots & a_{n-1} & b_n \end{bmatrix} \begin{bmatrix} 1 & 0 \\ 0 & 0 \\ \dots \\ 0 & 0 \\ 0 & 1 \end{bmatrix} = \begin{bmatrix} b_1 & 0 \\ a_1 & 0 \\ \dots \\ 0 & a_{n-1} \\ 0 & b_n \end{bmatrix}$$

and

$$[L_0, N]C = \begin{bmatrix} 0 & 0 \\ -a_1 & 0 \\ \dots \\ 0 & a_{n-1} \\ 0 & 0 \end{bmatrix} = -L_0 C \Sigma + C(C^T L_0 C)\Sigma$$

with

$$\Sigma = \begin{bmatrix} 1 & 0 \\ 0 & -1 \end{bmatrix}$$

Define $G_0(s)$ to be

$$G_0(s) = C^T(Is - L_0)^{-1}C$$

The Laurent expansion of $G_0(s)$ is

$$G_0(s) = \sum_{i=0}^{\infty} C^T L_0^i C s^{-i-1} = \tilde{G}_0 s^{-1} + \tilde{G}_1 s^{-2} + \dots$$

which implies that

$$C^T(Is - L_0)^{-1}L_0 C = sC^T(Is - L_0)^{-1}C - C^T C$$

**Lemma 1:** If $\dot{L}_0 = [L_0, [L_0, N]]$ with $L_0$ and $N$ as above then

$$\frac{d}{dt}G_0(s) = -s(G_0(s)\Sigma + \Sigma G_0(s)) + 2\Sigma + \Sigma \tilde{G}_1 G_0(s) + G_0(s)\tilde{G}_1 \Sigma$$

and in terms of the entries of $G$

$$G_0(s,t) = \left[ \begin{array}{cc} \tilde{g}_{11}(s) & \tilde{g}_{12}(s) \\ \tilde{g}_{12}(s) & \tilde{g}_{22}(s) \end{array} \right]$$

the equations of motion are

$$\frac{d}{dt}\tilde{g}_{11}(s,t) = -2s\tilde{g}_{11}(s,t) + 2 + 2b_1\tilde{g}_{11}(s,t)$$

$$\frac{d}{dt}\tilde{g}_{12}(s,t) = (b_1 - b_n)\tilde{g}_{12}(s,t)$$

$$\frac{d}{dt}\tilde{g}_{22}(s,t) = 2s\tilde{g}_{22}(s,t) - 2 - 2b_n\tilde{g}_{22}(s,t)$$

**Proof:** The evaluation of the derivative involves nothing more than the use of the identities given above. In considering these equations it is helpful to remember that $\tilde{g}_{11}(s)$ is of McMillian degree $n$ but of the special form (monic numerator)

$$\tilde{g}_{11}(s) = \frac{s^{n-1} + \tilde{p}_{n-2}s^{n-2} + \ldots + \tilde{p}_0}{s^n + \tilde{q}_{n-1}s^{n-1} + \ldots + \tilde{q}_0}$$

thus $2n - 1$ parameters are required for its specification. The entire problem is stated in terms of $L_0$ and $L_0$ only contains $2n - 1$ parameters. Let **Rat** $(2, n, n)$ denote the set of all real 2 by 2 rational matrices of McMillian degree $n$ and Cauchy index $n$. A short calculation shows that this space is a manifold of dimension $3n$. Given the parametrization of $L_0$, we see that $G_0$ flows on a $2n - 1$ dimensional submanifold of this space. The explicit characterization of this submanifold is a key step in our approach to solving these equations.

**Lemma 2:** If $L_0$ and $C$ are as given then the subset of **Rat** $(2, n, n)$ which is expressible as

$$G_0(s) = \left[ \begin{array}{cc} \frac{\tilde{p}_{11}(s)}{\tilde{q}(s)} & \frac{\tilde{p}_{12}(s)}{\tilde{q}(s)} \\ \frac{\tilde{p}_{12}(s)}{\tilde{q}(s)} & \frac{\tilde{p}_{22}(s)}{\tilde{q}(s)} \end{array} \right] = C^T(Is - L_0)^{-1}C$$

is completely characterized by the choice of a scalar rational function whose McMillian degree and Cauchy index are both n and whose numerator and denominator are monic. We may identify $\tilde{g}_{11}(s) = \tilde{p}_{11}(s)/\tilde{q}(s)$ with this function, in which case the remaining elements of $G_0(s)$ are determined by $\tilde{p}_{12}(s) = \beta$ with $\beta$ being determined to within a sign by

$$\beta^{-2} = \sum_{i=1}^{n} \frac{1}{\tilde{p}_{11}(\lambda_i)} \prod_{j \neq i}(\lambda_i - \lambda_j)$$

and

$$\frac{\tilde{p}_{22}(s)}{\tilde{q}(s)} = \sum_{i=1}^{n} \beta^2 \frac{(1/\tilde{p}_{11}(\lambda_i))}{(s+\lambda_i)} \prod_{j\neq i}(\lambda_i - \lambda_j)$$

Here $\{\lambda_i\}$ is the set of (necessarily real) zeroes of $\tilde{q}(s)$. Alternatively, we can characterize $G_0(s)$ has having real poles and a partial fraction expansion of the form

$$G_0(s) = \sum_{i=1}^{n} \frac{1}{s+\lambda_i} \begin{bmatrix} f_i & g_i \\ g_i & h_i \end{bmatrix}$$

with all the $f_i$ being positive, $\sum f_i = \sum h_i = 1$; $f_i h_i - g_i^2 = 0, i = 1, 2, ..., n$; $\sum g_i \lambda_i^k = 0; k = 0, 1, ..., n - 2$. Moreover, any $G$ of this form can be expressed as $C^T(Is - L_0)^{-1}C$ with $C$ and $L_0$ of the given form.

**Remark 1:** Notice that $L_0$ contains $2n - 1$ free parameters. The set of two by two rational symmetric matrices with a common denominator, monic of degree $n$, is a $4n$ parameter set. The conditions on the residues impose $2n + 1$ conditions on the rational functions.

**Proof:** Clearly the common denominator of the matrix elements is the polynomial $\tilde{q}(s) = \det(Is - L_0)$. To see that the off-diagonal elements of $G_0$ have constant numerators it is enough to observe that the off-diagonal elements of $C^T L_0^k C$ are zero for $k = 0, 1, ... n - 2$ and that $C^T L_0^{n-1} C$ has off-diagonal elements equal to the product $a_1 a_2 ... a_{n-1}$. This implies that the Laurent expansion of the off-diagonal terms begins with the term $a_1 a_2 ... a_{n-1} s^{-n}$. And thus, that the numerator is constant with the coefficient as given. The formula for the element $\tilde{g}_{22}$ follows from the fact that each residue matrix must be of rank one and must be positive semidefinite. The solution of this differential equation is to be sought in the space of rational functions whose denominator is the characteristic equation of $L_0$. The flow for $L_0$ is isospectral and so this is a fixed monic polynomial. To lay bare a completely decoupled set of equations we introduce $\{\mu_i\}$, the zeroes of $\tilde{p}_{11}$ and $\{\mu_i\}$, the zeroes of $\tilde{p}_{22}$. Differentiating the expressions $\tilde{g}_{11}(\mu_i) = 0$ and $\tilde{g}_{22}(\nu_i) = 0$ we get the differential equatation

$$\frac{d\mu_i}{dt} \tilde{g}'_{11}(\mu_i) = 2\mu_i \tilde{g}_{11}(\mu_i) - 2 - 2b_1 \tilde{g}_{11}(\mu_i)$$

Of course $\tilde{g}_{11}(\mu_i) = 0$ and so this simplifies to

$$\frac{d\mu_i}{dt} \tilde{g}'_{11}(\mu_i) = -2 \quad \text{or} \quad \dot{\mu}_i = \frac{-2}{\tilde{g}'_{11}(\mu_i)}$$

i.e., $n - 1$ decoupled differential equations, one for each $\mu_i$. The zero locations can, therefore, be considered as the angle variables of an action-angle system evolving independently of each other. The same effect can be had working with the zeroes of $\tilde{g}_{22}$. This confirms the complete integrability of the system. Although not directly relevant here, we remark that the results of [5] make it possible to write down a simple, closed-form, expression for the time evolution of $G$.

# 3 Periodic Toda Lattice

We now turn to the more interesting, and more difficult, periodic Toda lattice.
Again the equations take the form $\dot{L} = [L, [B(L)]]$ but now $L$ and $B(L)$ take the
form

$$
L = \begin{bmatrix} b_1 & a_1 & \ldots & 0 & a_n \\ a_1 & b_2 & \ldots & 0 & 0 \\ \ldots & \ldots & \ldots & \ldots & \ldots \\ a_n & 0 & \ldots & a_{n-1} & b_n \end{bmatrix} \;;\; B(L) = \begin{bmatrix} 0 & a_1 & \ldots & 0 & -a_n \\ -a_1 & 0 & \ldots & 0 & 0 \\ \ldots & \ldots & \ldots & \ldots & \ldots \\ a_n & 0 & \ldots & -a_{n-1} & 0 \end{bmatrix}
$$

If $C$ is a constant matrix with $n$ rows then from the definitions, including the
symmetry of $L$ and the skew-symmetry of $B[L]$, we see that

$$
\frac{d}{dt} C^T (Is - L(t))^{-1} C = C^T (Is - L(t))^{-1} B(L(t)) C + C^T B^T (L(t))(Is - L(t))^{-1} C
$$

As above, we let $C = [e_1, e_n]$ and adopt the notation $G(s) = C^T (Is - L)^{-1} C$. From
the definitions we have

$$
B(L)C = \begin{bmatrix} 0 & -a_n \\ -a_1 & 0 \\ \ldots & \ldots \\ 0 & a_{n-1} \\ a_n & 0 \end{bmatrix} \;;\; LC = \begin{bmatrix} b_1 & a_n \\ a_1 & 0 \\ \ldots & \ldots \\ 0 & a_{n-1} \\ a_n & b_n \end{bmatrix}
$$

One sees easily that

$$
C^T (Is - L)^{-1} C = \tilde{G}_0 s^{-1} + \tilde{G}_1 s^{-1} + \ldots = \begin{bmatrix} 1 & 0 \\ 0 & 1 \end{bmatrix} s^{-1} + \begin{bmatrix} b_1 & a_n \\ a_n & b_n \end{bmatrix} s^{-2} + \ldots
$$

and with $\Sigma$ as above

$$
B(L)C = -LC\Sigma + C\tilde{G}_1\Sigma + \frac{1}{2}C[\tilde{G}_1, \Sigma]
$$

We summarize these calculations as follows.

**Lemma 3:** If $L$, $C$, and $G(s)$ are as given above then

$$
\frac{d}{dt} G(s) = -s(G(s)\Sigma + \Sigma G(s)) + 2\Sigma + G(s)\tilde{G}_1\Sigma + \Sigma\tilde{G}_1 G(s) + \frac{1}{2}[G(s), [\tilde{G}_1, \Sigma]]
$$

and in terms of the entries of $G$

$$
G(s) = \begin{bmatrix} g_{11}(s) & g_{12}(s) \\ g_{12}(s) & g_{22}(s) \end{bmatrix}
$$

the equations of motion are

$$\frac{d}{dt}g_{11}(s,t) = -2sg_{11}(s,t) + 2 + 2b_1g_{11}(s,t) + 4a_ng_{12}(s,t)$$

$$\frac{d}{dt}g_{12}(s,t) = (b_1 - b_n)g_{12}(s,t) - 2a_n(g_{11}(s,t) - g_{22}(s,t))$$

$$\frac{d}{dt}g_{22}(s,t) = 2sg_{22}(s,t) - 2 - 2b_ng_{22}(s,t) - 4a_ng_{12}(s,t)$$

**Remark 2:** These equations differ from those of Lemma 1 by virtue of the additional double bracket term. This has the effect of coupling the $g_{11}$ equation with the $g_{12}$ equation, etc. In order to use the positions of the zeroes of $g_{11}$ as angle coordinates it will be necessary to reexpress $g_{12}(\mu_i)$ in terms $g_{11}(\mu_i)$.

**Remark 3:** As an aside, we observe that although the right-hand side of this equation is not quadratic in $G$ as written, it can be put in quadratic form by inserting factors of $G_0 = I$ as needed. For example, we can write

$$\frac{d}{dt}G(s) = -\Sigma\left(sG_0 + \tilde{G}_1\right)G(s) - G(s)\left(s\tilde{G}_0 + \tilde{G}_1\right)\Sigma + 2\tilde{G}_0\Sigma\tilde{G}_0 + G(s)\tilde{G}_1\Sigma$$
$$+\Sigma\tilde{G}_1G(s) + \frac{1}{2}[G(s),[\tilde{G}_1,\Sigma]]$$

As in the case of the aperiodic Toda lattice, $g_{11}(s)$ is of McMillian degree $n$ and of the special form

$$g_{11}(s) = \frac{s^{n-1} + p_{n-2}s^{n-2} + ... + p_0}{s^n + q_{n-1}s^{n-1} + ... + q_0}$$

Thus $2n - 1$ free parameters characterize it. However, the problem is stated in terms of $L$ and $L$ contains $2n$ parameters. Matters are clarified by the observation that the product of the $a$'s is constant along the flow. To see this, it is enough to note that

$$\dot{a}_i = (b_i - b_{i+1})a_i$$

with the variables $b_{n+1}$ and $b_1$ being identified. Clearly this implies that the derivative of the product of the $a$'s is zero.

**Lemma 4:** Let $C$ and $L$ be of the given form. The subset of **Rat** $(2,n,n)$ expressible as

$$G(s) = \begin{bmatrix} \frac{p_{11}(s)}{q(s)} & \frac{p_{12}(s)}{q(s)} \\ \frac{p_{12}(s)}{q(s)} & \frac{p_{22}(s)}{q(s)} \end{bmatrix} = C^T(Is - L)^{-1}C$$

is characterized by

$$G(s) = (I - G_0(s)K)^{-1}G_0(s)$$

with $G_0(s)$ being as in lemma 2 and $K$ being of the form

$$K = \begin{bmatrix} 0 & k \\ k & 0 \end{bmatrix} \quad ; \quad k = a_n$$

The elements of such $G$'s satisfy the relationship

$$(1 - kg_{12}(s))^2 - k^2 g_{11}(s)g_{22}(s) = 1 - 2\frac{(-1)^n a_1 a_2 ... a_n}{q(s)}$$

**Remark 4:** It is aconsequence of this lemma that the polynomial

$$q^2(s) + 2(-1)^n q(s) + a_n^2 p_{11}(s)p_{22}(s)$$

is a perfect square and

$$a_n p_{12}(s) = -q(s) \pm \sqrt{q^2(s) - 2(-1)^n q(s) + a_n^2 p_{11}(s)p_{22}(s)}$$

**Proof:** Notice that

$$CKC^T = \begin{bmatrix} 0 & 0 & ... & 0 & k \\ 0 & 0 & ... & 0 & 0 \\ ... & ... & ... & ... & ... \\ k & 0 & ... & 0 & 0 \end{bmatrix}$$

Thus the $L$ of the theorem statement can be expressed as $L_0 + CKC^T$ with $L_0$ as in the previous section and $k = a_n$. The feedback formula (now using $-K$) implies

$$G(s) = C^T(Is - L)^{-1}C = C^T(Is - L_0 - CKC^T)^{-1}C = (I - G_0K)^{-1}G_0$$

which establishes the first statement. Evaluating $(I - G_0K)^{-1}G_0$ we get

$$(I - G_0K)^{-1}G_0 = \begin{bmatrix} \dfrac{\tilde{q}\tilde{p}_{11}}{(\tilde{q}-a_n\tilde{p}_{12})^2 - a_n^2\tilde{p}_{11}\tilde{p}_{22}} & \dfrac{\tilde{q}\tilde{p}_{12} - a_n(\tilde{p}_{12}^2 - \tilde{p}_{11}\tilde{p}_{22})}{(\tilde{q}-a_n\tilde{p}_{12})^2 - a_n^2\tilde{p}_{11}\tilde{p}_{22}} \\ \dfrac{\tilde{q}\tilde{p}_{21} + a_n(\tilde{p}_{12}^2 - \tilde{p}_{11}\tilde{p}_{22})}{(\tilde{q}-a_n\tilde{p}_{12})^2 - a_n^2\tilde{p}_{11}\tilde{p}_{22}} & \dfrac{\tilde{q}\tilde{p}_{22}}{(\tilde{q}-a_n\tilde{p}_{12})^2 - a_n^2\tilde{p}_{11}\tilde{p}_{22}(s)} \end{bmatrix}$$

This calculation is facilitated by recognizing that the feedback term replaces $u_1$ by $u_1 + a_n y_2$ and $u_2$ by $u_2 + a_n y_1$. The result is the equation

$$\begin{bmatrix} \tilde{p}_{11} & \tilde{p}_{12} \\ \tilde{p}_{12} & \tilde{p}_{22} \end{bmatrix} \begin{bmatrix} u_1 \\ u_2 \end{bmatrix} = \begin{bmatrix} \tilde{q} - a_n\tilde{p}_{12} & -a_n\tilde{p}_{11} \\ -a_n\tilde{p}_{22} & \tilde{q} - a_n\tilde{p}_{12} \end{bmatrix} \begin{bmatrix} y_1 \\ y_2 \end{bmatrix}$$

To invert the coefficient of $y$ and solve for the new transfer function we evaluate the determinant

$$\det \begin{bmatrix} \tilde{q}(s) - a_n\tilde{p}_{12} & -a_n\tilde{p}_{11} \\ -a_n\tilde{p}_{22} & \tilde{q} - a_n\tilde{p}_{12} \end{bmatrix} = (\tilde{q} - a_n\tilde{p}_{12})^2 - a_n^2\tilde{p}_{11}\tilde{p}_{22}$$

Further manipulation gives the result

$$\begin{bmatrix} \tilde{q} - a_n\tilde{p}_{12} & -a_n\tilde{p}_{11} \\ -a_n\tilde{p}_{22} & \tilde{q} - a_n\tilde{p}_{12} \end{bmatrix}^{-1} \begin{bmatrix} \tilde{p}_{11} & \tilde{p}_{12} \\ \tilde{p}_{12} & \tilde{p}_{22} \end{bmatrix}$$

$$= \frac{1}{d} \begin{bmatrix} \tilde{q}\tilde{p}_{11} & \tilde{q}\tilde{p}_{12} - a_n(\tilde{p}_{12}^2 - \tilde{p}_{11}\tilde{p}_{22}) \\ \tilde{q}\tilde{p}_{21} - a_n(\tilde{p}_{12}^2 - \tilde{p}_{11}\tilde{p}_{22}) & \tilde{q}\tilde{p}_{22} \end{bmatrix}$$

from which the formula follows. The characteristic polynomial

$$\det(Is - L_0 - CKC^T) = \det \begin{bmatrix} s - b_1 & -a_1 & \ldots & 0 & k \\ -a_1 & s - b_2 & \ldots & 0 & 0 \\ \ldots & \ldots & \ldots & \ldots & \ldots \\ k & 0 & \ldots & -a_{n-1} & s - b_n \end{bmatrix}$$

has terms that are independent of $k$, linear in $k$ and terms that are quadratic in $k$, but no higher order dependence occurs. Thus for suitable choice of polynomials $r_0, r_1,$ and $r_2$, all independent of $k$, we have

$$\det(Is - L_0 + CKC^T) = r_0(s) + kr_1(s) + k^2 r_2(s)$$

An expansion by minors shows that $r_1$ is given by $r_1(s) =$

$$\det \begin{bmatrix} -a_1 & s - b_2 & \ldots & 0 & 0 \\ 0 & -a_2 & \ldots & 0 & 0 \\ \ldots & \ldots & \ldots & \ldots & \ldots \\ 0 & 0 & \ldots & 0 & -a_{n-1} \end{bmatrix} + \det \begin{bmatrix} -a_1 & 0 & \ldots & 0 & 0 \\ s - b_2 & -a_2 & \ldots & 0 & 0 \\ \ldots & \ldots & \ldots & \ldots & \ldots \\ 0 & 0 & \ldots & s - b_n & -a_{n-1} \end{bmatrix}$$

Thus $\det(Is - L_0 + CKC^T) = r_0(s) - 2(-1)^n a_1 \ldots a_n + a_n^2 r_2(s)$. Because the poles of $(I - G_0 K)^{-1} G_0$ must be zeroes of $(\tilde{q} - a_n\tilde{p}_{12})^2 - a_n^2 \tilde{p}_{11}\tilde{p}_{22}$ and the zeroes of the one-one element of the closed loop system correspond to the zeroes of the open loop system it follows that

$$(q(s) + a_n p_{12}(s))^2 - a_n^2 p_{11}(s)p_{22}(s) = q(s)(r_0(s) + a_n^2 r_2(s) - 2(-1)^n a_1 \ldots a_n))$$

from which the claim follows. In reference [3] the authors take the product of the $a$'s to be one and express matters in terms of the numerator of $g_{11}$ rather than $g_{11}$ itself.

**Theorem:** (Slightly modified from Kac and van Moerbeke [3]) If $G$ evolves according to the equations given in lemma 3 and if $p_{11}(\mu_i) = 0$ then

$$\frac{d\mu_i}{dt} g_{11}'(\mu_i) = -1 \pm \sqrt{1 + 2\frac{(-1)^n a_1 a_2 \ldots a_n}{q(\mu_i)}}$$

**Proof:** According to lemma 3 $g_{11}$ satisfies the equation

$$\frac{d}{dt} g_{11}(s,t) = -2sg_{11}(s,t) + 2 + 2b_1 g_{11}(s,t) + 4a_n g_{12}(s,t)$$

and so

$$\frac{d}{dt}\mu_i g'_{11}(\mu_i) = -2 + 4a_n g_{12}(\mu_i)$$

Using the conclusion of lemma 4

$$(1 + a_n g_{12}(s))^2 - a_n^2 g_{11}(s)g_{22}(s) = 1 - 2\frac{(-1)^{n-1}a_1 a_2...a_n}{q(s)}$$

which, when evaluated at $\mu_i$, gives an expression for $g_{12}(\mu_i)$ in terms of $q(\mu_i)$. Thus again, as in the non periodic case, the zeroes of the one-one element can be taken to be the angle variables. In this case the solution of the equations is more complex and involves hyperelliptic functions.

# 4 Recovering The Matrix L

The uniqueness part of the representation of $G(s)$ as $C(Is - A)^{-1}B$ asserts that two such representations, both minimal in the sense that the dimension of the "$A$" matrix is as small as possible, are related by a similarity transformation acting on $A$, $B$, $C$ according to the formulae $PAP^{-1}$, $PB$, and $CP^{-1}$. It is also known that two representations of the form $C^T(Is - A)^{-1}C$ with $A$ symmetric are necessarily related by an orthogonal transformation acting in the same way. An analysis of the action of the orthogonal matrices that fix $e_1$ and $e_n$ and send tridiagonal matrices into tridiagonal ones boils down to a study of diagonal orthogonal matrices. This means that the realization is determined to within a series of choices of signs. Kac and van Moerbeke explain the meaning of the signs in terms of the solutions of the differential equations.

# 5 References


1. Paul Fuhrmann, *Linear Systems and Operators in Hilbert Space*, McGraw-Hill, New York, 1981.

2. M. Kac and P. van Moerbeke, "A complete solution to the periodic Toda problem," *PNAS*, vol. 72, pp. 2879-2880, 1976

3. J. Moser, "Finitely many points on the line under the influence of an exponential potential– an integrable system," in *Dynamical Systems, Theory and Applications,* (J. Moser, Ed.), Lecture notes in Physics, Vol. 38, Springer-Verlag, 1975.

4. P. S. Krishnaprasad, "Symplectic mechanics and Rational Functions," *Ricerche di Automatica,* Vol. 10, (1979), pp. 107-135.

5. R. W. Brockett and Leonid Faybusovich, "Toda Flows, Inverse Spectral Transform and Realization Theory," *Systems and Control Letters*, Vol. 16, No. 2 (1991) pp. 79-88.

6. A. M. Bloch, R. W. Brockett and T. Ratiu, "A New Formulation of the Generalized Toda Lattice Equations and their Fixed Point Analysis via the Moment Map," *Bulletin of the American Mathematical Society*, Vol. 23, No 2 (1990) pp. 477-485.

Roger W. Brockett
Division of Applied Sciences
Harvard University
Cambridge, MA 02138, USA

# On Output-Stabilizability and Detectability of Discrete-Time LQ-Optimal Control Systems

Frank M. Callier
Facultés Universitaires Notre-Dame de la Paix
Namur, Belgium

**Abstract**: We handle the concept of optimizability in the context of infinite-horizon discrete-time Linear-Quadratic optimal control systems. It is shown that this concept is equivalent to: 1) the existence of a positive semi-definite (PSD) solution of the Algebraic Riccati Equation (ARE) , 2) optimizability using state feedback, and 3) output-stabilizability of the system, i.e. stabilizability of any observable part of the system. It is then found that the optimal minimal cost is defined by the minimal PSD-solution of the (ARE), which generates an optimal state feedback. The latter becomes stabilizing (i.e. the closed-loop state is bounded by a decreasing exponential) iff the system is output-stabilizable and detectable.

## 1  Introduction

The purpose of this paper is didactical. We handle the problem of optimizability (finite cost) of the discrete-time time-invariant infinite-horizon LQ(i.e. Linear-Quadratic)-optimal control problem, using first principle analytic arguments. We rediscover the importance of the concept of **output-stabilizability** of the system, as established by Geerts and Hautus [1] [2] for continuous-time systems using analysis, and by Wimmer [6] for the existence of a PSD solution of the (ARE) for discrete-time systems using algebra. We add next **detectability** and find then in a more direct way than in e.g. [3] [5], that the infinite-horizon LQ-problem is solvable by a unique (power-) stabilizing PSD solution of the (ARE). We are inspired here by some results of Wimmer [7].

## 2  Main Results

We consider a discrete-time linear time-invariant system $\Sigma = [A, B, C, 0]$ described, for $k \geq 0$ and $x_0 \in I\!\!R^n$ , by

$$\begin{cases} x_{k+1} &= Ax_k + Bu_k\,, \\ y_k &= Cx_k\,, \end{cases} \tag{1}$$

(where $x_k \in I\!\!R^n$ , $u_k \in I\!\!R^m$ , $y_k \in I\!\!R^p$ , $A \in I\!\!R^{n \times n}$ , $B \in I\!\!R^{n \times m}$ , $C \in I\!\!R^{p \times n}$ ), with associated finite-horizon and infinite-horizon (standard) cost given resp. by

$$V(x_0, N, u_k) \quad := \quad \sum_{k=0}^{N-1} (\|y_k\|^2 + \|u_k\|^2) \,, \tag{2}$$

and

$$V(x_0, \infty, u_k) \quad := \quad \sum_{k=0}^{\infty} (\|y_k\|^2 + \|u_k\|^2) \,. \tag{3}$$

The standard finite-horizon and infinite-horizon LQ-optimal control problems are then: for arbitrary $x_0 \in I\!\!R^n$

$$\text{find } V^0(x_0, N) := \min_{u_k} V(x_0, N, u_k) \,, \tag{LQ}$$

resp.

$$\text{find } V^0(x_0, \infty) := \min_{u_k} V(x_0, \infty, u_k) \,. \tag{LQ$^\infty$}$$

We introduce now some notation and notions. RHS, PSD, (ARE), (RRE) are abbreviations for "right-hand side", "(symmetric) positive semidefinite", "algebraic Riccati equation", "(backwards) recursive Riccati equation". $\mathbb{C}_<$ , $\mathbb{C}_=$ , $\mathbb{C}_>$ , $\mathbb{C}_\geq$ denote the sets of complex numbers $\lambda$ with $|\lambda| < 1$ , $|\lambda| = 1$ , $|\lambda| > 1$ , $|\lambda| \geq 1$ . For any square matrix $A \in I\!\!R^{n \times n}$ , $\sigma(A)$ denotes its spectrum ( set of eigenvalues ). $\mathcal{L}_<(A)$ denotes the $A$-invariant subspace of $I\!\!R^n$ spanned by the (generalized) eigenvectors of $A$ corresponding to its eigenvalues in $\mathbb{C}_<$ ; $\mathcal{L}_=(A)$ , $\mathcal{L}_>(A)$ and $\mathcal{L}_\geq(A)$ are similarly defined. $\mathcal{L}_<(A)$ is the (power-) stable subspace of $I\!\!R^n$ , i.e.

$$\mathcal{L}_<(A) = \{x \in I\!\!R^n : \lim_{k \to \infty} A^k x = 0\} \,.$$

$\mathcal{L}_\geq(A) = \mathcal{L}_=(A) \oplus \mathcal{L}_>(A)$ is the unstable subspace of $I\!\!R^n$ as the direct sum of the critical- and antistable subspace; one has

$$\mathcal{L}_\geq(A) = \{x \in I\!\!R^n : A^k x \text{ can be extended on } k < 0$$
$$\text{and } \|A^{-k}x\| \leq \pi(k) \text{ for } k \geq 0 \text{ with } \pi(k) \text{ a polynomial in k } \} \,.$$

A (symmetric) PSD matrix of $I\!\!R^{n \times n}$ is sometimes denoted $P = P^* \geq 0$ . For any pair of matrices $A \in I\!\!R^{n \times n}$ and $B \in I\!\!R^{n \times m}$ we denote by $R(A, B)$ and $S(A, B)$ resp. the $(A, B)$-reachable- and the $(A, B)$-stabilizable subspace of $I\!\!R^n$ , where

$$R(A, B) \quad := \quad \text{Im } [B; AB; \cdots ; A^{n-1}B] \,,$$

and

$$S(A, B) \quad := \quad R(A, B) + \mathcal{L}_<(A) \,. \tag{4}$$

$(A, B)$ or $\Sigma$ is said to be reachable, resp. stabilizable iff $R(A, B) = I\!\!R^n$ , resp. $S(A, B) = I\!\!R^n$ : the latter is equivalent to the existence of $K \in I\!\!R^{m \times n}$ such that

$A + BK$ is (power-) stable, i.e. $\sigma(A + BK) \subset \mathbb{C}_<$ i.e. $\mathcal{L}_{\geq}(A + BK) = \{0\}$ .
For any pair of matrices $A \in \mathbb{R}^{n \times n}$ and $C \in \mathbb{R}^{p \times n}$ , we denote by $NO(C, A)$
and $ND(C, A)$ resp. the $(C, A)$-unobservable- and $(C, A)$-undetectable subspace
of $\mathbb{R}^n$ , where

$$NO(C, A) \; := \; \{x \in \mathbb{R}^n : CA^k x = 0 \text{ for } k \geq 0\} \, ,$$

and

$$ND(C, A) \; := \; NO(C, A) \cap \mathcal{L}_{\geq}(A) \, . \tag{5}$$

$(C, A)$ or $\Sigma$ is said to be observable , resp. detectable iff $NO(C, A) = \{0\}$ resp.
$ND(C, A) = \{0\}$ . Finally $\ell^2$ denotes the Hilbert space of vector-valued sequences
$(f_k)_0^\infty$ ( $f_k \in \mathbb{R}^q$ ) that are square summable i.e. $\sum_{k=0}^\infty \|f_k\|^2 < \infty$ .
We start now by recalling the solution of problem (LQ) which we need for reference.
Upon setting $\mathcal{B} := BB^*$ and $\mathcal{C} := C^*C$ its backwards recursive Riccati equation
reads with $P_k = P_k^* \geq 0$ in $\mathbb{R}^{n \times n}$,

$$\begin{cases} P_k & = \; A^* P_{k+1}(I + \mathcal{B}P_{k+1})^{-1}A + \mathcal{C} \, , \\ P_N & = \; 0 \, , \end{cases} \tag{RRE}$$

where $k + 1 = N, N - 1, \cdots, 1$ . Sometimes the solution at $k$ for $P_N = 0$ is written
$P_k = P(k; N, 0)$ . Writing the closed-loop $A$-matrix and the state-feedback-matrix
defined by $P_{k+1}$ as

$$A_k \; := \; (I + \mathcal{B}P_{k+1})^{-1}A \, ,$$

resp.

$$K_k \; := \; -B^* P_{k+1} A_k \, ,$$

we find then the following result using the method of [5, Sec. 6.3.4].

**Theorem 1** *For a given horizon $N$ the optimal cost, optimal (state-feedback) control and optimal state trajectory of (LQ) read for $x_0 \in \mathbb{R}^n$ and $k = 0, 1, \cdots, N - 1$ :*

$$V^0(x_0, N) = x_0^* P_0 x_0 \, ,$$
$$u_k = K_k x_k \quad \text{with } x_{k+1} = A_k x_k \, .$$

We turn now to problem (LQ)$^\infty$ . Its Algebraic Riccati equation reads

$$P = A^* P(I + \mathcal{B}P)^{-1}A + \mathcal{C} \tag{ARE}$$

( with $\mathcal{B} := BB^*$ and $\mathcal{C} := C^*C$ ), where we are interested in PSD solutions $P \in$
$\mathbb{R}^{n \times n}$ . Every such $P$ defines a closed-loop $A$-matrix and a state-feedback matrix
given by resp.

$$A_P \; := \; (I + \mathcal{B}P)^{-1}A \, , \tag{6}$$
$$K_P \; := \; -B^* P A_P \, . \tag{7}$$

For efficiency we give now the following set of definitions.

**Definitions 1** Let $P$ be a PSD solution of the (ARE). We say that

a) $P$ solves (LQ)$^\infty$ if its optimal cost, -(state-feedback) control and -state trajectory are given, for $x_0 \in I\!\!R^n$ and $k \geq 0$ , by:

$$V^0(x_0, \infty) = x_0^* P x_0 \, ,$$
$$u_k = K_P x_k \quad \text{with } x_{k+1} = A_P x_k \, .$$

b) $P$ is (power-) stabilizing if

$$\sigma(A_P) \subset \mathbb{C}_< \quad \text{i.e.} \quad \mathcal{L}_\geq(A_P) = \{0\} \, .$$

c) $P$ solves (LQ)$^\infty$ with closed-loop stability if $P$ solves (LQ)$^\infty$ and $P$ is stabilizing.

The following well-known theorem can then be stated using information in e.g. [3] [4] and [5].

**Theorem 2** *Consider* $\Sigma = [A, B, C, 0]$ *with* $(A, B)$ *stabilizable and* $(C, A)$ *detectable. Then,*

*(i) the (ARE) has a unique PSD solution* $P = P^* \geq 0$ *and it is stabilizing.*

*(ii) $P$ solves (LQ)$^\infty$ with closed-loop stability.*

The proof of this classical result is usually done without considering first necessary and sufficient conditions for the existence of a solution of problem (LQ)$^\infty$ (i.e. finite cost). This however is an important question: before we consider optimality with closed-loop stability we should first know when optimality is possible. It is the purpose of this paper to follow this line of thought, which shall result in an improved version of Theorem 2.
It turns out that, as in [1] [2] and [6], the following two concepts and their interpretations are paramount.

**Definition 2** $\Sigma = [A, B, C, 0]$ is optimizable if for all $x_0 \in I\!\!R^n$ there exists a control sequence $(u_k)_0^\infty$ such that $V(x_0, \infty, u_k) < \infty$ , i.e. (see (3)) for all $x_0 \in I\!\!R^n$ there exists a control $(u_k)_0^\infty$ in $\ell^2$ such that the output $(y_k)_0^\infty$ is in $\ell^2$ .

**Definition 3** $\Sigma = [A, B, C, 0]$ is output-stabilizable if any observable part of $\Sigma$ is stabilizable i.e.

$$\frac{I\!\!R^n}{NO(C, A)} = \frac{S(A, B)}{NO(C, A)}$$

or equivalently

$$S(A, B) + NO(C, A) = I\!\!R^n \, . \tag{8}$$

**Remark:** Output-stabilizability can also be characterized in terms of the Kalman-decomposition of $\Sigma$ : to see this observe that (8) reads also

$$I\!R^n = \mathcal{L}_<(A) + (R(A,B) + NO(C,A))$$

or equivalently

$$\frac{I\!R^n}{R(A,B) + NO(C,A)} = \frac{\mathcal{L}_<(A)}{R(A,B) + NO(C,A)}$$

which means that the $A_{44}$-block of the Kalman-partitioned $A$-matrix must be stable, i.e. $\sigma(A_{44}) \subset \mathbb{C}_<$, see [7, Lemma 2.2].
The following two theorems concern optimizability of $(LQ)^\infty$ and optimizability with closed-loop stability. They are the main results of the paper.

**Theorem 3** *Consider $\Sigma = [A,B,C,0]$ and let $P(0;N,0)$ be the solution at 0 for $P_N = 0$ of the (RRE). Then the following are equivalent:*

*(i) $\Sigma$ is optimizable.*

*(ii) The (ARE) has a symmetric PSD solution.*

*(iii) $\Sigma$ is optimizable by state feedback control i.e. there exists $K \in I\!R^{m \times n}$ such that for all $x_0 \in I\!R^n$ there holds $V(x_0, \infty, u_k = Kx_k) < \infty$ .*

*(iv) $\Sigma$ is output-stabilizable.*

*Moreover under one of these conditions*

$$P_\mu := \lim_{N \to \infty} P(0;N,0) \text{ exists}, \tag{9}$$

*such that $P_\mu$ is the minimal PSD solution of the (ARE) and $P_\mu$ solves$(LQ)^\infty$.*

**Comment 1** The philosophy of Theorem 3 is mostly contained in [1] [2] and [6]. As it can be expected, the existence of the limit described by (9) is shown below to be crucial for beginning the proof of Theorem 3. It turns also out that in (iii) one may use any state feedback $K_P$ dictated by (7) by a PSD solution $P$ of the (ARE). Now, by Theorem 3, for optimality with closed-loop stability one should have that $P_\mu$ is stabilizing; it turns out that $\mathcal{L}_\geq(A_\mu) = ND(C,A)$ (with $A_\mu$ given by (6) for $P = P_\mu$ ). This and some other considerations result in the following improved version of Theorem 2.

**Theorem 4** *Consider $\Sigma = [A,B,C,0]$ and let $P_\mu$ denote the minimal PSD solution of the (ARE) of Theorem 3. Then the following are equivalent:*

*(i) $\Sigma$ is stabilizable and detectable.*

*(ii) $\Sigma$ is output-stabilizable and detectable.*

*(iii) The (ARE) has a PSD solution and $P_\mu$ is stabilizing.*

*(iv) $\Sigma$ is optimizable and $P_\mu$ solves $(LQ)^\infty$ with closed-loop stability.*

*Moreover then $P_\mu$ is the unique PSD solution of the (ARE).*

**Comment 2** (i) Theorem 4 is a more precise form of Theorem 2 where the last three equivalent statements illustrate precisely in consecutive order how for $(LQ)^\infty$ one realizes optimality with closed-loop stability. The first statement is classical but misleading in that it hides the second one which is more accurate for realizing optimality first without - and then with closed-loop stability.
(ii) Our proof of the uniqueness claim uses the facts that if $\Sigma$ is optimizable then 1) ([7, Sec.6]) every PSD solution of the (ARE) is (power-) stabilizing iff $\Sigma$ is detectable and 2) all stabilizing PSD solutions are equal.

The next sections of the paper contain sketches of the proofs of Theorems 3 and 4 and a conclusion.

# 3   Proof of Theorem 3

The proof of Theorem 3 starts by considering

**Lemma 1** *Let $\Sigma = [A, B, C, 0]$ be optimizable. Then*

*(i) $P_\mu := \lim_{N \to \infty} P(0; N, 0)$ exists and is a PSD solution of the (ARE).*

*(ii) For any PSD solution $P = P^* \geq 0$ of the (ARE) with $A_P$ and $K_P$ given by (6)-(7), the state feedback control $u_k = K_P x_k$ results in $x_{k+1} = A_P x_k$ , and for all $x_0 \in I\!\!R^n$ we have,*

$$0 \leq V(x_0, \infty, K_P x_k) \leq x_0^* P x_0 < \infty . \tag{10}$$

*Hence if $\Sigma$ is optimizable then $\Sigma$ is optimizable by state feedback dictated by any PSD solution of the (ARE).*

**Proof:**   (i) Since $\Sigma$ is optimizable we have that for all $x_0 \in I\!\!R^n$ there exists a control $(u_k)_0^\infty \in \ell^2$ such that the output $(y_k)_0^\infty \in \ell^2$ and $V(x_0, \infty, u_k) < \infty$ . Using Theorem 1 and the (RRE) we get

$$V^0(x_0, N) = x_0^* P(0; N, 0) x_0 \leq V(x_0, \infty, u_k)$$

where $V^0(x_0, N)$ is increasing as $N$ increases. Therefore, for arbitrary $x_0$, $p(x_0) = \lim_{N \to \infty} x_0^* P(0; N, 0) x_0$ exists, whence $P_\mu := \lim_{N \to \infty} P(0; N, 0)$ exists.

Finally considering the (RRE) at $k = 0$ with $P(1; N, 0) = P(0; N - 1, 0)$ shows that, as $N \to \infty$, $P_\mu = P_\mu^* \geq 0$ solves the (ARE).

(ii) We note first that $A_P = A + BK_P$ . We consider next the closed-loop (ARE) given by

$$P - A_P^* P A_P = K_P^* K_P + C^* C \tag{CARE}$$

One obtains with $x_k = A_P^k x_0$,

$$
\begin{aligned}
x_0^* P x_0 &= x_n^* P x_n + \sum_{k=0}^{n-1} (\|K_P x_k\|^2 + \|C x_k\|^2) \\
&= x_n^* P x_n + V(x_0, n, K_P x_k)
\end{aligned}
$$

with all terms positive. One gets (10) as $n \to \infty$ . ∎

One can be more precise.

**Lemma 2** *Let $\Sigma = [A, B, C, 0]$ be optimizable with $P_\mu$ the PSD solution of the (ARE) of Lemma 1. Let $P = P^* \geq 0$ be any PSD solution of the (ARE). Then*

*(i) $P_\mu$ solves $(LQ)^\infty$ .*

*(ii) $P_\mu \leq P$ , i.e. $P_\mu$ is the minimal PSD solution of the (ARE).*

**Proof:** In the sequel $A_\mu$ and $K_\mu$ denote $A_P$ and $K_P$ given by (6)-(7) for $P = P_\mu$ . (i) Denote by $V(x_0, \infty, u_k)$ any finite cost of $(LQ)^\infty$ and consider for $N < \infty$ an optimal finite-horizon control of Theorem 1. There results

$$x_0^* P(0; N, 0) x_0 \leq V(x_0, \infty, u_k)$$

such that by Lemma 1(i) as $N \to \infty$

$$x_0^* P_\mu x_0 \leq V(x_0, \infty, u_k) . \tag{11}$$

Now Lemma 1(ii) with $u_k = K_\mu x_k$ gives

$$V(x_0, \infty, K_\mu x_k) \leq x_0^* P_\mu x_0 < \infty \tag{12}$$

i.e. $K_\mu x_k$ generates a finite cost.
Hence by (11) and (12)

$$V(x_0, \infty, K_\mu x_k) = x_0^* P_\mu x_0 \leq V(x_0, \infty, u_k)$$

where the RHS is any (finite) cost of $(LQ)^\infty$ .
(ii) Using (i) and (10) one gets for all $x_0 \in I\!\!R^n$ ,

$$x_0^* P_\mu x_0 \leq V(x_0, \infty, K_P x_k) \leq x_0^* P x_0 . \qquad ∎$$

The following lemma is handy and easy.

**Lemma 3** *Consider first* $\Sigma = [A, B, C, 0]$ *and* $K \in I\!\!R^{m \times n}$ . *Consider also* $T :=$ $NO\left(\left[\begin{smallmatrix} C \\ K \end{smallmatrix}\right], A + BK\right)$ *and* $S := NO\left(\left[\begin{smallmatrix} C \\ K \end{smallmatrix}\right], A\right)$ . *Then*

*(i)* $T = S \subset NO(C, A)$ .

*(ii) on* $T$, $\quad (A + BK)^k = A^k$ *for* $k \geq 0$ .

**Proof of Theorem 3:** a) **Equivalences:** By Lemma 1 statements (i) (ii) and (iii) are equivalent. Hence it is sufficient to show that (i) is equivalent to (iv) , which we do next.

Consider therefore any observable part $\Sigma_2 = [A_2, B_2, C_2, 0]$ of $\Sigma = [A, B, C, 0]$ induced by the decomposition $I\!\!R^n = NO(C, A) \oplus S$ where $S$ is a complementary subspace. Then without loss of generality $x^T = (x_1^T, x_2^T)^T$ where $x_2 \in I\!\!R^{n_2}$ (equivalent to $\frac{I\!\!R^n}{NO(C,A)}$ ),

$$A = \begin{bmatrix} A_1 & A_{12} \\ 0 & A_2 \end{bmatrix} , \qquad B = \begin{bmatrix} B_1 \\ B_2 \end{bmatrix} , \qquad C = \begin{bmatrix} 0 & C_2 \end{bmatrix} , \qquad (13)$$

and $\Sigma_2$ is described by $x_{2(k+1)} = A_2 x_{2k} + B_2 u_k$ , $y_k = C_2 x_{2k}$ , with $NO(C_2, A_2)$ $= \{0\}$ . Observe that by (13) the cost $V(x_0, \infty, u_k)$ given by (3) and (1) depends only on $\Sigma_2$ and $x_{20} \in I\!\!R^{n_2}$ . Hence $\Sigma$ is optimizable iff $\Sigma_2$ is optimizable. Moreover since $\Sigma_2$ is an observable part of $\Sigma$ , $\Sigma$ will be output-stabilizable iff $\Sigma_2$ or $(A_2, B_2)$ is stabilizable. Hence, taking into account that (i) and (iii) are equivalent, we are reduced to show that $\Sigma_2$ is optimizable using state feedback if and only if $(A_2, B_2)$ is stabilizable.

Now sufficiency is a standard result, so we must show necessity. Consider therefore a state feedback which generates a finite cost for any $x_{20} \in I\!\!R^{n_2}$ , i.e. $u_k = K_2 x_{2k}$ , $K_2 \in I\!\!R^{m \times n_2}$ such that $x_{2k} = (A_2 + B_2 K_2)^k x_{20}$ , $y_k = C_2 x_{2k}$ and

$$\sum_{k=0}^{\infty} (\|y_k\|^2 + \|u_k\|^2) < \infty .$$

This implies for the unstable subspace of $A_2 + B_2 K_2$ that

$$\mathcal{L}_{\geq}(A_2 + B_2 K_2) \subset NO\left(\begin{bmatrix} C_2 \\ K_2 \end{bmatrix}, A_2 + B_2 K_2\right) ,$$

and using Lemma 3 we get

$$NO\left(\begin{bmatrix} C_2 \\ K_2 \end{bmatrix}, A_2 + B_2 K_2\right) \subset NO(C_2, A_2) = \{0\} .$$

Therefore $\mathcal{L}_{\geq}(A_2 + B_2 K_2) = \{0\}$ and $(A_2, B_2)$ is stabilizable.
b) **Additional information:** follows by Lemma 2. $\qquad\blacksquare$

# 4 Proof of Theorem 4

We start by two simple lemmas.

**Lemma 4** *Consider* $\Sigma = [A, B, C, 0]$ . *Then the following are equivalent:*

*(i)* $\Sigma$ *is stabilizable and detectable.*

*(ii)* $\Sigma$ *is output-stabilizable and detectable.*

**Proof:** Easy: use (4)-(5), (8) and

$$NO(C, A) \cap \mathcal{L}_<(A) \subset \mathcal{L}_<(A) \, . \qquad \blacksquare$$

**Lemma 5** *Let* $\Sigma = [A, B, C, 0]$ *be optimizable. Let* $P_1$ *and* $P_2$ *be two PSD solutions of the (ARE) that are stabilizing, i.e. with* $A_i = (I + BP_i)^{-1}A$ *for* $i = 1, 2$ , $\sigma(A_i) \subset \mathbb{C}_<$ . *Then* $P_1 = P_2$ .

**Proof:** Set $\Delta_{12} := P_1 - P_2$ . Some algebra gives

$$\Delta_{12} = A_1^* \Delta_{12} A_2 \, .$$

Since $A_1^*$ and $A_2$ are power-stable, $\Delta_{12} = 0$ . $\qquad \blacksquare$
The next lemma explains the role of the unobservable- and undetectable subspace of $\Sigma$ .

**Lemma 6** *Let* $\Sigma = [A, B, C, 0]$ *be optimizable and let* $P_\mu$ *be the minimal PSD solution of the (ARE) with* $A_\mu \in \mathbb{R}^{n \times n}$ *given by (6) for* $P = P_\mu$ . *Then*

*(i)* $\mathrm{Ker}\,(P_\mu) = NO(C, A)$ . $\qquad\qquad\qquad (14)$

*(ii) For any PSD solution* $P$ *of the (ARE), with* $A_P \in \mathbb{R}^{n \times n}$ *given by (6) ,*

$$\mathcal{L}_\geq(A_P) \subset ND(C, A) \, . \qquad\qquad (15)$$

*(iii)* $\mathcal{L}_\geq(A_\mu) = ND(C, A)$ . $\qquad\qquad\qquad (16)$

**Proof:** Let $K_\mu$ be given by (7) for $P = P_\mu$ .
(i) By Theorem 3, $P_\mu$ solves (LQ)$^\infty$, whence using $u_k = K_\mu x_k$ , $x_k = A_\mu^k x_0$ ( where $A_\mu = A + BK_\mu$ ),

$$x_0^* P_\mu x_0 = \sum_{k=0}^\infty (\|C x_k\|^2 + \|K_\mu x_k\|^2) \, .$$

Thus

$$\mathrm{Ker}\,(P_\mu) \subset NO\left(\begin{bmatrix} C \\ K_\mu \end{bmatrix}, A_\mu\right) \, . \qquad\qquad (17)$$

Hence, by Lemma 3, Ker $(P_\mu) \subset NO(C, A)$ . The converse holds because if $x_0 \in NO(C, A)$ then with zero control, $0 = V(x_0, \infty, 0) = V^0(x_0, \infty) = x_0^* P_\mu x_0$ .
(ii) Using Lemma 1 with $u_k = K_P x_k$ , $x_k = A_P^k x_0$ ( where $A_P = A + BK_P$ ), we get

$$\sum_{k=0}^{\infty} (\|Cx_k\|^2 + \|K_P x_k\|^2) \leq x_0^* P x_0 < \infty .$$

This and Lemma 3 give

$$\mathcal{L}_{\geq}(A_P) \subset NO\left(\begin{bmatrix} C \\ K_P \end{bmatrix}, A_P\right) \subset NO(C, A) \tag{18}$$

such that on $\mathcal{L}_{\geq}(A_P)$ , $A_P^k = A^k$ for $k \geq 0$ , with $A_P$ unstable. Thus $A$ is unstable on $\mathcal{L}_{\geq}(A_P)$ , i.e.

$$\mathcal{L}_{\geq}(A_P) \subset \mathcal{L}_{\geq}(A) .$$

Therefore this and (18) give (15).
(iii) For $P = P_\mu$ (15) gives $\mathcal{L}_{\geq}(A_\mu) \subset ND(C, A)$ . Hence we must show

$$ND(C, A) \subset \mathcal{L}_{\geq}(A_\mu) . \tag{19}$$

Now by (14) $ND(C, A) = $Ker $(P_\mu) \cap \mathcal{L}_{\geq}(A)$ . Thus by (17) and Lemma 3

$$ND(C, A) \subset NO\left(\begin{bmatrix} C \\ K_\mu \end{bmatrix}, A_\mu\right)$$

such that on $ND(C, A)$ , $A_\mu^k = A^k$ for $k \geq 0$ . Since $A$ is unstable on $ND(C, A)$ , $A_\mu$ is unstable on $ND(C, A)$ i.e. (19) holds. $\blacksquare$

**Proof of Theorem 4:** a) **Equivalences:** (i) and (ii) are equivalent by Lemma 4. (ii) and (iii) are equivalent by Theorem 3 and Lemma 6(iii). (iii) and (iv) are equivalent by Theorem 3 and Definition 1c).
b) **Additional information:** Since $(C, A)$ is detectable it follows by Lemma 6(ii) that any PSD solution of the (ARE) must be stabilizing. Therefore by Lemma 5 they are all equal to $P_\mu$ . $\blacksquare$

# 5 Conclusion

Theorems 3 and 4 are the core results for solving problem $(LQ)^\infty$ without and with closed-loop stability. Any well thaught course on the subject should contain their ideas. In that respect the notions of output-stabilizability and detectability are crucial and better than stabilizability and detectability, the former being only sufficient for optimizability. It is a fact that Theorem 3 and 4 have analogs in the continuous-time case using [1] [2] and [8]; some more research (if not already done) should lead to 1) infinite-dimensional state-space versions and 2) results concerning minimum error-variance state-estimators.

# References


[1] Geerts A.H.W., A necessary and sufficient condition for solvability of the linear quadratic control problem, Systems and Control Letters **11**, 1988, pp. 47–51.

[2] Geerts A.H.W. and M.L.J. Hautus, The output-stabilizable subspace and linear optimal control, in: Proceedings of the International Symposium MTNS-89 Vol. II, M.A. Kaashoek et al. (Eds.), Birkhäuser Verlag, Boston, 1990, pp. 113–120.

[3] Kucera V., Analysis and design of linear discrete control systems, Prentice Hall, 1992.

[4] Kucera V., The discrete Riccati equation of optimal control, Kybernetika **8**, 1972, pp. 430–447.

[5] Kwakernaak H. and Sivan R., Linear optimal control systems, Wiley-Interscience, New-York, 1972.

[6] Wimmer H.K., Existence of positive-definite and semi-definite solutions of discrete-time algebraic Riccati equations, International Journal of Control **59**, 1994, pp. 463–471.

[7] Wimmer H.K., The set of positive semi-definite solutions of the algebraic Riccati equation of discrete-time optimal control, IEEE Transactions on Automatic Control **41**, 1996, pp. 660–671.

[8] Wimmer H.K., Lattice properties of sets of semidefinite solutions of continuous-time algebraic Riccati equations, Automatica **31**, 1995, pp. 173–182.



Frank M. Callier
Department of Mathematics
Facultés Universitaires Notre-Dame de la Paix
Rempart de la Vierge, 8
B-5000 Namur, Belgium
Tel: (32)(81)72.49.32
Fax: (32)(81)72.49.14
Email: fcl@math.fundp.ac.be

# Matrix Pairs and 2D Systems Analysis

Ettore Fornasini, Giovanni Marchesini
and Maria Elena Valcher
Università di Padova, Padova, Italy

**Abstract**: Pairs of linear transformations on a finite dimensional vector space are of great relevance in the analysis of two-dimensional (2D) systems evolutions. In this paper, special properties of matrix pairs, such as finite memory, separability, property L and property P, as well as their dynamical interpretations, are investigated. Practical criteria for testing property L and property P in a finite number of steps are also presented.
The nonnegativity requirement on a matrix pair allows for much stronger characterizations of finite memory and separability, which in fact prove to be structural properties. Finally, the irreducibility and primitivity notions of positive matrix pairs are discussed, and connected with the dynamical behavior of the associated 2D systems.

## 1   Introduction

The theory of pairs and, more generally, $k$-tuples of linear transformations on a finite dimensional vector space dates from about the end of the last century. Nevertheless, in spite of its relatively long history, important general results, such as the invariant theory of $n \times n$ matrices (see [16] and the references therein) and a wealth of more specialized topics, recently appeared in mathematical journals, showing the permanent fertility of this field.

1D system theory has often taken advantage of these results (e.g. in Fliess' representation of bilinear systems) and, conversely, problems arising in modelling and control have provided a permanent source of new mathematical questions involving two or more linear transformations. Even more interestingly, system theoretic methods opened new vistas on purely algebraic problems on matrix sets, as in the case of matrix pairs with common eigenvectors [18].

In 2D system theory, virtually all the problems of modelling, realization and control involve a pair of linear transformations, whose spectral and combinatorial properties characterize the pattern of dynamical trajectories. Furthermore, 2D systems endowed with special structures, such as finite memory, separability, positivity etc., bring into focus special classes of matrix pairs, that deserve on their own a careful investigation. As several definitions and problems that one might hope to extend from 1D dynamical models do, in fact, carry over to 2D models, classical system theoretic concepts often suggest the guideline for tackling with pairs of linear transformations. It should be stressed, however, that the mathematical tools needed in the analysis of matrix pairs are substantially more difficult, and the mechanics of the proofs more involved, than most others occurring in Linear Algebra.

This paper deals with the circle of ideas that includes the theory of matrix pairs, two-dimensional signals and 2D state space models. In the next section the main properties of the characteristic polynomial and trace series of an unconstrained matrix pair are summarized. In section 3 some special classes of pairs, as well as their connections with the dynamical behavior of 2D state models, are investigated. Section 4 is devoted to nonnegative 2D systems and pairs. Detailed proofs of the results presented in sections 2 and 3 can be found in [6, 7], while for positive pairs we refer the interested reader to [8, 9]. The concluding section presents some open problems and further issues on matrix pairs arising from current researches.

## 2 Characteristic polynomials and trace series

2D systems theory connotes a large collection of problems and methods held together by a central theme: the analysis and control of processes and devices whose dynamics depends on two independent variables. In this note, however, we restrict our attention to the very special class of quarter plane causal unforced 2D state models, whose description essentially depends on a pair of square matrices associated with the shift operators along the coordinate axes [4]:

$$\mathbf{x}(h+1, k+1) = A_1\mathbf{x}(h, k+1) + A_2\mathbf{x}(h+1, k). \tag{1}$$

The *local states* $\mathbf{x}(h, k)$ are defined on (a suitable region of) the discrete plane $\mathbb{Z} \times \mathbb{Z}$ and take values in $\mathbb{R}^n$, $A_1$ and $A_2$ are real $n \times n$ matrices and the initial conditions are usually assigned by specifying all local state values $\mathbf{x}(-\ell, \ell)$ on the *separation set* $C_0 := \{(-\ell, \ell) : \ell \in \mathbb{Z}\}$. Upon resorting to formal power series, initial conditions are represented by the *global state* $\mathcal{X}_0 := \sum_{\ell=-\infty}^{+\infty} \mathbf{x}(-\ell, \ell)z_1^{-\ell}z_2^{\ell}$ and the induced sequence of local states in the half-plane $\{(h, k) : h + k \geq 0\}$ is described by

$$X(z_1, z_2) = \sum_{h,k}\mathbf{x}(h, k)z_1^h z_2^k = (I - A_1z_1 - A_2z_2)^{-1}\mathcal{X}_0 = \left[\sum_{i,j=0}^{\infty} A_1{}^i\!\sqcup\!{}^j A_2\ z_1^i z_2^j\right]\mathcal{X}_0.$$

where the *Hurwitz products* $A_1{}^i\!\sqcup\!{}^j A_2$, $i, j \in \mathbb{N}$, are the matrix coefficients of the power series expansion of $(I - A_1z_1 - A_2z_2)^{-1}$. Each Hurwitz product decomposes as $A_1{}^i\!\sqcup\!{}^j A_2 = \sum_{\nu_1,\nu_2,\dots,\nu_{i+j}} A_{\nu_1} A_{\nu_2} \dots A_{\nu_{i+j}}$, the summation being extended to all products that include the factors $A_1$ and $A_2$, $i$ and $j$ times respectively. In particular, when assuming $\mathbf{x}(-\ell, \ell) = 0$ for every $\ell \neq 0$, the state in $(h, k)$ is given by

$$\mathbf{x}(h, k) = \sum_{\nu_1,\nu_2,\dots,\nu_{i+j}} A_{\nu_1} A_{\nu_2} \dots A_{\nu_{i+j}}\mathbf{x}(0, 0), \tag{2}$$

and it can be interpreted as the sum of the elementary contributions along all paths connecting $(0,0)$ to $(h,k)$ in the two-dimensional grid.

Most analytic and combinatorial properties of the pair $(A_1, A_2)$ obviously reflect into the behavior of (1), and it is often useful to translate an abstract question on the pair into a dynamical problem for the corresponding state model. This is well illustrated by the *characteristic polynomial* and the *trace series* of $(A_1, A_2)$, defined as

$$\Delta_{A_1,A_2}(z_1, z_2) := \det(I - A_1 z_1 - A_2 z_2)$$

and

$$T_{A_1,A_2}(z_1, z_2) := \sum_{h=1}^{\infty} \Big( \sum_{i+j=h} \operatorname{tr}(A_1{}^i \sqcup{}^j A_2) z_1^i z_2^j \Big),$$

respectively. Like the characteristic polynomial of a single matrix, which in general does not capture its Jordan structure, $\Delta_{A_1,A_2}$ does not identify the similarity orbit of the pair and hence does not provide a complete information on the underlying linear transformations. Nevertheless, important aspects of the 2D motion completely rely on it. There is, first of all, the internal stability of system (1), which depends only on the variety of the zeros of $\Delta_{A_1,A_2}$. Additional insights into the structure of 2D systems come from the factorization of the characteristic polynomial. Actually, properties like finite memory and separability, and interesting features of the spectrum of $\alpha A_1 + \beta A_2$, such as property L, can be restated as conditions on the factors of $\Delta_{A_1,A_2}$.

We shall not dwell here with stability issues, and concentrate instead on the remaining topics.

**Proposition 2.1** *Let* $\Delta_{A_1,A_2}(z_1, z_2) = 1 - \sum_{h=1}^{n} \delta_h(z_1, z_2)$ *and* $T_{A_1,A_2}(z_1, z_2) = \sum_{h=1}^{\infty} \tau_h(z_1, z_2)$, *with* $\delta_h(z_1, z_2)$ *and* $\tau_h(z_1, z_2)$ *homogeneous forms of degree* $h$, *be the characteristic polynomial and trace series, respectively, of the* $n \times n$ *matrix pair* $(A_1, A_2)$. *Then*

*i)  the homogeneous components* $\delta_h(z_1, z_2)$ *and* $\tau_h(z_1, z_2)$ *satisfy*

$$\begin{aligned}
&\tau_1(z_1, z_2) - \delta_1(z_1, z_2) = 0 \\
&\tau_2(z_1, z_2) - \delta_1(z_1, z_2)\tau_1(z_1, z_2) - 2\delta_2(z_1, z_2) = 0 \\
&\qquad \cdots \\
&\tau_n(z_1, z_2) - \delta_1(z_1, z_2)\tau_{n-1}(z_1, z_2) - \ldots - n\delta_n(z_1, z_2) = 0
\end{aligned} \tag{3}$$

*and, for all* $k > 0$,

$$\tau_{n+k}(z_1, z_2) - \sum_{i=1}^{n} \tau_{n+k-i}(z_1, z_2)\delta_i(z_1, z_2) = 0; \tag{4}$$

*ii)  the traces of* $A_1{}^i \sqcup{}^j A_2$ *and the coefficients* $d_{ij}$ *of* $\Delta_{A_1,A_2}(z_1, z_2)$ *satisfy*

$$\operatorname{tr}(A_1{}^i \sqcup{}^j A_2) = \sum_{0 < r+s < i+j} d_{rs}\operatorname{tr}(A_1{}^{i-r} \sqcup{}^{j-s} A_2) + (i+j)d_{ij}, \tag{5}$$

*where $d_{rs} = 0$ for $r + s > n$, and $A_1{}^r ш^s A_2$ is the zero matrix whenever $r$ or $s$ is negative.*

Equation (5) has some simple, but useful consequences. First of all, it provides an algorithm for recursively computing the traces of $A_1{}^i ш^j A_2$ from the coefficients of the characteristic polynomial. On the other hand, once the traces are given, also the converse, i.e. the computation of the coefficients of $\Delta_{A_1,A_2}(z_1, z_2)$, is made possible. Actually, if an upper bound $\bar{n}$ on the degree of $\Delta_{A_1,A_2}(z_1, z_2)$ is known, then assigning $\text{tr}(A_1{}^i ш^j A_2)$ for $i + j \leq \bar{n}$ allows to recover both $\Delta_{A_1,A_2}(z_1, z_2)$ and the traces of $A_1{}^i ш^j A_2$ for $i + j > \bar{n}$. Finally, it is easy to realize that $T_{A_1,A_2}$ is a rational power series, since its coefficients satisfy the recursive relation (5). Further connections of $T_{A_1,A_2}$ with the structure of the characteristic polynomial are enlightened by the following proposition.

**Proposition 2.2** *Let $\Delta_{A_1,A_2}(z_1, z_2) = 1 - \sum_{h=1}^{n} \delta_h(z_1, z_2)$ be the characteristic polynomial of the matrix pair $(A_1, A_2)$. The corresponding trace series $T_{A_1,A_2}$ can be expressed as*

$$T_{A_1,A_2}(z_1, z_2) = \frac{\delta_1(z_1, z_2) + 2\delta_2(z_1, z_2) + \ldots + n\delta_n(z_1, z_2)}{\Delta(z_1, z_2)}. \tag{6}$$

*Moreover, if $\Delta_{A_1,A_2}$ factorizes as $\Delta(z_1, z_2) = \prod_{i=1}^{t} \Delta_i(z_1, z_2)^{\nu_i}$, with $\Delta_i(z_1, z_2) = 1 - \sum_{j=1}^{r_i} \delta_j^{(i)}(z_1, z_2)$ irreducible distinct factors, $\nu_i \in \mathbb{N}$ and $\delta_j^{(i)}(z_1, z_2), i = 1, 2, \ldots, t$, homogeneous polynomials of degree $j$, then*

$$T_{A_1,A_2}(z_1, z_2) = \sum_{i=1}^{t} \nu_i \frac{\sum_{j=1}^{r_i} j\, \delta_j^{(i)}(z_1, z_2)}{\Delta_i(z_1, z_2)}. \tag{7}$$

In (7) the trace series $T_{A_1,A_2}(z_1, z_2)$ is expressed as a partial fraction expansion, whose $i$-th term is the trace series of the irreducible factor $\Delta_i(z_1, z_2)$, weighted with the corresponding multiplicity $\nu_i$. Thus the denominator of every irreducible rational function which represents a trace series factorizes into irreducible factors with multiplicity one. On the other hand, when an irreducible rational function $T(z_1, z_2)$ has been given, (7) suggests a way to check whether $T(z_1, z_2)$ can be expanded into a trace series.

Finally, consider the set of all matrix pairs $\mathcal{M} = \{(A_1, A_2) : A_1, A_2 \in \mathbb{C}^{n \times n}, n \in \mathbb{N}\}$, and introduce in $\mathcal{M}$ the equivalence relation

$$(A_1, A_2) \sim (\hat{A}_1, \hat{A}_2) \quad \Leftrightarrow \quad \Delta_{A_1 A_2}(z_1, z_2) = \Delta_{\hat{A}_1 \hat{A}_2}(z_1, z_2).$$

Basing on the previous results, useful sets of complete invariants for the equivalence relation $\sim$ are easily derived, involving Hurwitz products and linear combinations of the matrix pairs, as illustrated in the following proposition.

**Proposition 2.3** *Let $(A_1, A_2)$ and $(\hat{A}_1, \hat{A}_2)$ be two (complex-valued) matrix pairs, possibly of different size. The following statements are equivalent:*

*i)* $\Delta_{A_1,A_2}(z_1, z_2) = \Delta_{\hat{A}_1,\hat{A}_2}(z_1, z_2)$;

*ii) for all $\alpha, \beta \in \mathbb{C}$, $\Lambda_0(\alpha A_1 + \beta A_2) = \Lambda_0(\alpha \hat{A}_1 + \beta \hat{A}_2)$, where $\Lambda_0(A)$ denotes the set of nonzero eigenvalues of the matrix $A$, each of them counted according with the corresponding algebraic multiplicity;*

*iii) for all $\alpha, \beta \in \mathbb{C}$ and $k \in \mathbb{N}_+$, $\mathrm{tr}(\alpha A_1 + \beta A_2)^k = \mathrm{tr}(\alpha \hat{A}_1 + \beta \hat{A}_2)^k$;*

*iv) for all $(i, j) \neq (0, 0)$, $\mathrm{tr}(A_1{}^i \sqcup^j A_2) = \mathrm{tr}(\hat{A}_1{}^i \sqcup^j \hat{A}_2)$.*

*v)* $T_{A_1,A_2}(z_1, z_2) = T_{\hat{A}_1,\hat{A}_2}(z_1, z_2)$;

# 3  Pairs of matrices with special structure

In this section we aim to specifically focus on matrix pairs endowed with property L or property P. Pairs with property L occur quite frequently in the applications: indeed, the important classes of finite memory and separable 2D systems are described by pairs with this property. A pair of $n \times n$ matrices with entries in $\mathbb{C}$, $(A_1, A_2)$, has *property L* if the eigenvalues of $A_1$ and $A_2$ can be ordered into two $n$-tuples

$$\Lambda(A_1) = (\lambda_1, \lambda_2, \dots, \lambda_n) \quad \text{and} \quad \Lambda(A_2) = (\mu_1, \mu_2, \dots, \mu_n) \tag{8}$$

such that, for all $\alpha, \beta$ in $\mathbb{C}$, the spectrum of $\Lambda(\alpha A_1 + \beta A_2)$ is given by

$$\Lambda(\alpha A_1 + \beta A_2) = (\alpha \lambda_1 + \beta \mu_1, \dots, \alpha \lambda_n + \beta \mu_n). \tag{9}$$

Clearly matrices $\hat{A}_1 = \mathrm{diag}\{\lambda_1, \lambda_2, \dots, \lambda_n\}$ and $\hat{A}_2 = \mathrm{diag}\{\mu_1, \mu_2, \dots, \mu_n\}$ constitute a pair with property L. Any other pair $(A_1, A_2)$ of the same dimension, with property L w.r.t. the orderings (8), satisfies, by Proposition 2.3, $\Lambda(\alpha A_1 + \beta A_2) = \Lambda(\alpha \hat{A}_1 + \beta \hat{A}_2)$. A straightforward consequence is the chain of equivalences listed in the following proposition.

**Proposition 3.1** *Let $A_1$ and $A_2$ be in $\mathbb{C}^{n \times n}$. The following statements are equivalent:*

*L)* $(A_1, A_2)$ *has property L w.r.t. the orderings (8) of their spectra;*

*$L_1$)* $\Delta_{A_1,A_2}(z_1, z_2) = \prod_{i=1}^{n}(1 - \lambda_i z_1 - \mu_i z_2)$;

*$L_2$) for all $\alpha, \beta \in \mathbb{C}$ and $k \in \mathbb{N}$, $\mathrm{tr}(\alpha A_1 + \beta A_2)^k = \sum_{i=1}^{n}(\alpha \lambda_i + \beta \mu_i)^k$;*

*$L_3$) for every $(h, k) \in \mathbb{N} \times \mathbb{N}$, $\mathrm{tr}(A_1{}^h \sqcup^k A_2) = \binom{h+k}{h} \sum_{i=1}^{n} \lambda_i^h \mu_i^k$;*

*$L_4$)* $T_{A_1,A_2}(z_1, z_2) = \sum_{h+k>0} \mathrm{tr}(A_1{}^h \sqcup^k A_2) z_1^h z_2^k = \sum_{i=1}^{n} \dfrac{\lambda_i z_1 + \mu_i z_2}{1 - \lambda_i z_1 - \mu_i z_2}$.

Hankel matrix theory provides a direct method to check, in a finite number of steps, whether a given pair $(A_1, A_2)$ is endowed with property L. To reach this goal, we introduce [3] the infinite Hankel matrix $\mathcal{H}(s)$ of a formal power series $s$ in the commuting variables $z_1$ and $z_2$. If $s := \sum_{h,k} \langle s, z_1^h z_2^k \rangle z_1^h z_2^k$, with $\langle s, z_1^h z_2^k \rangle$

denoting the coefficients of the monomial $z_1^h z_2^k$ in $s$, we have

$$
\mathcal{H}(s) := \begin{bmatrix}
\langle s,1 \rangle & \langle s,z_1 \rangle & \langle s,z_2 \rangle & \langle s,z_1^2 \rangle & \langle s,z_1 z_2 \rangle & \langle s,z_2^2 \rangle & \cdots \\
\langle s,z_1 \rangle & \langle s,z_1^2 \rangle & \langle s,z_1 z_2 \rangle & \langle s,z_1^3 \rangle & \langle s,z_1^2 z_2 \rangle & \cdots & \cdots \\
\langle s,z_2 \rangle & \langle s,z_1 z_2 \rangle & \langle s,z_2^2 \rangle & \cdots & \cdots & \cdots & \cdots \\
\langle s,z_1^2 \rangle & \cdots & \cdots & \cdots & \cdots & \cdots & \cdots \\
\cdots & \cdots & \cdots & \cdots & \cdots & \cdots & \cdots
\end{bmatrix}.
$$

Rows and columns of $\mathcal{H}(s)$ take indices in the multiplicative monoid of the terms $\mathcal{T} := \{z_1^h z_2^k : h, k \in \mathbf{N}\}$. For all $M', M'' \in \mathbf{N}$, we denote by $\mathcal{H}_{M' \times M''}(s)$ the submatrix, appearing in the upper left corner of $\mathcal{H}(s)$, whose rows (columns) are indexed by the terms of homogeneus degree not greater than $M'$ ($M''$).
We are now in a position to state the following criterion.

**Proposition 3.2** *Let $(A_1, A_2)$ be a pair of matrices in $\mathbb{C}^{n \times n}$ and consider the power series*

$$
R_{A_1,A_2}(z_1, z_2) := \sum_{i,j=0}^{\infty} \operatorname{tr}(A_1{}^i \sqcup^j A_2) \binom{i+j}{i}^{-1} z_1^i z_2^j.
$$

*$(A_1, A_2)$ has property L if and only if*

$$
\operatorname{rank} \mathcal{H}_{(n-1) \times (n-1)}(R_{A_1,A_2}) = \operatorname{rank} \mathcal{H}_{n \times n}(R_{A_1,A_2}) \leq n.
$$

The results of Proposition 3.1 provide a convenient framework for understanding the internal dynamics of finite memory and separable 2D state models. A 2D system (1) that reaches the zero (global) state within a finite number of steps, independently of its initial conditions, is called *finite memory* [5, 6]. It is almost immediate to show that the finite memory (FM) property corresponds to assume that $(I - A_1 z_1 - A_2 z_2)^{-1}$ is a polynomial matrix or, equivalently, that the characteristic polynomial of the pair $(A_1, A_2)$ is unitary. The following proposition summarizes a useful set of different characterizations of the FM property.

**Proposition 3.3** *Let $A_1$ and $A_2$ be matrices in $\mathbb{C}^{n \times n}$. The following facts are equivalent*

$FM_1)$ $\Delta_{A_1,A_2}(z_1, z_2) = 1;$

$FM_2)$ $\Lambda(\alpha A_1 + \beta A_2) = (0,0,...,0), \ \forall \alpha, \beta \in \mathbb{C};$

$FM_3)$ $\Lambda(\nu A_1 + A_2) = \Lambda(A_1 + \nu A_2) = (0,0,...,0), \ \text{for every } \nu = 1,...,n+1;$

$FM_4)$ $\operatorname{tr}(A_1{}^i \sqcup^j A_2) = 0, \ \forall \ (i,j) \neq (0,0);$

$FM_5)$ $A_1{}^i \sqcup^j A_2 = 0, \ \text{for } i+j \geq n.$

2D systems whose characteristic polynomials factorize into the product of a polynomial in $z_1$ and a polynomial in $z_2$ are called *separable* [5, 6], and are usually thought of as the simplest examples of 2D systems with infinite impulse response. Actually, many properties one may hope to extrapolate from an understanding of 1D systems carry over to separable systems, and just the knowledge that the system is separable allows one to make fairly strong statements about its behaviour. In particular, internal stability can be quickly deduced from the general theory of discrete time 1D systems, as the long term performance of separable systems is determined by the eigenvalues of $A_1$ and $A_2$.

**Proposition 3.4** *Let $A_1$ and $A_2$ be matrices in $\mathbb{C}^{n \times n}$. The following statements are equivalent*

$S_1)$ $\Delta_{A_1,A_2}(z_1, z_2) = r(z_1)s(z_2)$;

$S_2)$ *$A_1$ and $A_2$ satisfy property L w.r.t. the orderings of the spectra $\Lambda(A_1) = (\lambda_1, ..., \lambda_\rho, 0, ..., 0, 0, ..., 0)$, $\Lambda(A_2) = (0, ..., 0, \mu_1, ..., \mu_\sigma, 0, ..., 0)$, so that $\Lambda(\alpha A_1 + \beta A_2) = (\alpha\lambda_1, ..., \alpha\lambda_\rho, \beta\mu_1, ..., \beta\mu_\sigma, 0, ..., 0)$ for every $\alpha, \beta \in \mathbb{C}$;*

$S_3)$ *$\mathrm{tr}(A_1{}^i \omega^j A_2) = 0$ whenever $i$ and $j$ are both nonzero;*

$S_4)$ *$\mathrm{tr}(\alpha A_1 + \beta A_2)^k = \mathrm{tr}(\alpha A_1)^k + \mathrm{tr}(\beta A_2)^k$, $\forall \alpha, \beta \in \mathbb{C}, k \in \mathbb{N}_+$.*

Before turning to the discussion of property P, we have to introduce some background notations on polynomials and series in noncommutative variables. Given an alphabet $\Xi = \{\xi_1, \xi_2\}$, the free monoid $\Xi^*$ with base $\Xi$ is the set of all words $w = \xi_{i_1}\xi_{i_2} \cdots \xi_{i_m}$, $m \in \mathbb{N}$, $\xi_{i_h} \in \Xi$. The integer $m$ is called the length of $w$ and denoted by $|w|$, while $|w|_i$ represents the number of occurencies of $\xi_i$ in $w$, $i = 1, 2$. If $v = \xi_{j_1}\xi_{j_2} \cdots \xi_{j_p}$ is another element of $\Xi^*$, the product is defined by concatenation $wv = \xi_{i_1}\xi_{i_2} \cdots \xi_{i_m}\xi_{j_1}\xi_{j_2} \cdots \xi_{j_p}$. This produces a monoid with $1 = \emptyset$, the empty word, as unit element. Clearly, $|wv| = |v| + |w|$ and $|1| = 0$.

$\mathbb{C}\langle \xi_1, \xi_2 \rangle$ and $\mathbb{C}\langle\langle \xi_1, \xi_2 \rangle\rangle$ are the algebras of polynomials and formal power series respectively in the noncommuting indeterminates $\xi_1$ and $\xi_2$. For each pair of matrices $(A_1, A_2)$ in $\mathbb{C}^{n \times n}$, the map $\psi$ defined on $\{1, \xi_1, \xi_2\}$ by the assignments $\psi(1) = I_n$ and $\psi(\xi_i) = A_i$, $i = 1, 2$, uniquely extends to an algebra morphism of $\mathbb{C}\langle \xi_1, \xi_2 \rangle$ into $\mathbb{C}^{n \times n}$. The $\psi$-image of a polynomial $\mathcal{P}(\xi_1, \xi_2) \in \mathbb{C}\langle \xi_1, \xi_2 \rangle$ is denoted by $\mathcal{P}(A_1, A_2)$.

A pair of $n \times n$ matrices $(A_1, A_2)$ with elements in $\mathbb{C}$ is said to have *property P* if the eigenvalues of $A_1$ and $A_2$ can be ordered into two $n$-tuples

$$\Lambda(A_1) = (\lambda_1, \lambda_2, \ldots \lambda_n), \qquad \Lambda(A_2) = (\mu_1, \mu_2, \ldots \mu_n), \tag{10}$$

such that, for every polynomial $\mathcal{P}(\xi_1, \xi_2) \in \mathbb{C}\langle \xi_1, \xi_2 \rangle$,

$$\Lambda(\mathcal{P}(A_1, A_2)) = (\mathcal{P}(\lambda_1, \mu_1), \mathcal{P}(\lambda_2, \mu_2), \ldots, \mathcal{P}(\lambda_n, \mu_n)). \tag{11}$$

According to a celebrated result of Mc Coy [12], property P is equivalent to simultaneous triangularizability, a feature that allows for good insights into the geometry of system (1), which can be viewed as a cascade of 2D systems having dimension one. Moreover, when considering state models with inputs and outputs, triangular matrix pairs provide a class of 2D systems large enough for realizing all transfer functions $p(z_1, z_2)/q(z_1, z_2)$ with denominators of the form $q(z_1, z_2) = \prod_j (1 - \lambda_j z_1 - \mu_j z_2)$ and, in particular, all transfer functions with separable denominators [2]. It should be stressed that the same is not true, however, if we consider only commutative matrix pairs.

As property P trivially implies property L, while examples can be given [14, 15] disproving the converse, the set of all pairs with property P is properly included in the set of pairs with property L. On the other hand, we have seen in Proposition 3.1 that a pair is endowed with property L if and only if its characteristic polynomial factorizes into a product of linear terms, and it is obvious that any such polynomial corresponds also to some matrix pair with property P. Consequently, no possibility is left of describing property P basing only on the characteristic polynomial. The appropriate tools are, instead, certain noncommutative polynomials and power series associated with the pair, as well as the corresponding Hankel matrices.

**Proposition 3.5** *Let $A_1$ and $A_2$ be $n \times n$ matrices with entries in $\mathbb{C}$, and consider the orderings of their spectra given in (10). The following statements are equivalent:*

$P_1$) *$(A_1, A_2)$ has property P w.r.t. the orderings (10);*

$P_2$) *for any $w \in \Xi^*$, with $|w|_1 = h$ and $|w|_2 = k$, $\operatorname{tr}(w(A_1, A_2)) = \sum_{i=1}^{n} \lambda_i^h \mu_i^k$;*

$P_3$) *the noncommutative power series $\mathcal{N} = \sum_{w \in \Xi^*} \operatorname{tr}(w(A_1, A_2))w$ can be represented as $\mathcal{N} = \sum_{i=1}^{n}(1 - \lambda_i \xi_1 - \mu_i \xi_2)^{-1}$, and hence is recognizable [3];*

$P_4$) *for any $w \in \Xi^*$, with $|w|_1 = h$ and $|w|_2 = k$, $\det(zI - w(A_1, A_2)) = \prod_{i=1}^{n}(z - \lambda_i^h \mu_i^k)$.*

An effective method for testing property P depends on the Hankel matrix [3, 17] $\mathcal{H}(\mathcal{N})$ of the noncommutative power series $\mathcal{N}$, i.e. the infinite matrix whose rows and columns are indexed by the words of $\Xi^*$, and whose element with indexes $u$ and $v$ is equal to $\langle \mathcal{N}, uv \rangle$. The words in $\Xi^*$, and consequently the row and column indexes in $\mathcal{H}(\mathcal{N})$, are ordered according to their length, while the lexicographical order is adopted for words of the same length. For all $M', M'' \in \mathbb{N}$, we shall denote by $\mathcal{H}_{M' \times M''}(\mathcal{N})$ the submatrix appearing in the upper left corner of $\mathcal{H}(\mathcal{N})$, whose rows (columns) are indexed by words of length not greater than $M'$ ($M''$).

**Proposition 3.6** *Let $A_1$ and $A_2$ be $n \times n$ matrices with entries in $\mathbb{C}$, and consider the associated noncommutative power series $\mathcal{N} = \sum_{w \in \Xi^*} \operatorname{tr}(w(A_1, A_2))w$. The following statements are equivalent:*

*i) $(A_1, A_2)$ has property P;*

*ii) rank $\mathcal{H}_{(n^2-1) \times (n^2-1)}(\mathcal{N}) = \bar{n} \leq n$ and, for all pairs of words $w, \bar{w}$ with length not greater than $2\bar{n}$,*

$$|w|_i = |\bar{w}|_i, \quad i = 1, 2 \quad \Rightarrow \quad \operatorname{tr}(w(A_1, A_2)) = \operatorname{tr}(\bar{w}(A_1, A_2)); \tag{12}$$

*iii) (12) holds for all pairs of words $w, \bar{w}$ with length not greater than $2n^2$.*

## 4  Nonnegative matrix pairs

The interest in nonnegative matrix pairs is largely motivated, apart from theoretical reasons, by their possible applications in building discrete models of dynamical systems that involve only nonnegative variables, such as pressures, concentrations, levels of population etc.[10, 11, 19].

The results so far obtained have extended to positive pairs the available characterizations of finite memory, separability and property L. Furthermore, some basic notions of positive matrix theory, like irreducibility, primitivity and the Perron-Frobenius theorem, particularly meaningful for their relevance in (1D) positive system dynamics, have been generalized to the 2D case.

In this section the following notation will be adopted: given a matrix $A = [a_{ij}]$, we will write $A \gg 0$ ($A$ *strictly positive*), if $a_{ij} > 0$ for all $i, j$; $A > 0$ ($A$ *positive*), if $a_{ij} \geq 0$ for all $i, j$, and $a_{hk} > 0$ for at least one pair $(h, k)$; $A \geq 0$ ($A$ *nonnegative*), if $a_{ij} \geq 0$ for all $i, j$.

## 4.1   Nonnegative pairs with special properties

Introducing the nonnegativity assumption allows to obtain more penetrating characterizations of finite memory and separable matrix pairs. In particular, finite memory and separability turn out to be "structural properties" of a pair, in the sense that they depend only on the zero patterns of the matrices and not on the specific values their nonzero entries take.

This is a quite interesting feature. Indeed, in many cases the information available on the physical process one aims to model allows to assume that no interaction exists among certain variables, and, consequently, that some entries of the matrices $A_1$ and $A_2$ are exactly 0, whereas the others can be assumed nonnegative, and known with some level of uncertainty. This is always the case of compartmental models [10], where nonzero entries correspond to the existence of flows between different compartments, and physical or biological reasons guarantee that some pairs of compartments have no direct interaction at all.

**Proposition 4.1**   *For a pair of $n \times n$ nonnegative matrices $(A_1, A_2)$, the following statements are equivalent*

*i) $\Delta_{A_1, A_2}(z_1, z_2) = 1$;*

*ii) $A_1 + A_2$ is a nilpotent (and, a fortiori, a reducible) matrix;*

*iii) $A_1{}^i \sqcup{}^j A_2$ is nilpotent, for all $(i, j) \neq (0, 0)$;*

*iv) $w(A_1, A_2)$ is nilpotent, for all $w \in \Xi^* \setminus \{1\}$;*

*v) there exists a permutation matrix $P$ such that $P^T(A_1 + A_2)P$ is upper triangular with zero diagonal entries.*

Notice that in the general case, when the matrix entries assume both positive and negative values, condition $ii)$ is necessary, but not sufficient, for guaranteeing the finite memory property, which depends on the nilpotency of all linear combinations $\alpha A + \beta A_2$, $\alpha, \beta \in \mathbb{C}$. On the contrary, examples can be given [8] showing that conditions $iii)$ and $iv)$ are sufficient, but not necessary, for the finite memory property. Moreover, while for a general finite memory pair $(A_1, A_2)$ we can only guarantee that the Hurwitz products $A^i \sqcup{}^j A_2$ are zero when $i + j \geq n$, in the nonnegative case this property extends to all matrix products $w(A_1, A_2), w \in \Xi^*$ and $|w| \geq n$.

In analyzing nonnegative separable pairs we end up with some results that strictly parallel those obtained in the finite memory case. A fairly complete spectral characterization of separability is summarized in the following proposition.

**Proposition 4.2** *For a nonnegative matrix pair* $(A_1, A_2)$, *the following statements are equivalent:*

*i)* $\Delta_{A_1, A_2}(z_1, z_2) = r(z_1)s(z_2)$;

*ii)* $\det[I - (A_1 + A_2)z] = \det[I - A_1 z]\det[I - A_2 z]$;

*iii)* $A_1{}^i \sqcup{}^j A_2$ *is nilpotent for all* $(i, j)$ *with* $i, j > 0$;

*iv)* $w(A_1, A_2)$ *is nilpotent, for all* $w \in \Xi^*$ *such that* $|w|_i > 0$, $i = 1, 2$;

*v) there exists a permutation matrix* $P$ *such that* $P^T A_1 P$ *and* $P^T A_2 P$ *are conformably partitioned into block triangular matrices*

$$P^T A_1 P = \begin{bmatrix} [A_1]_{11} & * & * & * \\ & [A_1]_{22} & * & * \\ & & \ddots & * \\ & & & [A_1]_{tt} \end{bmatrix} \quad P^T A_2 P = \begin{bmatrix} [A_2]_{11} & * & * & * \\ & [A_2]_{22} & * & * \\ & & \ddots & * \\ & & & [A_2]_{tt} \end{bmatrix}, \tag{13}$$

*with* $[A_1]_{ii} \neq 0$ *implying* $[A_2]_{ii} = 0$;

*vi) there exists a nonsingular matrix* $T \in \mathbb{C}^{n \times n}$ *such that* $\hat{A}_1 = T^{-1}A_1 T$ *and* $\hat{A}_2 = T^{-1}A_2 T$ *are upper triangular matrices, and the Hadamar product* $A_1 * A_2$ *has zero diagonal entries.*

The characterizations given in points v) of Propositions 4.1 and 4.2 have a combinatorial nature and make it clear that finite memory and separability are preserved under all perturbations of the positive entries. Property L and property P, instead, are not structural ones, as examples can be given [8] showing that they depend on the specific values assumed by the nonzero entries.

## 4.2   Irreducibility and primitivity

In positive matrix theory, irreducibility property of a matrix $A$ can be introduced in combinatorial terms: indeed, irreducible matrices are those that cannot be reduced, by means of cogredience transformations, to block-triangular form [13]. Several equivalent descriptions are also available: algebraic ones, which connect irreducibility to the zero-patterns of the powers of $A$, and graph-theoretic and system theoretic characterizations, which relate it to the structure of the corresponding directed graph $\mathcal{D}(A)$ and to the behavior of the associated state model. More precisely, a positive matrix $A \in \mathbb{R}^{n \times n}$ is irreducible if and only if positive integers $h$ and $T$ can be found, such that

$$\sum_{i=t+1}^{t+h} A^i \gg 0, \qquad \text{for every } t \geq T, \tag{14}$$

or equivalently, if and only if for every pair of vertices $i$ and $j$ in $\mathcal{D}(A)$ there exists a path connecting $i$ to $j$ or, finally, if and only if for every positive initial condition $\mathbf{x}(0) > 0$ the dynamical model $\mathbf{x}(t+1) = A\mathbf{x}(t), t = 0, 1, \ldots$, produces state vectors satisfying

$$\sum_{i=t+1}^{t+h} \mathbf{x}(i) \gg \mathbf{0}, \tag{15}$$

for sufficiently large values of $t$.

Searching for a natural extension of the irreducibility definition to positive matrix pairs, we can try to generalize anyone of the above characterizations. For instance, we can look for a two-dimensional extension of (15), and refer to the state evolution of the 2D system (1), corresponding to an arbitrary set $\mathcal{X}_0$ of nonnegative local states. Working on the discrete grid $\mathbb{Z} \times \mathbb{Z}$, it seems reasonable to replace the interval $[t+1, t+h]$ appearing in (15) with some finite set $\mathcal{F} \subset \mathbb{Z} \times \mathbb{Z}$ and to define irreducible any nonnegative pair $(A_1, A_2)$ for which a finite set $\mathcal{F} \subset \mathbb{Z} \times \mathbb{Z}$ can be found such that for every nonnegative $\mathcal{X}_0$ the following condition

$$\sum_{[i \ \ j] \in [h \ \ k] + \mathcal{F}} \mathbf{x}(i, j) \gg 0, \qquad \forall\, [h \ \ k] \in \mathbb{Z} \times \mathbb{Z} \text{ s.t. } h + k \geq T, \tag{16}$$

holds true for some suitable positive integer $T$. If $\mathcal{X}_0$ consists of a finite number of nonzero local states, however, or no upper bound exists on the distance between consecutive nonzero local states on $\mathcal{C}_0$, condition (16) can be satisfied only for $T \to +\infty$. So, in order to make our definition more consistent, we confine ourselves to *admissible* sets of initial conditions, namely to nonnegative sequences $\mathcal{X}_0$ such that $\sum_{\ell=h}^{h+N} \mathbf{x}(\ell, -\ell) > 0$ holds true for some $N > 0$ and every $h \in \mathbb{Z}$. Irreducibility can now be characterized as follows.

**Definition** A pair $(A_1, A_2)$ of $n \times n$ positive matrices is *irreducible* if there is a finite set $\mathcal{F} \subset \mathbb{Z} \times \mathbb{Z}$ such that for every admissible set of initial conditions $\mathcal{X}_0$ a positive integer $T$ can be found such that

$$\sum_{[i \ \ j] \in [h \ \ k] + \mathcal{F}} \mathbf{x}(i, j) \gg 0, \qquad \forall\, [h \ \ k] \in \mathbb{Z} \times \mathbb{Z} \text{ s.t. } h + k \geq T. \tag{17}$$

An alternative description of irreducible matrix pairs can be obtained by replacing in (14) the power matrices with the Hurwitz products, and the half-line $[T, +\infty)$ with a suitable solid convex cone. In fact, intuitively speaking, a positive matrix pair $(A_1, A_2)$ should be thought of as irreducible if there are a finite "window" and a solid convex cone such that, independently of the window position within the cone, the sum of all Hurwitz products $A_1{}^r \sqcup{}^s A_2$ corresponding to integer pairs in the window, is strictly positive.

**Proposition 4.5** *A pair of $n \times n$ positive matrices $(A_1, A_2)$ is irreducible if and only if there are a solid convex cone $\mathcal{K}^* \subset \mathbb{R}_+^2$ and a finite set $\mathcal{F} \subset \mathbb{N}^2$ such* *that*

$$\sum_{[r\ \ s]\in[h\ \ k]+\mathcal{F}} A_1{}^r \sqcup^s A_2 \gg 0, \qquad \forall\, [h\ \ k] \in \mathbf{N}^2 \text{ s.t. } [h\ \ k] + \mathcal{F} \subset \mathcal{K}^*. \tag{18}$$

The minimal value of the cardinality of a set $\mathcal{F}$ for which anyone of the above equivalent conditions, (17) or (18), holds true is called the *imprimitivity index* of the pair $(A_1, A_2)$. This index can be related to the number of points in which the variety $\mathcal{V}(\Delta_{A_1,A_2})$ of the characteristic polynomial $\Delta_{A_1,A_2}(z_1, z_2)$ intersects the polydisc $\mathcal{P}_{r^{-1}} := \{(z_1, z_2) \in \mathbb{C}^2 : |z_1| \le r^{-1}, |z_2| \le r^{-1}\}$, with $r = \rho(A_1 + A_2)$ the spectral radius of $A_1 + A_2$. More precisely, it has be shown [9] that $\mathcal{V}(\Delta_{A_1,A_2})$ intersects $\mathcal{P}_{r^{-1}}$ only in $(r^{-1}, r^{-1})$, and in a finite number of points $(r^{-1}e^{i\theta_1}, r^{-1}e^{i\theta_2})$ of its distinguished boundary $\mathcal{T}_{r^{-1}} := \{(z_1, z_2) \in \mathbb{C}^2 : |z_1| = r^{-1}, |z_2| = r^{-1}\}$, where the pairs $(\theta_1, \theta_2)$ satisfy the congruence relations $h\theta_1 + k\theta_2 \equiv 0 \pmod{2\pi}$, for all $(h, k) \in \operatorname{supp}(\Delta_{A,B})$. This result can be viewed as the two-dimensional analogue of the Perron-Frobenius theorem.

For sake of completeness, we mention that irreducibility property of a matrix pair admits also a graph-theoretic interpretation, based on the notion of 2D-directed graph, i.e. a directed graph with two classes of arcs, corresponding to the nonzero entries of $A_1$ and $A_2$, respectively. For more details, the interested reader is referred to [9].

A *primitive* pair of positive matrices is an irreducible pair with unitary imprimitivity index. The following proposition illustrates an interesting set of equivalent primitivity conditions that motivate the above definition.

**Proposition 4.7** *Let $(A_1, A_2)$ be an irreducible pair of $n \times n$ positive matrices, with $\rho(A_1 + A_2) = r$. The following facts are equivalent:*

*i) $(A_1, A_2)$ is primitive;*

*ii) there exists a strictly positive Hurwitz product;*

*iii) there is a solid convex cone $\mathcal{K}^*$ in $\mathbb{R}_+^2$ such that for all $(h, k) \in \mathbf{N}^2 \cap \mathcal{K}^*$ the Hurwitz product $A_1{}^h \sqcup^k A_2$ is strictly positive;*

*iv) for every admissible set of initial conditions there is a positive integer $T$ such that $\mathbf{x}(h, k) \gg 0$ for all $(h, k) \in \mathbf{N}^2$, $h + k \ge T$;*

*v) the variety $\mathcal{V}(\Delta_{A_1,A_2})$ intersects the polydisk $\mathcal{P}_{r^{-1}}$ only in $(r^{-1}, r^{-1})$.*

# 5  Concluding remarks

In this paper we have presented an outline of some results about matrix pairs, in particular nonnegative ones, and of their relevance in the analysis of 2D systems. Further reasearch will, it is to be hoped, better clarify the connections between the spectral and combinatorial properties of a pair and the dynamics of the associated state model. An interesting question, for instance, concerns the existence of a dominating vector representing the asymptotic direction of all local states $\mathbf{x}(h, k)$

as $h + k$ goes to infinity. Preliminary results have been obtained in the scalar case, in the case of 2D Markov chains and when $(A_1, A_2)$ is a $k$-commuting pair.

The positive realization problem for 2D rational functions provides a wide set of research issues involving positive matrix pairs. In particular, basing on the 1D analogue, it is reasonable to expect that the singularities of a positively realizable transfer function fulfil some regularity constraints, inherited from the Perron-Frobenius structure of the peripheral spectrum of a positive pair.

Finally, an interesting topic of research is the investigation of the structure of matrix pairs over finite fields, which arise in the synthesis of encoders and decoders for 2D convolutional codes.

# References


[1] A. Berman and R.J. Plemmons, Nonnegative matrices in the mathematical sciences, Academic Press, New York (NY), (1979).

[2] M. Bisiacco, E. Fornasini, and G. Marchesini, 2D partial fraction expansions and minimal commutative realizations, IEEE Trans. Circ. Sys. **35** (1988), pp.1533–1538.

[3] M. Fliess, Un outil algébrique: Les séries formelles non commutatives, Springer Lect.Notes in Econ. Math. Sys. **131**, (1975), pp.122–149.

[4] E. Fornasini and G. Marchesini, Doubly indexed dynamical systems, Math. Sys. Theory **12**, (1978), pp.59–72.

[5] E. Fornasini and G. Marchesini, Properties of pairs of matrices and state-models for 2D systems, In C.R.Rao, editor, Multivariate Analysis: Future Directions. North Holland Series in Probability and Statistics **5**, (1993), pages 131–80.

[6] E. Fornasini, G. Marchesini, and M.E. Valcher, On the structure of finite memory and separable 2D systems, Automatica **29**, (1994), pp.347–350.

[7] E. Fornasini and M.E. Valcher, Matrix pairs in 2D systems: an approach based on trace series and Hankel matrices, SIAM J. Contr. Optim. **33**, 4 (1995), pp.1127–1150.

[8] E. Fornasini and M.E. Valcher, On the spectral and combinatorial structure of 2D positive systems, Lin. Alg. Appl. **245**, (1996), pp.223–258.

[9] E. Fornasini and M.E. Valcher, Directed graphs, 2D state models and characteristic polynomial of irreducible matrix pairs, to appear in Lin. Alg. Appl., (1997).

[10] E. Fornasini and M.E. Valcher, Recent developments in 2D positive system theory , to appear in J. of Applied Mathematics, (1997).

[11] W.P. Heath. Self-tuning control for two-dimensional processes, J.Wiley & Sons, (1994).

[12] N.H. McCoy, On the characteristic roots of matrix polynomials, Bull. Amer.Math.Soc. **42**, (1936), pp.592–600.

[13] H. Minc, Nonnegative Matrices, J.Wiley & Sons, New York, (1988).

[14] T.S. Motzkin and O. Taussky, Pairs of matrices with property L, Trans. Amer. Math.Soc. **73**, (1952), pp.108–114.

[15] T.S. Motzkin and O. Taussky, Pairs of matrices with property L (II), Trans. Amer. Math.Soc. **80**, (1955), pp.387–401.

[16] C. Procesi, The invariant theory of $n \times n$ matrices. Advances in Methematics **19**, (1976), pp.306–381.

[17] A. Salomaa and M. Soittola, Automata theoretic aspects of formal power series, Springer-Verlag, (1978).

[18] D. Shemesh, Common eigenvectors of two matrices, Linear Algebra Appl. **62**, (1984), pp. 11-18.

[19] S. Vomiero, Un'applicazione dei sistemi 2D alla modellistica dello scambio sangue-tessuto, PhD thesis, (in Italian), Univ.di Padova, Italy, (1992).



Ettore Fornasini
Giovanni Marchesini
Maria Elena Valcher
Dipartimento di Elettronica ed Informatica, Università di Padova,
via Gradenigo 6/A, 35131 Padova, Italy
phones: + 39-49-827-7605 (7610-7795)
fax: + 39-49-827-7699
e-mails: `fornasini@dei.unipd.it`, `meme@dei.unipd.it`

# Spectral Minimality of $J$-Positive Linear Systems of Finite Order

Aurelian Gheondea
Mathematical Institute, Bucharest, Romania

Raimund J. Ober
University of Texas at Dallas, Richardson, USA

*Dedicated to Paul Fuhrmann on the occasion of his $60^{th}$ birthday*

**Abstract**: It is shown that reachable completely $J$-positive linear systems of finite order are spectrally minimal.

## 1   Introduction

It is well-known (see e.g. [4],[7]) that a minimal finite dimensional linear system $(A, B, C, D)$ is *spectrally minimal*, i.e.

$$\sigma(A) = \sigma(G),$$

where $\sigma(A)$ denotes the set of eigenvalues of $A$ and $\sigma(G)$ denotes the set of singularities of the rational transfer function $G(z) = C(zI - A)^{-1}B + D$. This fact is one of the key relationships between the transfer function description and the state space description of system theory. Unfortunately, for infinite dimensional systems spectral minimality does not hold in general. The issue is very subtle and is, for example, related to the problem that the state space isomorphism theorem does not hold in general for infinite dimensional systems (see e.g. [4]), i.e. two reachable and observable infinite dimensional realization of a nonrational transfer function need not be equivalent.

The work by Fuhrmann and co-workers has been fundamental for the current understanding of the issue of spectral minimality for infinite dimensional systems (see [4] for an exposition). Through this work it is clear that spectral minimality results can still be obtained if certain restrictions are imposed on the problem. In ([4],[5]) it was shown that restricting the class of transfer functions to those which are strictly non-cyclic, spectral minimality can be established for reachable and observable realizations of such transfer functions. This work relies very heavily on the functional calculus for shifts restricted to invariant subspaces. Another mathematical tool, the spectral theorem and functional calculus for self-adjoint and, respectively, normal operators, provides the techniques underlying the second set of results. In these results ([1],[2]) spectral minimality is shown for infinite dimensional reachable and observable systems $(A, B, C, D)$ such that $A$ is self-adjoint and, respectively, normal. In a paper by Feintuch ([3]) further spectral minimality results were obtained for the case when $A$ is compact or a spectral operator.

In this paper we will study systems with symmetry properties in an indefinite metric. We will examine a class of systems which we consider to be a prototype of a more general situation, more precisely, the class of completely $J$-positive systems of finite order. For these systems, our main result, cf. Theorem 5, shows that observable and reachable realizations are also spectrally minimal.

## 2   Preliminaries

Let $\mathcal{H}$ be a Hilbert space with the scalar product denoted by $\langle \cdot, \cdot \rangle$ and let $J$ be a fixed symmetry on $\mathcal{H}$, that is $J^* = J = J^{-1}$. Then one can introduce an indefinite inner product on $\mathcal{H}$ denoted $[\cdot, \cdot]$

$$[x, y] = \langle Jx, y \rangle, \quad x, y \in \mathcal{H}.$$

The Hilbert space $\mathcal{H}$ endowed with such an indefinite inner product $[\cdot, \cdot]$ is called a *Kreĭn space*. Most often one does not fix the positive definite inner product (there are infinitely many and all of them produce the same strong topology) of a Kreĭn space, but even though this point of view is the most natural, we will not follow this approach since we would have to introduce too much Kreĭn space terminology.

A bounded operator $A \in \mathcal{L}(\mathcal{H})$ is called *$J$-selfadjoint* if $JA = A^*J$. It is clear that the operator $A$ is $J$-selfadjoint if and only if the operator $JA$ is selfadjoint in the Hilbert space $\mathcal{H}$. A $J$-selfadjoint operator $A$ on $\mathcal{H}$ is called *$J$-positive of order $n$* if $JA^n \geq 0$. Similarly one defines $J$-negative operators of order $n$. A $J$-positive operator of order 1 is called simply a $J$-positive operator.

We briefly review some of the main results on the spectral theory for $J$-positive operators which will be of importance in this paper. In the following we denote by $\mathcal{R}_0$ the Boole algebra generated by intervals $\Delta$ in $\mathbb{R}$ such that its boundary $\partial \Delta$ does not contain the point 0. We recall now a particular case of a celebrated theorem of H. Langer and some of its consequences, cf. [8], [9].

**Theorem 1** *Let $A \in \mathcal{L}(\mathcal{H})$ be a $J$-positive operator of order $n$. Then $\sigma(A) \subset \mathbb{R}$ and there exists a mapping $E \colon \mathcal{R}_0 \to \mathcal{L}(\mathcal{H})$, uniquely determined with the following properties:*

(1) $E(\Delta)$ *is $J$-selfadjoint for all $\Delta \in \mathcal{R}_0$.*

(2) *E is a Boole algebra morphism, that is, it is additive and multiplicative.*

(3) $E(\mathbb{R}) = I$.

(4) *For all $\Delta \in \mathcal{R}_0$ such that the polynomial $t^n$ is positive (negative) on $\Delta$, the operator $E(\Delta)$ is J-positive (J-negative).*

(5) *For all $\Delta \in \mathcal{R}_0$ the operator $E(\Delta)$ is in $\{A\}''$ (the bicommutant of the algebra generated by the operator $A$).*

(6) *For all $\Delta \in \mathcal{R}_0$ we have $\sigma(A|E(\Delta)\mathcal{H}) \subseteq \overline{\Delta}$.*

The mapping $E$ which is uniquely associated to the $J$-positive operator $A$ of some finite order $n$ is called *the spectral function* of $A$. As a consequence of Theorem 1, the spectral function has also the following properties.

**Corollary 1** *With the notation as in Theorem 1 let $\Delta \in \mathcal{R}_0$ be closed and such that $0 \notin \Delta$. Then:*

(a) *The function $E_\Delta$ defined by*

$$E_\Delta(\Lambda) = E(\Delta \cap \Lambda), \quad \Lambda \in \mathcal{R}_0,$$

*can be extended uniquely to a bounded measure with $\operatorname{supp} E_\Delta \subseteq \Delta$.*

(b) *The operator $AE(\Delta)$ is similar with a selfadjoint operator on a Hilbert space, in particular it has spectral measure.*

(c) *$E_\Delta$ is the spectral measure of the operator $AE(\Delta)$, in particular*

$$AE(\Delta) = \int_\Delta t\, d\, E(t).$$

Corollary 1 shows that the spectral function $E$ of a $J$-positive operator of some finite order $n$ can be regarded as a, generally unbounded, spectral measure on $\mathbb{R} \setminus \{0\}$.

According to the general theory of a selfadjoint definitizable operator $A$ in a Kreĭn space $\mathcal{K}$ [8], [9], the non-real points in the spectrum of $A$ as well as the real points in the spectrum $\sigma(A)$ where the spectral function $E$ is not defined, the so-called *critical points*, have some additional properties. For example, an isolated point in the spectrum of $A$ is necessarily an eigenvalue of finite geometric multiplicity (that is, the maximal length of Jordan chains). Also, if a critical point is an eigenvalue then its geometric multiplicity is finite. In particular, in the case of a $J$-positive operator of order $n$, its only possible critical point is 0 and, if it is an eigenvalue, then its geometric multiplicity is less than or equal to $n + 1$.

In this paper we consider *linear systems*, i.e. quadruples $(A, B, C, D)$ where $A \in \mathcal{L}(\mathcal{H})$ is a contraction, $B \in \mathcal{L}(\mathcal{U}, \mathcal{H})$, $C \in \mathcal{L}(\mathcal{H}, \mathcal{Y})$, and $D \in \mathcal{L}(\mathcal{U}, \mathcal{Y})$ and $\mathcal{H}, \mathcal{U}$, and $\mathcal{Y}$ are Hilbert spaces. Usually the spaces $\mathcal{U}, \mathcal{H}$, and $\mathcal{Y}$ are called, respectively, the *input space*, the *state space* and the *output space*. Also, the operators $A$, $B$, $C$, and $D$ are called, respectively, the *main operator*, the *input operator*, the *output operator*, and the *feedthrough operator*.

With every linear system $(A, B, C, D)$ there is associated its *transfer function* $G \colon \rho(A) \to \mathcal{L}(\mathcal{U}, \mathcal{Y})$ as follows

$$G(\lambda) = D + C(\lambda I - A)^{-1} B, \quad \lambda \in \rho(A). \tag{1}$$

Since the main operator $A$ is assumed contractive, the transfer function is defined and analytic for all $|\lambda| > 1$.

Let us assume that $\mathcal{U} = \mathcal{Y}$ and that on $\mathcal{H}$ there is fixed a symmetry $J$ (and hence the associated Kreĭn space $(\mathcal{H}, [\cdot, \cdot])$). A linear system $(A, B, C, D)$ is called *completely $J$-symmetric* if the operator $A$ is $J$-selfadjoint, $C = JB^*$, and $D = D^*$. The completely $J$-symmetric system is called *completely $J$-positive of order $n$* if the operator $A$ is $J$-positive of order $n$.

In [6] an extensive analysis was carried out for $J$-positive systems of finite order, but the question of spectral minimality was only partially resolved. It is the topic of this paper to establish spectral minimality for reachable and observable $J$-positive systems of finite order.

The transfer function of a completely $J$-symmetric system is given in the following theorem.

**Theorem 2 ([6])** *Let $(A,B,C,D)$ be a linear system which is completely $J$-positive of order $n$, such that $\mathcal{U} = \mathcal{Y} = \mathbb{C}^m$, and consider its transfer function $G$ as in (1). Then, there exist a $J$-positive operator $N \in \mathcal{L}(\mathcal{H})$, such that $N^2 = AN = 0$, and a symmetric matrix valued Borel measure $d\nu$ on $[-1, 1] \setminus \{0\}$ such that*

$$G(\lambda) = D + \sum_{k=1}^{n} \frac{1}{\lambda^k} B^* J A^{k-1} B + \frac{1}{\lambda^{n+1}} B^* J N B + \frac{1}{\lambda^n} \int_{[-1,1]\setminus\{0\}} \frac{t^n}{(\lambda - t)} d\nu(t). \tag{2}$$

*The measure $d\nu$ has also the following two properties:*

*(a) $t^n d\nu(t)$ is a positive matrix valued finite Borel measure on $[-1, 1]$;*

*(b) The function*

$$g(z) = \frac{1}{z}\left(G(\frac{1}{z}) - D\right) = \sum_{k>0} a_k z^k, \tag{3}$$

*which is analytic for $|z| < 1$, has its Taylor coefficients*

$$a_k = \begin{cases} B^* J A^k B, & 1 \le k \le n-1; \\[2mm] B^* J N B + \int\limits_{[-1,1]\setminus\{0\}} t^n \mathrm{d}\,\nu(t), & k = n; \\[2mm] \int\limits_{[-1,1]\setminus\{0\}} t^k \mathrm{d}\,\nu(t), & k \ge n+1. \end{cases} \tag{4}$$

*The measure $\mathrm{d}\,\nu$ is uniquely determined by these two properties, more precisely, if $E$ denotes the spectral function of $A$ we have $\mathrm{d}\,\nu(t) = \mathrm{d}\,B^* J E(t) B$, and the operator $N$ can be chosen*

$$N = A^n - \int\limits_{\mathbb{R}\setminus\{0\}} t^n \mathrm{d}\,E(t). \tag{5}$$

*If, in addition, $\pm 1 \notin \sigma_p(A)$ then $\mathrm{d}\,\nu(\{-1,1\}) = 0$ and $\lim\limits_{k\to\infty} \|a_k\| = 0$.*

The matrix valued measure $\mathrm{d}\,\nu$ as in Theorem 2 is called the *defining measure* of the system $(A, B, C, D)$. Under the assumptions of Theorem 2 and as a consequence of the representation (2) it follows that the transfer function $G$ has analytic continuation onto $\mathbb{C} \setminus \mathrm{supp}\,(\mathrm{d}\,\nu)$.

Let $\mathcal{U}$ be a Hilbert space and assume that $G\colon \{z \in \mathbb{C} \mid |z| > 1\} \to \mathcal{L}(\mathcal{U})$ is an operator valued function which is analytic everywhere on its domain of definition and at infinity. One can define an operator valued analytic function $g\colon \mathbb{D} \to \mathcal{L}(\mathcal{U})$ by

$$g(z) = \frac{1}{z}\left(G(\frac{1}{z}) - G(\infty)\right), \quad |z| < 1.$$

Then $g$ has the Taylor expansion

$$g(z) = \sum_{k \ge 0} S_k z^k, \quad |z| < 1.$$

Associated with the function $g$ one can consider the block-operator Hankel matrix

$$H = \begin{bmatrix} S_0 & S_1 & S_2 & \ldots & S_k & \ldots \\ S_1 & S_2 & S_3 & \ldots & S_{k+1} & \ldots \\ S_2 & S_3 & \ldots & & & \\ \vdots & & & & & \\ S_k & & & & & \\ \vdots & & & & & \end{bmatrix}. \tag{6}$$

We consider again a completely $J$-positive system $(A, B, B^* J, D)$ of order $n$, where $J$ is a fixed symmetry on the state space $\mathcal{H}$, and let $\mathcal{U}$ denote the input/output space. Following the general theory we consider $O\colon \mathcal{D}(O)(\subseteq \mathcal{H}) \to \ell^2_{\mathcal{U}}$, the *observability operator* defined by

$$\mathcal{D}(O) = \{h \in \mathcal{H} \mid \sum_{k \geq 0} \|B^* J A^k h\|^2 < \infty\},$$

$$Oh = (B^* J A^k h)_{k \geq 0}, \quad h \in \mathcal{D}(O).$$

By duality one introduces the *reachability operator* $R\colon \mathcal{D}(R)(\subseteq \ell^2_{\mathcal{U}}) \to \mathcal{H}$

$$\mathcal{D}(R) = \{(x_k)_{k \geq 0} \in \ell^2_{\mathcal{U}} \mid \sum_{k \geq 0} \|A^k B x_k\|^2 < \infty\},$$

$$R((x_k)_{k \geq 0}) = \sum_{k \geq 0} A^k B x_k, \quad (x_k)_{k \geq 0} \in \mathcal{D}(R).$$

Note that the domain of $R$ is dense in $\ell^2_{\mathcal{U}}$ since it contains all the sequences with finite support.

In the following results the observability and reachability of $J$-positive systems of finite order are exmanined.

**Theorem 3 ([6])** *If $\pm 1 \notin \sigma_p(A)$ then $\mathcal{D}(O)$ is dense in $\mathcal{H}$, $O = R^* J$ and $R = J O^*$, in particular both operators $O$ and $R$ are closed.*

**Corollary 2 ([6])** *Assume that $\pm 1 \notin \sigma_p(A)$. Then the following assertions are equivalent:*

(i) *the observability operator $O$ is bounded;*

(ii) *the reachability operator $R$ is bounded.*

Recall that a system $(A, B, C, D)$ is called *observable* if the observability operator $O$ is bounded and injective. The system is called *reachable* if the reachability operator $R$ is bounded and has dense range. Note that, as a consequence of Theorem 3, the completely $J$-positive system $(A, B, B^* J, D)$ of order $n$, such that $\pm 1 \notin \sigma_p(A)$, is observable if and only if it is reachable. The kernel of the observability operator is characterized as follows.

**Proposition 1 ([6])** *Let $E$ be the spectral function and let $N$ be the nilpotent operator associated with the main operator $A$ of the completely $J$-positive system $(A, B, B^* J, D)$ of order $n$. Then*

$$\ker(O) = \bigcap_{k=0}^{n} \ker(B^* J A^k) \cap \bigcap_{\Delta \in \mathcal{R}_0} \ker(B^* J E(\Delta)) \cap \ker(B^* J N).$$

Following N.J. Young [10], we say that a system $(A, B, C, D)$ is *parbalanced* if the corresponding observability and reachability operators $O$ and, respectively, $R$ are bounded and the observability gramian $O^*O$ coincide with the reachability gramian $RR^*$.

The following realization result was also established in [6]

**Theorem 4** *Let $\mathcal{U}$ be a Hilbert space and let $G: \{z \in \mathbb{C} \mid |z| > 1\} \to \mathcal{L}(\mathcal{U})$ be an operator valued function which is analytic on its domain and at infinity, such that $G$ is symmetric, that is,*

$$G(\bar{z}) = G(z)^*, \quad |z| > 1.$$

*If the Hankel block-operator matrix in (6) defines a bounded operator in $\ell^2_{\mathcal{U}}$ then there exists a Kreĭn state space $\mathcal{K}$ with a specified fundamental symmetry $J$ on $\mathcal{K}$ such that $G$ is realized by a completely $J$-symmetric linear system $(A, B, B^*J, D)$ which is observable, reachable and parbalanced.*

*If, in addition, for some $n \geq 0$ and all $|\lambda| > 1$ the function $G$ has the representation*

$$G(\lambda) = D + \sum_{k=1}^{n} \frac{1}{\lambda^k} S_{k-1} + \frac{1}{\lambda^{n+1}} \Gamma + \frac{1}{\lambda^n} \int\limits_{[-1,1]\backslash\{0\}} \frac{t^n}{(\lambda - t)} \, d\nu(t),$$

*where $\{S_k\}_{k=0}^{n-1}$ is a family of bounded selfadjoint operators on $\mathcal{U}$, $D \in \mathcal{L}(\mathcal{U})$, $D = D^*$, $\Gamma \in \mathcal{L}(\mathcal{U})$, $\Gamma \geq 0$, and $\nu$ is a hermitian $\mathcal{L}(\mathcal{U})$-valued measure on $[-1,1] \backslash \{0\}$ such that $t^n d\nu(t)$ is a finite and positive measure, then the realization $(A, B, C, D)$ constructed above is completely $J$-positive of order $n$.*

# 3   Spectral Minimality

So far we have considered the transfer function of a discrete-time system $(A,B,C,D)$ as a function defined outside the closed disk in the complex plane with center 0 and radius larger than $\|A\|$. In order to study the question of spectral minimality one needs to extend this definition. It is clear that the transfer function can be defined as an analytic function on the resolvent set of $A$. Let now $G$ be the maximal analytic continuation of this function. The set of singularities $\sigma(G)$ of $G$ are then defined to be the complement of the points of analyticity of $G$. Clearly $\sigma(G)$ is a closed set. The system $(A, B, C, D)$ is called *spectrally minimal* if $\sigma(G) = \sigma(A)$.

**Proposition 2** *Let $(A_i, B_i, C_i)$ be a discrete-time system with bounded system operators and state space $\mathcal{H}_i$ such that the corresponding observability (reachability) operator $O_i$ $(R_i)$ is bounded $i = 1, 2$. Assume that $\sigma(A_1) \cap \sigma(A_2) = \emptyset$. Also assume* *that the resolvent sets of $A_1$ and $A_2$ have only one component. Then the system $(A, B, C)$ given by*

$$A = \begin{pmatrix} A_1 & 0 \\ 0 & A_2 \end{pmatrix}, \qquad B = \begin{pmatrix} B_1 \\ B_2 \end{pmatrix}, \qquad C = \begin{pmatrix} C_1 & C_2 \end{pmatrix},$$

*with state space $\mathcal{H} := \mathcal{H}_1 \oplus \mathcal{H}_2$ is observable (reachable) if and only if the systems $(A_1, B_1, C_1)$ and $(A_2, B_2, C_2)$ are obervable (reachable).*

**Proof:** Let $\mathcal{H}_i$ be the state space of the system $(A_i, B_i, C_i)$, $i = 1, 2$. Then $\mathcal{H} = \mathcal{H}_1 \oplus \mathcal{H}_2$ is the state space of $(A, B, C)$ and $O : \mathcal{H} \to l_y^2$, $x = (x_1, x_2) \mapsto O_1 x_1 + O_2 x_2$ is the observability operator of $(A, B, C)$. Clearly $O$ is bounded if and only if $O_1$ and $O_2$ are bounded. Moreover the injectivity of $O$ implies injectivity of $O_1$ and $O_2$. Therefore the observability of $(A, B, C)$ implies the observability of $(A_1, B_1, C_1)$ and $(A_2, B_2, C_2)$.

Now assume that $(A_1, B_1, C_1)$ and $(A_2, B_2, C_2)$ are observable but that $(A, B, C)$ is not observable. Then there exists $x = (x_1, x_2) \in \mathcal{H} = \mathcal{H}_1 \oplus \mathcal{H}_2$ such that $0 = Ox = O_1 x_1 + O_2 x_2$. Let $G_i$ be the transfer function of the discrete-time system $(A_i, x_i, C_i)$. Note that for $|z| > \|A_i\|$, $i = 1, 2$,

$$G_i(z) = C_i(zI - A_i)^{-1} x_i = \sum_{n=0}^{\infty} \frac{1}{z^{n+1}} C_i A_i^n x_i.$$

Since $(C_1 A_1^n x_1)_{n \geq 1} = O_1 x_1 = -O_2 x_2 = -(C_2 A_2^n x_2)_{n \geq 0}$, we have that $G_1(z) = -G_2(z)$, $|z| > \max\{\|A_1\|, \|A_2\|\}$. Since by assumption the resolvent sets of $A_1$ and $A_2$ only have one component, this implies that $G_1 = -G_2$ and $\sigma(G_1) = \sigma(-G_2) = \sigma(G_2)$.

Clearly $\sigma(G_i) \subseteq \sigma(A_i)$, $i = 1, 2$. Since by assumption $\sigma(A_1) \cap \sigma(A_2) = \emptyset$ we have that $\sigma(G_1) \cap \sigma(G_2) = \emptyset$. As $\sigma(G_1) = \sigma(G_2)$, this implies that $\sigma(G_1) = \sigma(G_2) = \emptyset$. Hence $G_1$ and $G_2$ are analytic on $\mathbb{C}$. But $\lim_{|s| \to \infty} \|G_i(s)\| = 0$, $i = 1, 2$. Hence $G_1 = G_2 = 0$ on $\mathbb{C}$. Therefore $O_1 x_1 = -O_2 x_2 = 0$. By the injectivity of $O_1$ and $O_2$ we have that $x_1 = 0$, $x_2 = 0$ and hence $x = (x_1, x_2) = 0$. Therefore $O$ is injective and hence $(A, B, C)$ is observable.

The statements concerning reachability follow by duality. $\blacksquare$

We can now determine the spectral minimality of a reachable and observable completely $J$-positive realization of finite order. The part of the proof which deals with the non-zero spectrum is inspired by the proof of the spectral minimality for continuous-time symmetric systems with bounded operators in ([1]).

**Theorem 5** *Let $(A, B, B^* J, D)$ be a completely $J$-positive realization of finite order of the transfer function $G$ such that the reachability operator $R$ has dense range. Then the system is spectrally minimal, i.e. $\sigma(A) = \sigma(G)$.*

**Proof:** Note that the regions of analyticity of $G$ and the resolvent of $A$ have only one connected component. By the earlier remarks we have that $\sigma(G) \subseteq \sigma(A)$. We therefore need to consider the reverse set inclusion.

Let $\Delta$ be a compact real interval which does not contain 0. Then either $\Delta \subset (-\infty, 0)$ or $\Delta \subset (0, +\infty)$. To make a choice, assume that $\Delta \subset (0, +\infty)$.

From the construction of the spectral function $A$ of a $J$-positive operator of finite order as in [9] we have

$$E(\Delta) = \lim_{\varepsilon \to 0} \lim_{\delta \to 0} \frac{1}{2\pi i} \int_{C^\delta_{\Delta_\varepsilon}} (zI - A)^{-1} \mathrm{d}\, z, \tag{7}$$

where $\Delta_\varepsilon = [a - \varepsilon, b + \varepsilon]$, assuming that $\Delta = [a, b]$, $C^\delta_{\Delta_\varepsilon}$ is a rectangle symmetric with respect to the real axis constructed around the interval $\Delta_\varepsilon$ from which we remove two segments of length $2\delta$ around the points of coordinates $(-\varepsilon + a, 0)$ and $(b + \varepsilon, 0)$.

Let us assume now that $G$ has analytic continuation in a neighbourhood of $\Delta$. Then, from (7) and taking into account that the system $(A, B, B^*J, D)$ is a realization of $G$, applying the Cauchy formula we get

$$B^*JE(\Delta)B = \lim_{\varepsilon \to 0} \lim_{\delta \to 0} \frac{1}{2\pi i} \int_{C^\delta_{\Delta_\varepsilon}} (G(z) - D) \mathrm{d}\, z = 0.$$

Taking into account of Theorem 1 it follows that $E(\Delta)\mathcal{H}$ is a uniformly positive subspace of $\mathcal{H}$ and hence, for arbitrary $u$ in the input/output space $\mathcal{U}$ we have

$$0 = \langle B^*JE(\Delta)Bu, u \rangle = [E(\Delta)Bu, E(\Delta)Bu] \geq \alpha \|E(\Delta)Bu\|^2,$$

for some $\alpha > 0$. Then $E(\Delta)B = 0$ follows and since the spectral function $E$ commutes with the main operator $A$ we obtain

$$0 = A^k E(\Delta)B = E(\Delta)A^k B, \quad k \geq 0.$$

From here we obtain that $E(\Delta)|\mathcal{R}(R) = 0$ and since it is assumed that the reachability operator $R$ has dense range, it follows that $E(\Delta) = 0$ and hence $\Delta \cap \sigma(A) = \emptyset$.

We have therefore shown that $\sigma(A) \setminus \{0\} \subseteq \sigma(G)$ and hence that $\sigma(A) \setminus \{0\} = \sigma(G) \setminus \{0\}$.

Now we need to clarify the role of the point 0. If $0 \in \sigma(G)$ then the proof is completed. Now assume that $0 \notin \sigma(G)$. Since $\sigma(G)$ is closed this implies that there exists a $\varepsilon > 0$ such that $[-\varepsilon, \varepsilon] \cap \sigma(G) = \emptyset$. If $0 \notin \sigma(A)$ the proof is also completed. It therefore remains to exclude the case that $0 \in \sigma(A)$. To do this assume that $0 \in \sigma(A)$. Note that since $\sigma(A) \setminus \{0\} = \sigma(G) \setminus \{0\}$, we have that 0 is an isolated spectral point of $A$ as by assumption $[-\varepsilon, \varepsilon] \cap \sigma(G) = \emptyset$. But by [9] this implies that 0 is an eigenvalue of $A$.

Now consider $\Delta_r := ]-\infty, -\varepsilon[ \cup ]\varepsilon, \infty[$ and $\Delta_0 = [-\varepsilon, \varepsilon]$. Let

$$(A_r, B_r, C_r, D) := (E(\Delta_r)AE(\Delta_r), E(\Delta_r)B, CE(\Delta_r), D),$$

$$(A_0, B_0, C_0, 0) := (E(\Delta_0)AE(\Delta_0), E(\Delta_0)B, CE(\Delta_0), 0)$$

and define the corresponding state spaces $H_r = E(\Delta_r)H$ and $H_0 = E(\Delta_0)H$. Let $G_r$ and $G_0$ be the corresponding transfer functions. Note that since $0 \in \sigma_p(A)$ the system $(A_0, B_0, C_0, 0)$ is not the zero system.

We have that the system $(A, B, C, D)$ is the sum of these two systems. Since $(A, B, C, D)$ is reachable and observable and the spectra of $A_r$ and $A_0$ are disjoint these two systems are also reachable and observable by the previous Proposition. Also note that $\sigma(A_r) = \sigma(G) \cap \Delta_r$ and $\sigma(A_0) = \sigma_p(A_0) = \{0\}$, where $\sigma_p(A_0)$ denotes the point spectrum of $A_0$.

Since $H_r$ is a uniformly positive subspace the system $(A_r, B_r, C_r, D)$ is completely $J$-symmetric with $J = I$. By the spectral minimality result of the first part of the proof (or the spectral minimality result for self-adjoint systems in [1]) it follows that

$$\sigma(G_r) = \sigma(A_r) = \sigma(G) \cap \Delta_r.$$

Since $G = G_r + G_0$ and the sets of singularities of the two functions $G_r$ and $G_0$ are disjoint, we have that

$$\sigma(G) = \sigma(G_r) \cup \sigma(G_0) = (\sigma(G) \cap \Delta_r) \cup \sigma(G_0) = \sigma(G) \cup \sigma(G_0).$$

Since $\sigma(G) \cap \{0\} = \emptyset$ and $\sigma(G_0) = \{0\}$, this implies that $\sigma(G_0) = \emptyset$.

On the other hand $A_0$ is a nilpotent operator of finite order, more precisely, $A_0^{n+1} = 0$, cf. [8], [9]. But then for $z \neq 0$,

$$G_0(z) = C_0(zI - A_0)^{-1}B_0 = \sum_{n=0}^{\infty} \frac{1}{z^{n+1}} C_0 A_0^n B_0 = \sum_{n=0}^{N} \frac{1}{z^{n+1}} C_0 A_0^n B_0.$$

Therefore $G_0$ is a rational function. Since the system $(A_0, B_0, C_0, 0)$ is minimal this implies that this system is finite dimensional. By the spectral minimality of minimal finite dimensional systems we have that

$$\{0\} = \sigma(A_0) = \sigma(G_0).$$

This is a contradiction to $\sigma(G_0) = \emptyset$. Hence $0 \notin \sigma(A)$ if $0 \notin \sigma(G)$ and therefore $\sigma(A) = \sigma(G)$. $\blacksquare$

# 4   Acknowledgements

This research was supported in part by NSF grant DMS-9501223.

# References

[1] J.S. BARAS, R.W. BROCKETT, P.A. FUHRMANN: State-space models for infinite-dimensional systems, *IEEE Trans. Autom. Control*, **19**(1974), pp. 693–700.

[2] R.W. BROCKETT, P.A. FUHRMANN: Normal symmetric dynamical systems, *SIAM Journal on Control and Optimization*, **14**(1976), 107–119.

[3] A. FEINTUCH: Spectral minimality for infinite-dimensional linear systems, *SIAM Journal on Control and Optimization*, **14**(1976), 945–950.

[4] P.A. FUHRMANN: *Linear Systems and Operators in Hilbert Space*, McGraw-Hill, 1981.

[5] P.A. FUHRMANN: On spectral minimality and fine structure of the shift realization, in *Distributed parameter systems: Modelling and Identification, Proceedings of the IFIP working conference*, Lecture Notes In Control and Information Sciences, no 1, Springer Verlag, 1976.

[6] A. GHEONDEA, R.J. OBER: Completely $J$-positive linear systems of finite order, *Mathematische Nachrichten*, in press.

[7] T. KAILATH: *Linear Systems*, Prentice-Hall, 1980.

[8] H. LANGER: *Spektraltheorie linearer Operatoren in J-Räumen und einige Anwendungen auf die Schar $L(\lambda) = \lambda^2 I + \lambda B + C$*, Habilitationsschrift, Dresden 1965.

[9] H. LANGER: Spectral functions of definitizable operators in Kreĭn spaces, in *Functional Analysis*, Lecture Notes in Mathematics, vol. 948, Springer-Verlag, Berlin 1982, pp. 1–46.

[10] N.J. YOUNG: Balanced realizations in infinite dimensions, in *Operator Theory: Advances and Applications*, Vol. **19**, Birkhäuser Verlag, Basel 1986, pp. 449–471.

Aurelian Gheondea
Institutul de Matematică
al Academiei Române
C.P. 1-764, 70700 Bucureşti
România
e-mail: gheondea@imar.ro

Raimund J. Ober
Center for Engineering Mathematics EC35
University of Texas at Dallas
Richardson, Texas 75083-0688
USA
e-mail: ober@utdallas.edu

# State Feedback in General Discrete-Time Systems


M. L. J. Hautus
Eindhoven University of Technology
Eindhoven, The Netherlands



**Abstract:** A characterization is given for the class of systems obtainable by static state feedback in a general discrete-time system. To this extent, an input output description of such systems is given. A definition is given of feedback and the problem of the determining the feedback class, and a definition of state response maps. The concept of the Nerode map and equivalence enables us to give an algebraic treatment of the problem.


## 1 Introduction

An obvious question in system and control theory is as how we can change the system dynamics of a plant using feedback, in particular, using static feedback, i.e. feedback without dynamics. This problem can be stated more precisely as follows: 'Determine the feedback class of a system, that is, the class of systems obtainable by static feedback.'

This question has been investigated in detail in literature, in particular for linear systems. It has become apparent that useful results can only be expected in the case of state feedback. The research in linear systems resulted in the discovery of a complete set of invariants by P. Brunovsky (see [1]). It turned out that these feedback invariants have significance not only for problems directly related to feedback but also for realization theory, in particular, for the description of canonical forms of realizations (see e.g. [4]). The connection between realization theory and the feedback problem has been revealed in a more direct way in [2], where the feedback class of a plant has been characterized, not with the aid of canonical forms, but utilizing the fundamental treatment of realization theory by R.E. Kalman given in Chapter 10 of [3].

The method used in [2] is of a fundamental nature, and it is not surprising that it can be generalized. It is the purpose of this paper to describe the essential ideas of this method in the setting of 'general system theory', that is, on the set theoretic level. Of course one cannot expect deep mathematical ideas in this setup. However, it is hoped that this paper will give an indication of the method of treatment and the type of result to be expected in the case of systems with an algebraic or topological structure.

In order to simplify the description, we discuss only discrete-time systems.

Figure 1: A system

## 2 General discrete-time systems

We start by introducing some notation and terminology for (finite) sequences. If $S$ is a given set, we consider strings, i.e. sequences $\sigma = s_1 \cdots s_n$ of elements of $S$. The set of such sequences is called the **free semigroup generated by** $S$ and denoted with $S^+$. Multiplication in this semigroup is defined to be concatenation, i.e., if $\sigma = s_1 \cdots s_n$ and $\tau = t_1 \cdots t_m$, then $\sigma\tau := s_1 \cdots s_n t_1 \cdots t_m$. We obtain the **free monoid** $S^*$ **generated by** $S$ by adjoining the empty sequence $e$ to $S^+$, and extending the product defined in $S^+$ to $S^*$ by $ee = e$ and $e\sigma = \sigma e = \sigma$ for $\sigma \in S^+$. In addition, we introduce the following notation:

- The **length** $|\sigma|$ of $\sigma = s_1 \cdots s_n$ equals $n$. Also $|e| := 0$. Obviously, $|\sigma\tau| = |\sigma| + |\tau|$.

- The **last element map** $\ell_S : S^+ \to S$ is defined by $\ell_S(s_1 \cdots s_n) := s_n$. If there is no danger of confusion, we write simply $\ell$ instead of $\ell_S$.

The set $S$ is imbedded in $S^+$ and in $S^*$ in a natural way: the elements of $S$ represent sequences of length 1.

Next we introduce systems. We start with a set $U$, called the **inputalphabet**, and a set $Y$, the **output alphabet**. We do not require these sets to be finite. We will describe a system in two ways. The simplest description of the system is given by the **response map** $f : U^+ \to Y$.

The interpretation of this map is as follows: The string $\omega = u_1 \cdots u_n$ is offered as input to the system, that is, the elements of the string, $u_k$ for $k = 1, \ldots, n$ are entered consecutively as inputs (See Figure 1). Then the output after the application of the string will be $y = f(\omega)$. It is assumed that the system yields the output $y$ at the same time instant at which $u_n = \ell(\omega)$ is applied, so that there is not necessarily a delay in the system.

The set of response maps will be denoted $\Gamma(U, Y)$. Note that the map $\ell : U^+ \to U$ is an example of a response map with $Y = U$.

A different characterization of the system is provided by the **i/o map (input / output map)**. Here it is assumed again that a string $\omega = u_1 \cdots u_n$ is offered to the system. But now the output is read each time an element is entered. This results in an output sequence $\gamma = y_1 \cdots y_n =: \bar{f}(\omega)$. It is convenient to say that the empty string input yields an empty output. The relation between $\bar{f}$ and $f$ is straightforward. We have $\bar{f}(e) = e, \bar{f}(\omega u) = \bar{f}(\omega)f(\omega u)$ $(\omega \in U^*, u \in U)$, which defines $\bar{f}$ recursively, based on $f$, and $f(\omega) = \ell_Y \circ \bar{f}(\omega)$ $(\omega \in U^+)$, hence, $f = \ell_Y \circ \bar{f}$. We will always use the notation $\bar{f}$ for the i/o map corresponding to the response map $f$. Note that $\bar{\ell} : U^* \to U^*$ is the identity map. Not every map $F : U^* \to Y^*$ corresponds to a response map this way. In fact, we have the following result:

PROPOSITION 2.1 $F : U^* \to Y^*$ *is an i/o-map (i.e., there exists* $f : U^+ \to Y$ *such that* $F = \bar{f}$*) if and only if*

- *$F$ is **length preserving**, that is, $|F(\omega)| = |\omega|$ for all $\omega \in U^*$.*

- *$F$ is **causal**, that is, if $F(\omega_1\omega_2) = \gamma_1\gamma_2$ and $|\gamma_1| = |\omega_1|$ then $F(\omega_1) = \gamma_1$.*

The proof is straightforward. Henceforth we will identify systems by either the response map or the i/o-map. We define some special classes of response maps:

- $f : U^+ \to Y$ is called **static** (or **memoryless**) if $f(\omega)$ depends only on $\ell(\omega)$ (hence, if $\ell(\omega_1) = \ell(\omega_2)$ implies that $f(\omega_1) = f(\omega_2)$).

- $f : U^+ \to Y$ is called **strictly causal** if $f(u_1 \cdots u_{n-1}u_n)$ depends only on $u_1 \cdots u_{n-1}$ and not on $u_n$. In particular, if $u \in U$, and $f$ is strictly causal, $f(u)$ is independent of $u$.

So, a response map is static if it can be factored as $f := F \circ \ell_U$, where $F : U \to Y$ is an arbitrary map. Often, when there is no danger of confusion, we will use the notation $F$ or $F : U^* \to Y^*$ for the i/o-map $\overline{F \circ \ell_U}$.

If $f$ is a strictly causal map, we define $f_- : U^* \to Y$ by $f_-(\omega) := f(\omega u)$, for any $u$. In particular, $f_-(e) = f(u)$, where $u \in U$ is arbitrary. The map $f_-$ is called the **strict response map** corresponding to the strictly causal map $f$. For every map $h : U^* \to Y$, there exists exactly one strictly causal map $f$ such that $f_- = h$.

# 3 Feedback

Our objective is to study feedback in the general setting of the previous section. We want to use the output of a system $f$ to be feedback to the input. In addition, we assume that there is another input $u$ available. So, we assume that the actual input $v$ of $f$ is the output of a system having as input both $u$, the original input, and $y$, the output of the system.

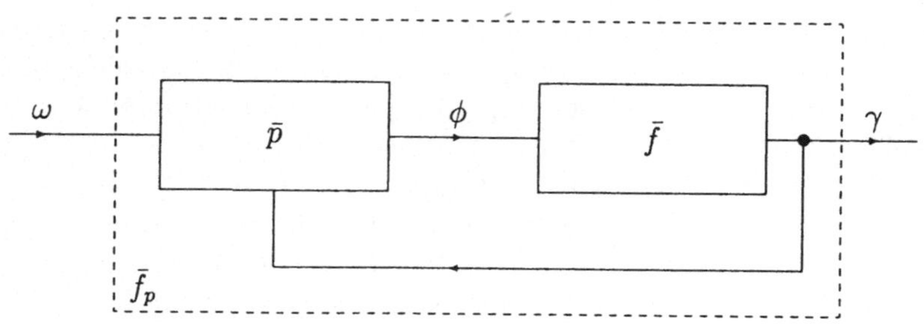

Figure 2: Feedback

Let us suppose we are given input aphabets $U$ and $V$, and an output aphabet $Y$. Let

$$f : V^+ \to Y, \quad p : (Y \times U)^+ \to V$$

be response maps. We say that $p$ defines a **feedback connection** for $f$, if there exists a unique response map $f_p : U^+ \to Y$ such that for every $\omega \in U^+$, we have

$$\gamma = \bar{f}(\phi), \tag{1}$$

where

$$\gamma := \bar{f}_p(\omega), \quad \phi := \bar{p}(\gamma, \omega). \tag{2}$$

Figure 2 gives a symbolic representation of the equations (1) and (2). If we use the notation $\omega = u_1 \cdots u_n, \phi = v_1 \cdots v_n, \gamma = y_1 \cdots y_n$, then these equations read as follows:

$$y_k = f(v_1 \cdots v_k), \quad v_k = p(y_1 \cdots y_k, u_1 \cdots u_k)$$

for $k = 1, \ldots, n$. These equations show immediately:

LEMMA 3.1 *If $f : V^+ \to Y$ is strictly causal then every $p \in \Gamma(Y \times U, V)$ defines a feedback connection.*

For convenience, we will, when applying feedback to a system with response map $f$, restrict ourselves to the case where $f$ is strictly causal.

If $\mathcal{P} \subseteq \Gamma(Y \times U, V)$ is given, then we want to investigate the following problem:

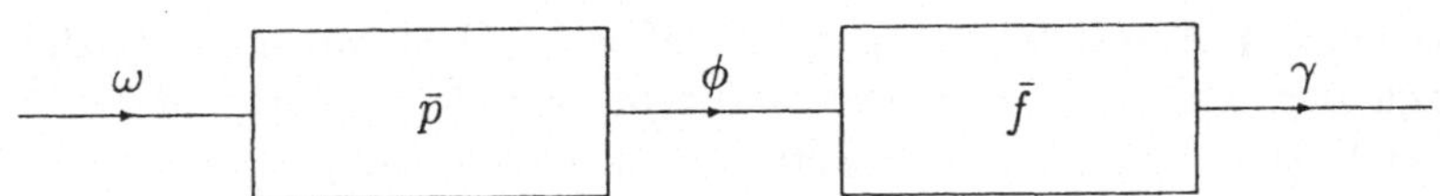

Figure 3: A cascade connection

PROBLEM 3.2 *Given a strictly causal map $f : U^+ \to Y$ and $\mathcal{P}$, determine the $\mathcal{P}$-feedback class of $f$, i.e., $\{f_p | p \in \mathcal{P}\}$.*

It will be seen in a moment that the problem is rather trivial in the case $\mathcal{P} = \Gamma(Y \times U, V)$. A more interesting class $\mathcal{P}$ will be the set of all *static* $p \in \Gamma(Y \times U, V)$.

It follows from equations (1) and (2) that

$$\phi = \bar{p}(\bar{f}(\phi), \omega). \tag{3}$$

By the strict causality of $f$, we may solve $\phi$ from this equation. This yields:

$$\phi = \bar{q}(\omega)$$

for some response map $q : U^+ \to Y$. Hence the feedback connection results in the same i/o-map as the cascade connection

$$\gamma = \bar{f} \circ \bar{q}(\omega),$$

that is, $f_p = f \circ \bar{q}$. It follows that every response map obtainable by a feedback connection is also obtainable by a cascade connection (See Figure 3). The problem of the determination of the $p$-feedback class of $f$ can be rephrased as

PROBLEM 3.3 *For what response maps $q$ does there exist $p \in \mathcal{P}$ such that $f_p = f \circ \bar{q}$?*

In other words, what cascade connections can be implemented by feedback? If $\mathcal{P} = \Gamma(Y \times U, V)$, every cascade connection is obtainable by feedback. A trivial possibility is $p(\gamma, \omega) = q(\omega)$. We will restrict ourselves to a special class of feedback connections:

DEFINITION 3.4 *A response map $p \in \Gamma(Y \times U, V)$ is called a **regular static feedback** if $p$ is a static response map represented by a map of the form $P : Y \times U \to V$ (i.e., $p = P \circ \ell$, see section 2) satisfying the property that the map $u \mapsto P(y, u)$ is invertible for every $y \in Y$. We denote the class of regular static feedbacks by $\mathcal{P}$.*

The invertibility condition serves to guarantee that every feedback connection in $\mathcal{P}$ can be 'undone' by another feedback (cf. Section 7). Note that in the rest of the paper, we will use the notation $\mathcal{P}$ exclusively for regular static feedback maps in $\Gamma(Y \times U, V)$.

The equations of the system with feedback read

$$y_k = f(v_1 \cdots v_k), \quad v_k = P(y_k, u_k).$$

A first necessary condition for Problem 3.3 can be given easily: Let $p \in \mathcal{P}$. We denote the inverse of $u \mapsto P(y, u)$ by $v \mapsto R(y, v)$. Now, it follows from equation (3) that $\bar{q}$ is invertible, and that the inverse is given by

$$\bar{q}^{-1} = \bar{r}, \tag{4}$$

where

$$r(\phi) := R(f(\phi), \ell(\phi)). \tag{5}$$

The invertibility of $\bar{q}$ is a necessary but not a sufficient condition for $\bar{q}$ to be obtainable by regular static feedback. In order to formulate a sufficient condition, we introduce a new concept:

DEFINITION 3.5 *Let $f : U^+ \to X, r : U^+ \to Y$ and $H : X \times U \to Y$. We say that $(f, H)$ is a **model** of $r$ if*

$$r(\omega) = H(f(\omega), \ell(\omega)) \quad (\omega \in U^+).$$

*If $f : U^+ \to Y, r : U^+ \to Z$, we say that $f$ is a **semimodel** of $r$ if there exists $H$ such that $(f, H)$ is a model of $r$.*

The following is an immediate consequence of (4) and (5):

PROPOSITION 3.6 *Given a strictly causal response map $f : V^+ \to Y$ and an invertible i/o-map $\bar{q} : U^* \to V^*$, there exists $p \in \mathcal{P}$ such that $f_p = f \circ \bar{q}$ if and only if $f$ is a semimodel of $r$, where $\bar{r} := \bar{q}^{-1}$.*

Thus our problem is reduced to the following problem:

PROBLEM 3.7 *Given a strictly causal map $f : V^+ \to Y$, of which response maps $r : V^+ \to U$ is $f$ a semimodel?*

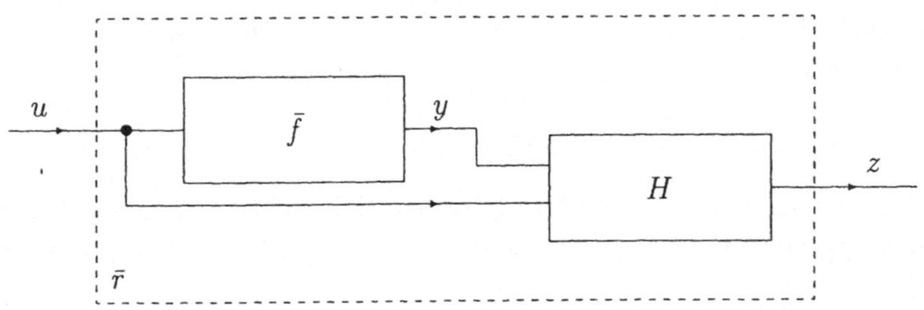

Figure 4: A model

# 4 State response maps and realizations

We will solve Problem 3.7 for the case where $f$ is a state response map.

DEFINITION 4.1 *Let $X$ be a set. A strictly causal map $f : U^+ \to X$ is called a **state response map** if $f_-(\omega)$ only depends on $f(\omega)$ and $\ell(\omega)$. A **reachable (state response) map** is a surjective state response map.*

That is, for a state response map, there exists $x^0 \in X$ and $F : X \times U \to X$ such that for every $\omega = u_1 \cdots u_n \in U^*$, we have

$$x_1 = x^0, \quad x_{k+1} = F(x_k, u_k) \quad (k = 1, \ldots, n-1),$$

where $x_1 \cdots x_n := \bar{f}(\omega)$ (note that $x^0 = f(e)$).

DEFINITION 4.2 *Let $r : U^+ \to Y$ be a response map. A pair $(f, H)$, where $f : U^+ \to X$ and $H : X \times U \to Y$, is called a **realization** of $r$ if $f$ is a state response map and $(f, H)$ is a model of $r$. A state response map $f : U^+ \to X$ is called a **semirealization** of $r$ if $f$ is a semimodel of $r$.*

It follows that $(f, H)$ is a realization of $r$ if and only if $r_-(\omega) = r(\omega u) = H(f_-(\omega), u)$ for all $\omega \in U^*, u \in U$. Hence, if for $\omega = u_1 \cdots u_n$, we define $x_1 \cdots x_n := \bar{f}(\omega)$, and $y_1 \cdots y_n = \bar{r}(\omega)$, we have the relations $x_1 = x^0$ and

$$x_{k+1} = F(x_k, u_k) \text{ and } y_k = H(x_k, u_k) \quad (k = 1, \ldots, n-1).$$

For reachable state response maps $f$, a condition on $r$ for $f$ to be a semirealization of $r$ can be given in terms of Nerode equivalence. In order to facilitate the formulation of this condition, we introduce some algebraic terminology.

# 5 U*-modules

In this section, $U$ will be a fixed set, called the input alphabet.

DEFINITION 5.1 *Let $S$ be set and $\phi : S \times U^* \to S$. The pair $(S, \phi)$ is called a (**right**) $U^*$-**module** if*

$$\phi(s, e) = s, \quad \phi(s, \omega_1 \omega_2) = \phi(\phi(s, \omega_1), \omega_2).$$

Usually, we speak of the $U^*$-module $S$ rather than $(S, \phi)$. Also we denote the action $\phi$ by juxtaposition: $\phi(s, \omega) = s\omega$. Consequently, the defining properties of the module action are

$$se = s, \quad s(\omega_1 \omega_2) = (s\omega_1)\omega_2.$$

EXAMPLE 5.2

1. $U^*$ itself is a $U^*$-module, where, if $s = \omega$, we define $s\omega_1 := \omega\omega_1$. Here the juxtaposition in the right-hand side denotes concatenation.

2. If $X$ is a set and $F : X \times U \to X$ a map, then the formulas

$$
\begin{aligned}
xe \quad &:= x, \\
xu \quad &:= F(x, u) \quad &&(u \in U) \\
x(\omega u) \quad &:= (x\omega)u \quad &&(\omega \in U^*, u \in U),
\end{aligned}
$$

   applied recursively, define a $U^*$-module.

3. If $Y$ is a set, then $\Gamma(U, Y)$ is a $U^*$-module with $f\omega$ defined by $f\omega(\omega_1) := f(\omega\omega_1)$.

$\square$

The usual algebraic concepts can be defined for $U^*$-modules: homomorphisms, submodules, congruences and quotient modules. We omit the details.

As a consequence of example 5.2, state response maps can be characterized as follows:

PROPOSITION 5.3 *A response map $f : U^+ \to X$ is a reachable state response map if and only if it is strictly causal and $X$ can be endowed with a $U^*$-module structure such that $f_- : U^* \to X$ is a epimorphism.*

# 6  Nerode maps

In this section, $U^*$ and $\Gamma(U, Y)$ will be considered $U^*$-modules as described in Example 5.2.

DEFINITION 6.1 *Let $f \in \Gamma(U, Y)$. Then*

$$\tilde{f} : U^* \to \Gamma(U, Y) : \omega \mapsto f\omega$$

*is called the* **Nerode map** *of $f$. Here, $f\omega$ denotes the result of the right $U^*$-module action of $\omega$ on $f$.*

This means that we have:

$$\tilde{f}(\omega)(\omega_1) = f(\omega\omega_1) \quad (\omega \in U^*, \omega_1 \in U^*).$$

If $S$ and $T$ are sets and $F : S \to T$ is a map, we denote by $\ker F$ the equivalence relation on $S$ induced by $F$, i.e.,

$$\ker F := \{(s_1, s_2) \in S \times S \mid f(s_1) = F(s_2)\}.$$

DEFINITION 6.2 *If $f : U^+ \to Y$ is a response map, then $\ker \tilde{f}$ is called the* **Nerode equivalence relation** *of $f$.*

Accordingly, two input sequences $\omega_1$ and $\omega_2$ are Nerode equivalent if and only if $f(\omega_1\omega) = f(\omega_2\omega)$ for all $\omega \in U^*$.

PROPOSITION 6.3 *If $f \in \Gamma(U, Y)$, then $\tilde{f} : U^* \to \Gamma(U, Y)$ is a $U^*$-homomorphism and $\ker \tilde{f}$ is a congruence on $U^*$.*

We have the following characterization of state response maps:

THEOREM 6.4 *Let $f : U^* \to X$ be a strictly causal reponse map. Then $f$ is a reachable state response map if and only if*

$$\mathrm{im} f_- = X, \quad \ker f_- = \ker \tilde{f}.$$

PROOF The first condition is equivalent to the surjectivity of $f$. It is easily verified that the inclusion $\ker \tilde{f} \subseteq \ker f_-$ is always valid (i.e., also for strictly causal maps that are not necessarily state response maps). Suppose that $f$ is a reachable state response map. Then $X$ is a $U^*$-module and $f_-$ is a $U^*$-epimorphism. If $f_-(\omega_1) = f_-(\omega_2)$ for some $\omega_1, \omega_2 \in U^*$, then we have for all $\omega \in U^*$:

$$f_-(\omega_1\omega) = f_-(\omega_1)\omega = f_-(\omega_2)\omega = f_-(\omega_2\omega),$$

where we have used the module action on $X$. Hence $f_-(\omega_1\omega) = f_-(\omega_2\omega)$ for all $\omega \in U^+$, that is, $\tilde{f}(\omega_1) = \tilde{f}(\omega_2)$.

Conversely, assume that $\ker \tilde{f} = \ker f_-$ and that $f_-$ is surjective. Then there exists an isomorphism $\phi : X \to U^*/\ker f_- = U^*/\ker \tilde{f}$ such that the diagram

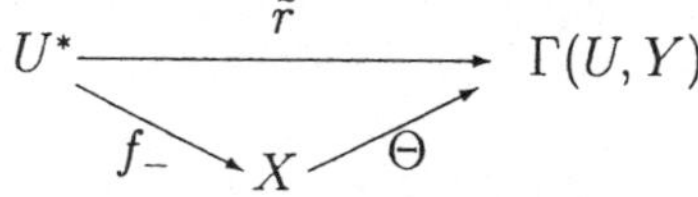

commutes. Here $\pi$ denotes the canonical projection. The isomorphism $\phi$ defines a $U^*$-module structure on $X$, and $f_- = \phi^{-1} \circ \pi$ is a homomorphism. The result now follows from Proposition 5.3. $\qquad\square$

The next result gives a necessary and sufficient condition for a state response map to be a semirealization of an arbitrary response map.

THEOREM 6.5 *Let* $r : U^+ \to Y$ *be a response map and* $f : U^+ \to X$ *be a reachable state response map. Then the following conditions are equivalent:*

1. $f$ *is a semirealization of* $r$,

2. $\ker f_- = \ker \tilde{f} \subseteq \ker \tilde{r}$,

3. *There exists a homomorphism* $\Theta$ *such that the following diagram commutes:*

$$
\begin{array}{ccc}
U^* & \xrightarrow{\;\;\tilde{r}\;\;} & \Gamma(U, Y) \\
 & {}_{f_-}\searrow \quad {}_{\Theta}\nearrow & \\
 & X &
\end{array}
$$

PROOF

$1 \Rightarrow 2$  Let $(f, H)$ be a realization of $r$. Then $r(\omega) = H(f(\omega), \ell(\omega))$ for $\omega \in U^+$. Suppose that for some $\omega_1, \omega_2 \in U^*$, we have $\tilde{f}(\omega_1) = \tilde{f}(\omega_2)$. Then $f(\omega_1 \omega) = f(\omega_2 \omega)$ for all $\omega \in U^+$, and hence

$$
r(\omega_1 \omega) = H(f(\omega_1 \omega), \ell(\omega)) = H(f(\omega_2 \omega), \ell(\omega)) = r(\omega_2 \omega).
$$

It follows that $\tilde{r}(\omega_1) = \tilde{r}(\omega_2)$.

$2 \Rightarrow 3$  This is a well-known situation: For $x \in X$, choose any $\omega \in U^*$ with $f_-(\omega) = x$, and set $\Theta(x) := \tilde{r}(\omega)$. Condition 2 guarantees that $\Theta$ is a well-defined homomorphism.

$3 \Rightarrow 1$  We define $H : X \times U \to Y$ by $H(x, u) := \Theta(x)(u)$, and show that $(f, H)$ is a (reachable) realization of $r$. That is, we show that for each $\omega \in U^+, u \in U$, the equality $r(\omega u) = H(f_-(\omega), u)$ holds. In fact, we have

$$
H(f_-(\omega), u) = \Theta(f_-(\omega))(u) = \tilde{r}(\omega)(u) = r(\omega u).
$$

$\qquad\square$

REMARK 6.6 We see that for a state response map $f$, the condition $\ker \tilde{f} \subseteq \ker \tilde{r}$ is necessary and sufficient for $f$ to be a semirealization of $r$. In fact, this property is characteristic for reachable state response maps:

THEOREM 6.7 *Let* $f : U^+ \to X$ *be a surjective strictly causal response map. Then* $f$ *is a state response map if and only if* $f$ *is a semimodel of every response map* $r : U^+ \to X$ *satisfying* $\ker \tilde{f} \subseteq \ker \tilde{r}$.

PROOF The 'if' part has been shown in the previous theorem. Assume that $f$ is a surjective strictly causal response map which is a semimodel of every response map $r : U^+ \to Y$ with $\ker \tilde{f} \subseteq \ker \tilde{r}$. Define $S := U^*/\ker \tilde{f}$, and let $r : U^+ \to S$ be the strictly causal response map such that $r_-$ is the canonical projection $U^* \to S$. Then $r$ is a state response map (see Proposition 5.3). Hence $\ker r_- = \ker \tilde{r}$. $\quad\square$

On the other hand, since $r_-$ is the canonical projection, we have $\ker r_- = \ker \tilde{f}$. Consequently, $\ker \tilde{r} = \ker \tilde{f}$. By assumption, $f$ is a semimodel of $r$, i.e., there exists $H : X \times U \to S$ such that for every $u \in U, \omega \in U^*$:

$$r_-(\omega) = r(\omega u) = H(f_-(\omega), u).$$

This equality implies $\ker f_- \subseteq \ker r_- = \ker \tilde{r}$. Now the result follows from Theorem 6.4. $\quad\square$

# 7 Feedback equivalence

Using the results of the preceding section, we can give a solution to Problem 3.3, that is, the determination of all response maps $q$ such that $f_p = f \circ \bar{q}$ for some $p \in \mathcal{P}$.

THEOREM 7.1 *If* $f : V^+ \to X$ *is a reachable state response map, and* $q : U^+ \to V$, *then there exists* $p \in \mathcal{P}$ *such that* $f_p = f \circ \bar{q}$ *if and only if the following conditions are satisfied*

  1. $\bar{q} : U^* \to V^*$ *is invertible,*

  2. $\ker \tilde{f} \subseteq \ker \tilde{r}$, *where* $r := \bar{q}^{-1}$.

In the case of a linear system, this condition reduces to the condition given in [2].

In the remaining part of this section, we pay attention to the $\mathcal{P}$-feedback class of $f$, that is, to the set of response maps $f_p$, where $p \in \mathcal{P}$.

THEOREM 7.2 *Let $f$ be a reachable state response map, and let $p \in \mathcal{P}$. Then $f_p$ is a reachable state response map.*

PROOF This result follows easily from the definition: If the state transition is given by $x_{k+1} = F(x_k, v_k)$, then the state feedback $v_k \stackrel{.}{=} p(x_k, u_k)$ yields $x_{k+1} = F_p(x_k, u_k)$, where $F_p(x, u) := F(x, p(x, u))$. The reachability of $f_p$ is a straightforward property. $\qquad\square$

In the following, we assume that $V = U$. In view of our invertibility assumption on $p$, this is not a real restriction. Then we can define a product relation on $\mathcal{P}$ by:

$$p \star q(x, y) := p(x, q(x, u)).$$

It is easily seen that with this multiplication, $\mathcal{P}$ is group, with identity $\mathrm{id}(x, u) := u$. This group acts on the set of reachable state response maps, and the group action satisfies:

$$f_{p \star q} = (f_p)_q, \quad f_{\mathrm{id}} = f.$$

For this reason, we can speak of *feedback equivalent state response maps*, and we might investigate invariance properties. There are two reasons to expect that such a direct approach might be difficult:

- In the case of a linear system given by an equation of the form $x_{k+1} = Fx_k + Gu_k$, the feedback would naturally be of the form $v = Lx + Vu$, where $V$ is invertible. It turns out (see [1]) that a simple description of the feedback invariants can only be given if, in addition to $\mathcal{P}$, one also has the group of state-space isomorphisms $x \mapsto Ax : X \to X$ (with $A$ nonsingular) acting on $g$. Accordingly, in the general case, one must expect that simple results are only obtainable if one admits the group $\mathcal{A} : X \to X$ of bijections, in addition to $\mathcal{P}$. Every element $A$ of this group $\mathcal{A}$ acts on $f$ in the following way:

$$f \mapsto f_A : x \mapsto A(f(A^{-1}(x), u)).$$

  The group obtained by combining the action of $\mathcal{P}$ and $\mathcal{A}$ is rather conplicated.

- We have to consider two state response maps simultaneously. But then, the algebraic terminology introduced in section 5 cannot be used, unless we allow two different $U^*$-modules on $X$. It seems unlikely that this is a convenient description.

For this reason, we will use a different approach to the description of feedback invariance, inspired by [2].

We say that two state response maps $f_i : U^+ \to X_i$ $(i = 1, 2)$ are **isomorphic** if there exists a $U^*$-isomorphism $\alpha : X_1 \to X_2$ such that the diagram

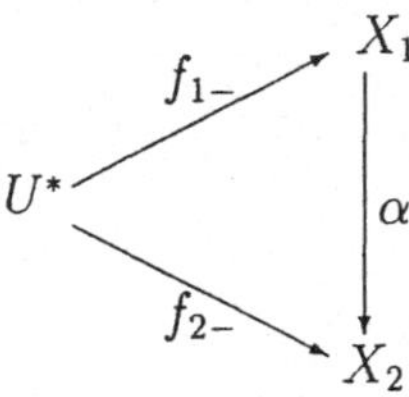

commutes. We allow (but do not require) the $U^*$-modules $X_1$ and $X_2$ to have the same carrier set. Rather than dealing with state space $X$ itself, we use the Nerode equivalences of the the response maps. This approach is based on the following result.

PROPOSITION 7.3

1. *If $C$ is a congruence on $U^*$, there exists state space $X$ and a (reachable) state response map $f : U^* \to X$ such that $\ker f_- = C$.*

2. *If $f_i : U^+ \to X_i$ for $i = 1, 2$ are reachable response maps satisfying $\ker f_{1-} = \ker f_{2-}$, then $f_1$ and $f_2$ are isomorphic.*

PROOF

- Define $X := X_C := U^*/C$, and let $\pi : U^* \to X$ be the canonical projection. Let $f := f_C$ be the strictly causal response map satisfying $(f_C)_- = \pi$. Then $f$ is a reachable state response map an d $\ker f_- = C$.

- We note that every reachable state response map $f : U^+ \to X$ with $\ker f_- = C$ is isomorphic to $f_C$. This follows from the existence of a $U^*$-isomorphism $\alpha$ such that the following diagram commutes:

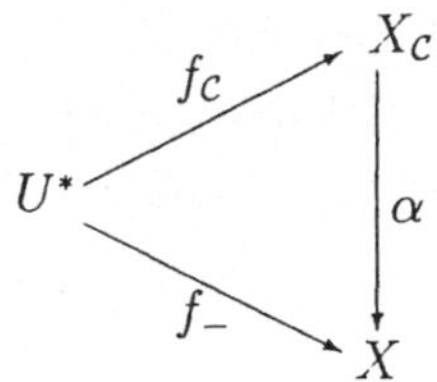

□

It follows that, if we work with Nerode equivalence classes instead of states, then the state-space isomorphisms mentioned before are built in. In addition, all Nerode equivalence classes to be considered are submodules of one $U^*$-module, viz., $U^* \times U^*$, so that we do not encounter the difficulty mentioned before.

DEFINITION 7.4 *Two congruences $C_1$ and $C_1$ are called **feedback equivalent** if there exist feedback equivalent state-response maps $f_i : U^+ \to X_i$ for $i = 1, 2$ such that $\ker f_{1-} = C_1, \ker f_{2-} = C_2$.*

We introduce the following notation: If $f : S \to T$ is a map, then $f^2 : S \times S \to T \times T$ is defined by

$$f^2(s_1, s_2) := (f(s_1), f(s_2)).$$

Now we can state the following result:

THEOREM 7.5 *Two congruences $C_1$ and $C_2$ are feedback equivalent if and only if there exists a map $r : U^+ \to U$ such that*

1. *$\bar{r} : U^* \to U^*$ is invertible,*

2. *$\bar{r}^2(C_1) = C_2$,*

3. *$C_1 \subseteq \ker \bar{r}$.*

PROOF Suppose that there exists a map $r$ satisfying conditions 1,2 and 3. Define $X := U^*/C_1$, and let $f_1 : U^* \to X$ be a response map such that $f_{1-} : U^* \to X$ is the canonical projection. Then $f_1$ is a reachable state response map satisfying $\ker f_1 = C_1$ (see Proposition 7.3). Define $q : U^+ \to U$ by $\bar{q} := \bar{r}^{-1}$. Condition 3 guarantees that $f_2 := f_1 \circ \bar{q}$ is a reachable state response map obtainable from $f_1$ by regular static feedback. It remains to be shown that $\ker f_{2-} = C_2$. Now, .

$$(\omega_1, \omega_2) \in \ker f_{2-} \Leftrightarrow f_{2-}(\omega_1) = f_{2-}(\omega_2) \Leftrightarrow f_{1-}(\bar{q}(\omega_1)) = f_{1-}(\bar{q}(\omega_2))$$

$$\Leftrightarrow \bar{q}^2(\omega_1, \omega_2) = (\bar{q}(\omega_1), \bar{q}(\omega_2)) \in \ker f_{1-} = C_1 \Leftrightarrow (\omega_1, \omega_2) \in C_2.$$

The converse statement is immediate. $\qquad\square$

Finally, we give a result about the existence of a static feedback which is not necessarily regular. Here we drop the assumption $U = V$.

THEOREM 7.6 *Let $f_- : V^+ \to X$ be a reachable state response map, let $q : U^+ \to V$ and define $f_1 := f \circ q$. There exists a static feedback $P : X \times U \to X$ such that $f_P = f_1$ if and only if*

- *$f_1$ is a state response map,*

- *$\ker \tilde{f}_1 \subseteq \ker \tilde{q}$.*

The proof is similar to the proof of theorem 7.1. It will be omitted.

# References

[1] BRUNOVSKY P., "A classification of linear systems", *Kybernetika* 44, 1970, pp. 173-188

[2] HAUTUS M.L.J. & HEYMANN MICHAEL, "Feedback- An algebraic approach", *SIAM J. Control* 16, 1978, pp. 83-105

[3] KALMAN R.E., FALB P.L., ARBIB M.A., "Topics in mathematical Systems Theory", McGraw-Hill, New York, 1969

[4] WOLOVICH W.A., "Linear Multivariable Control" , *Lecture Notes in Economics and Mathematical Systems* 101, Springer Verlag, Berlin, 1974

Malo Hautus
Department of Mathematics and Computing Science
Eindhoven University of Technology
P.O. Box 513, 5600 MB Eindhoven
The Netherlands

# Nonblocking Supervisory Control of Nondeterministic Systems[1]


Michael Heymann[2]
Technion, Isreal Institute of Technology, Haifa, Israel

Feng Lin
Wayne State University, Detroit, USA



**Abstract**: In this paper we extend the theory of supervisory control of nondeterministic discrete-event systems, subject to nondeterministic specification, developed in [9]. We focus our attention on nonblocking and liveness considerations and develop algorithms for nonblocking-supervisor synthesis.


## 1 A personal perspective (of the first author)

The recognition in the late 1950s that linear filtering can be accomplished recursively, inspired the development of the Kalman filter and its dual, the linear quadratic optimal controller. The concepts of controllability and observability, the cornerstones of Algebraic Systems Theory, were not far behind. Early research on Algebraic Systems Theory focused on these concepts and issues related to canonical forms, canonical (minimal) realizations, system structural equivalences and invariants. These and related questions occupied much of the Algebraic Systems Theory research agenda of the 1960s. In the second half of that decade, the discovery of the pole shifting theorem provided the first insights into the connection between controllability and state-feedback. This later led to the emergence of the "geometric" theory of linear systems where the connection between pole shifting and controllability (and their obsevational duals) played major roles. At around the same time the fundamental feedback invariants of linear systems were first discovered.

I became familiar with Paul Fuhrmann's work on systems theory when I first came accross his classical paper "Algebraic System Theory - an Analyst's Point of View". At the time, I was working on the paper "Linear Feedback - An Algebraic Approach" which dealt with the algebraic stucture of linear state-feedback. (The


---

[1]This research is supported in part by the National Science Foundation under grant ECS-9315344 and in part by the Technion Fund for Promotion of Research.

[2]The work by this author was completed while he was a Senior NRC Research Associate at NASA Ames Research Center, Moffett Field, CA 94035.

two papers have many points of contact). This first exposure and many interactions thereafter led to my continuing friendship with Paul for which I am very grateful.

In the early 1980s I reached the (personal) conclusion that a major obstacle to the practical "implementability" of continuous (especially linear) control theory, is the fact that many practical systems possess discontinuities, discrete interactions with their environment, parametric and structural uncertainty, nondeterminism and the like. A major component in the puzzle of how to control complex systems appeared to be missing!

This led to my embarkation on a new path - exploring discrete event control systems - where attention is focused exclusively on the logical (or discrete) aspects of system dynamics. With the recent maturing of the discrete event control theory and with many of the discrete aspects becoming clarified, it is progressively becoming feasible to begin an assault on "hybrid" systems in which the continuous and discrete aspects of system dynamics fully interact and coexist. It is not unlikely that in the not too distant future algebraic aspects of hybrid systems will become important, and an "Algebraic " theory of hybrid systems will emerge.

The present paper deals with the control of nondeterministic discrete event systems which, hopefully, sometime in the not too distant future will find its logical connection to my "algebraic" origins.

## 2 Introduction

Most of the published research on control of discrete-event systems (DES) has focused on systems that are modeled as deterministic finite state machines. For such systems, an extensive theory has been developed [23]. A great deal of attention was also given to the control of partially observed discrete-event systems [15], in which only a subset of the system's events are available for external observation. For such systems, necessary and sufficient conditions for existence of supervisors [15] [22] [23], algorithms for supervisor synthesis [2] [15] [16], for off-line as well as on-line implementation [2] [7], have been obtained, and a wide variety of related questions have been investigated.

Partially observed systems frequently exhibit nondeterministic behavior. There are, however, situations in which the system's model is nondeterministic not because of partial observation but, rather, because either the system is inherently nondeterministic, or because only a partial model of the system is available and some or all of its internal activities are unmodeled.

In contrast to deterministic discrete-event systems, whose behaviors are fully specified by their generated language, nondeterministic systems exhibit behaviors whose description requires much more refinement and detail. Further, while in the deterministic case, legal behavior of a system can be adequately expressed in terms of a language specification, this is clearly not always true when the system is nondeterministic. Indeed, to formally capture and specify legal behavior of the controlled system, it may be necessary to state, in addition to the permitted language, also the degree of nondeterminism that the controlled system is allowed to retain. Various semantic formalisms have been introduced over the years for modeling and specification of nondeterministic behaviors. These differ from each other, among other things, in the degree of nondeterministic detail that they capture and distinguish. These formalisms include CSP [11] and the associated *failures* semantics, bisimulation semantics [19] and labeled transition systems [4]. In [5] and [6] the *trajectory model* formalism was introduced as a semantic framework for modeling and specification of nondeterministic behaviors with specific focus on discrete event control. It was shown there that this semantic is a *language congruence* that adequately captures nondeterministic behaviors that one might wish to discriminate and distinguish by discrete-event control. Thus, for control purposes, nondeterministic discrete-event systems can be modeled either as nondeterministic automata (with $\epsilon$-transitions) or as trajectory models.

In recent years, there has been increasing interest in supervisory control of nondeterministic systems as reported, e.g., in [3] [12] [20] [21] [25]. However, while some existence conditions for control of nondeterministic systems have been derived, only limited progress on development of algorithms for supervisor synthesis has been reported (see e.g. [20] where a synthesis algorithm based on failures semantics is presented). Indeed, the direct supervisor synthesis for nondeterministic systems seems to be quite a difficult task (and, as will be shown below, unnecessary).

Motivated by this observation, we began an investigation, [8] [9], of the connection between the supervisory control problem for general nondeterministic systems and the corresponding problem for partially observed deterministic systems. Our work led us to develop an approach to synthesis of supervisors for nondeterministic systems wherein direct advantage is taken of the existing theory for control under partial observation.

In [9] we considered the supervisory control problem of nondeterministic discrete-event systems subject to trajectory-model specifications. Our approach to the supervisor synthesis was based on the following basic idea: We first synthesized from the given system, by adding to it hypothetical transitions and hypothetical uncontrollable and unobservable events, a deterministic system whose partially observed image is the original nondeterministic system (in the sense that the hypothetical events are obviously not observed). We called this procedure *lifting*. Before performing the lifting, the legal (trajectory model) specification was embedded in the original nondeterministic system model so that it can readily be dealt with in the corresponding lifted deterministic system. The next step of the synthesis was to construct a supervisor for the lifted system subject to the (obvious) condition that the artificially added events are neither observable nor controllable. Such a supervisor can easily be constructed using the well known theory and algorithms for supervisory control of partially observed systems. It is self evident, and we showed it formally, that a supervisor synthesized in this way is applicable for the original nondeterministic system and satisfies the specifications. Moreover, we showed that if the supervisor designed using this approach is optimal for the lifted system, it is also the optimal supervisor for the original system. Thus, since control under partial observation is well understood, we only had to, ultimately focus on the auxiliary steps of model lifting and specification embedding.

The present paper is a continuation of this research. In [8] [9] we focused our attention only on safety specifications, without consideration of liveness issues. We did not worry about questions related to task completion, nor about the problem of possible blocking. We extend here the results of [9] to include nonblocking issues and liveness considerations. This generalization which, in spirit, is very similar to the parallel situation in the deterministic case, introduces several additional complexities to the theory, that have to be examined in detail. We develop the theory and the associated synthesis algorithms for nonblocking supervisory control by examining the so called, *static case*, where a subset of target (or marked) states and a subset of forbidden states of the system are specified. The control objective is then to disable the smallest subset of transitions such that, in the controlled system, no path leads to a forbidden state and every path can be extended to a target state. It can be shown that the more general *dynamic case*, where the specification is given by a trajectory model (or as a nondeterministic automaton), is transformed into the simpler static setting, in which the supervisor is then synthesized. Detailed algorithms for optimal supervisor synthesis are provided. We also briefly address the problem of control under partial observation (where some of the actual events in the modeled system are unobservable) and the problem of decentralized control.

Due to space limitation, some details are omitted, which can be found in [10].

# 3 Nonblocking supervisors

We model a nondeterministic discrete-event system by the trajectory model introduced in [6], whose notations are adopted in this paper. For the purpose of specification, we often represent a trajectory model by a nondeterministic automaton. Similar to the language model and deterministic automaton used in modeling deterministic systems, the trajectory model representation and the automaton representation of nondeterministic systems are interchangeable: For each nondeterministic automaton, there exists a unique trajectory generated by the automaton; and for each trajectory model, we can construct an automaton generating the trajectory model. In particular, for each path $p$ in the automaton, we can find its corresponding trajectory $t_p$. To simplify the notation, we will use the same symbol to denote both the nondeterministic automaton and its associated trajectory model.

Since we are interested in the blocking and liveness issues in this paper, in addition to the usual elements of an automaton, we specify a set of marked states

$Q_m$ that represent, for example, task completions. To specify the desired behavior of the controlled system, we can use either a "static" specification or a "dynamic" specification. Similar to the algorithm developed in [9], we can always transform a dynamic specification into a static specification, where a subset $Q_b \subseteq Q$ of forbidden states that the system is not allowed to visit is specified. Thus, our system model can be written as

$$\mathcal{P} = (\Sigma \cup \{\epsilon\}, Q, \delta, q_0, Q_m, Q_b).$$

As in the deterministic case, we assume that $\mathcal{P}$ is trim (i.e., both accessible and co-accessible).

We define the set of marked trajectories of $\mathcal{P}$ as

$$\mathcal{P}_m = \{t_p : p \text{ ends in a marked state}\}.$$

The supervisory control problem is to synthesize a supervisor $\gamma$, (defined as a function $\gamma : L(\mathcal{P}) \to 2^{\Sigma_c}$ that after each observed string $s \in L(\mathcal{P})$ of executed transitions, disables a subset $\gamma(s) \subseteq \Sigma_c$ of controllable events,) such that the supervised system satisfies the state restrictions in that each path of the supervised system is a *legal* path; that is, each path ends at a target state (in $Q_m$) without ever entering a forbidden state (in $Q_b$). When such a supervisor exists, we would like to find, among all possible solutions, a least restrictive one; that is, a solution that disables as few as possible transitions.

For a supervisor $\gamma$, the language generated by the supervised system $\gamma/\mathcal{P}$ is given inductively as [9]

1. $\epsilon \in L(\gamma/\mathcal{P})$; and

2. $(\forall s \in L(\gamma/\mathcal{P}))(\forall \sigma \in \Sigma)s\sigma \in L(\gamma/\mathcal{P}) \Leftrightarrow s\sigma \in L(\mathcal{P}) \wedge \sigma \notin \gamma(s).$

The supervised system is then given by

$$\gamma/\mathcal{P} = \mathcal{P} \| det(L(\gamma/\mathcal{P}))$$

where $\|$ denotes the strict synchronous (parallel) composition and $det(L(\gamma/\mathcal{P}))$ the deterministic process generating language $L(\gamma/\mathcal{P})$ (as defined in [9]).

In principle, our goal is to design a supervisor $\gamma$ such that

$$\gamma/\mathcal{P} = \mathcal{P}_s$$

where $\mathcal{P}_s$ is (the trajectory model of) the (largest) trim subautomaton of

$$\overline{\mathcal{P}_s} = (\Sigma \cup \{\epsilon\}, Q_s, \delta_s, q_0, Q_{sm}).$$

where $Q_s = Q - Q_b$, $\delta_s = \delta|_{Q_s}$, ($\delta|_{Q_s}$ being the restriction of $\delta$ to $Q_s$), and $Q_{sm} = Q_s \cap Q_m$. Without loss of generality we shall assume that $\mathcal{P}_s = \overline{\mathcal{P}_s}$.

Such a supervisor is nonblocking in the sense that every trajectory enabled by the supervisor is a prefix of a trajectory that ends at a marked state.

As we shall see, such a supervisor does not always exist, and when it does not, we shall seek its best nonblocking approximation, as will be discussed below.

To obtain the desired supervisor, we proceed, just as in [9], by first transforming $\mathcal{P}$ to a deterministic automaton

$$\tilde{\mathcal{P}} = (\Sigma \cup \Sigma', \tilde{Q}, \tilde{\delta}, \tilde{q}_0, \tilde{Q}_m, \tilde{Q}_b)$$

using the procedure "Extend" given below.

**Procedure Extend**

    **Input:** $\qquad\qquad \mathcal{P} = (\Sigma \cup \{\epsilon\}, Q, \delta, q_0, Q_m, Q_b)$.

    **Output:** $\qquad\quad \tilde{\mathcal{P}} = (\Sigma \cup \Sigma', \tilde{Q}, \tilde{\delta}, \tilde{q}_0, \tilde{Q}_m, \tilde{Q}_b)$.

1. $\tilde{Q} := Q$;

2. For each $q \in \tilde{Q}$ and $\sigma \in \Sigma$

    If $|\delta(q, \sigma)| > 1$, add one more state, $q'$

    and add $\epsilon$−transitions as follows:

$$\tilde{Q} := \tilde{Q} \cup \{q'\};$$
$$\tilde{\delta}(q, \sigma) := \{q'\};$$
$$\tilde{\delta}(q', \epsilon) := \delta(q, \sigma);$$

    else set

$$\tilde{\delta}(q, \sigma) := \delta(q, \sigma);$$

3. For each $q \in \tilde{Q}$

    replace the $\epsilon$−transitions by transitions labeled $\tau_1, \tau_2, \ldots$ as follows:

    If $\tilde{\delta}(q, \epsilon) = \{q_1, \ldots, q_n\}$, then set

$$\tilde{\delta}(q, \tau_1) := \{q_1\};$$
$$\ldots$$
$$\tilde{\delta}(q, \tau_n) := \{q_n\};$$

4. Set

$$\Sigma' := \{\tau_1, \tau_2, \ldots\},$$
$$\tilde{Q}_m := Q_m,$$
$$\tilde{Q}_b := Q_b \cup \{\tilde{q} \in \tilde{Q} - Q : \delta(\tilde{q}, \epsilon) \subseteq Q_b\}.$$

5. End of algorithm                ■

We now define the following languages:

$$L(\tilde{\mathcal{P}}) := \{s\in\Sigma^* : \tilde{\delta}(\tilde{q}_0, s) \text{ is defined}\},$$
$$L_m(\tilde{\mathcal{P}}) := \{s\in L(\tilde{\mathcal{P}}) : \tilde{\delta}(\tilde{q}_0, s)\in\tilde{Q}_m\},$$
$$E := \{s\in L_m(\tilde{\mathcal{P}}) : (\forall s'\le s)\tilde{\delta}(\tilde{q}_0, s')\in\tilde{Q} - \tilde{Q}_b\}.$$

From the definition of $E$ it is clear that

$$E = L_m(\tilde{\mathcal{P}}) \cap \overline{E},$$

that is, $E$ is $L_m(\tilde{\mathcal{P}})$-closed [22]. From Proposition 7 of [9] it follows that the projection of $\tilde{\mathcal{P}}$ on $\Sigma$ is $\mathcal{P}$, that is,

$$\tilde{\mathcal{P}}\backslash_{\Sigma'} = \mathcal{P}.$$

We call a path marked if it ends in a marked state of $Q_m = \tilde{Q}_m$. We can prove the following

**Proposition 1** A marked path $p$ of the system $\mathcal{P}$ is legal (that is, is a path in $\mathcal{P}_s$) if and only if it is the projection of a path associated with a string $s\in E$ in $\tilde{\mathcal{P}}$.

**Proof**

Consider a marked path $p = (q_0, ..., \sigma_i, q_i, ..., \sigma_k, q_k)$ of $\mathcal{P}$ that visits a state $q_b \in Q_b$. The corresponding path in $\tilde{\mathcal{P}}$ has the same form with possible insertions of pairs (executions) $\sigma', q'$, where $\sigma' \in \Sigma'$ and $q' \in \tilde{Q} - Q$. Hence the corresponding path $\tilde{p}$ in $\tilde{\mathcal{P}}$ also visits $q_b \in Q_b \subseteq \tilde{Q}_b$. Conversely, let $\tilde{p} = (q_0, ..., \sigma_i, q_i, ..., \sigma_k, q_k)$ be a marked path in $\tilde{\mathcal{P}}$ and assumes it visits a state $q_b\in\tilde{Q}_b$. If $q_b \in Q_b$, then the projected path in $\mathcal{P}$ also visits the state $q_b$. If $q_b \in \tilde{Q}_b - Q_b$, then $q_b \neq q_k$ and by the definition of $\tilde{Q}_b$, the next state visited by the path in $\tilde{\mathcal{P}}$ must be in $Q_b$. This bad state will be visited also by the projected path in $\mathcal{P}$. Thus, a marked path in $\tilde{\mathcal{P}}$ visits only states in $\tilde{Q} - \tilde{Q}_b$ if and only if the corresponding marked path in $\mathcal{P}$ visits only states in $Q - Q_b$.

∎

We can now state the main result of this section that summarizes the conditions for existence of the desired supervisor.

**Theorem 1** There exists a nonblocking supervisor $\gamma$ such that $\gamma/\mathcal{P} = \mathcal{P}_s$ if and only if $E$ is controllable and observable with respect to $L(\tilde{\mathcal{P}})$.

**Proof**

By the results of [15], there exists a nonblocking supervisor $\gamma : PL(\tilde{\mathcal{P}}) = L(\mathcal{P}) \to 2^{\Sigma_c}$ such that $L_m(\gamma/\tilde{\mathcal{P}}) = E$ if and only if $E$ is controllable, observable, and $L_m(\tilde{\mathcal{P}})$-closed.

Since $E$ is $L_m(\tilde{\mathcal{P}})$-closed by definition, the result follows from Proposition 1.

∎

If $E$ is not controllable and observable, we will synthesize an optimal supervisor $\gamma$ (under partial observation) for $\tilde{P}$ such that $L_m(\gamma/\tilde{P}) = supC\mathcal{N}(E)$, the supremal controllable and normal sublanguage of $E$. The reason that we can replace here the requirement of observability by normality, is due to the fact that in the lifted system all unobservable events $\Sigma'$ are also uncontrollable, in which case a language is controllable and observable if and only if it is controllable and normal [18].

The synthesis is discussed in the next section.

# 4   Supervisor synthesis

Our objective is to design a nonblocking supervisor $\gamma$ for $\tilde{P}$ such that

$$L_m(\gamma/\tilde{P}) = supC\mathcal{N}(E).$$

This supervisor tracks only the events of $\Sigma$, and hence can be applied directly to $P$. It will be least restrictive in the sense that it allows the system $P$ to visit as many states in $Q_s$ as possible (see [9]).

Such a supervisor can be designed with or without the lifting procedure, as outlined in the two ensuing algorithms.

**Algorithm 1** *(Synthesis by lifting)*

1. Lift $P$ to $\tilde{P}$ using Procedure Extend:

$$\tilde{P} = (\Sigma \cup \Sigma', \tilde{Q}, \tilde{\delta}, \tilde{q}_0, \tilde{Q}_m, \tilde{Q}_b).$$

2. Compute the sublanguage $supC\mathcal{N}(E)$, that is, the supremal controllable and normal sublanguage of $E$.

3. Compute the projection $P(supC\mathcal{N}(E))$ of the language $supC\mathcal{N}(E)$ on $\Sigma$ and let the supervisor $\gamma$ be defined by

$$(\forall s \in \overline{PsupC\mathcal{N}(E)})\gamma(s) := \{\sigma \in \Sigma : s\sigma \notin \overline{PsupC\mathcal{N}(E)}\}.$$

∎

In the above algorithm, Step 1 is described in Section 3. Steps 2 and 3 are standard elements in the design of supervisors under partial observation [15].

The correctness of the above algorithm is obvious (see also [9]) and is stated in the following

**Theorem 2** The supervisor synthesized using Algorithm 1 is nonblocking and satisfies

$$L_m(\gamma/\tilde{P}) = supC\mathcal{N}(E).$$

**Proof**

Elementary.

$\blacksquare$

An alternate procedure for supervisor synthesis, that does not require the lifting of $\mathcal{P}$, is described in the next algorithm, where $Acc(.)$ denotes the accessible part of an automaton.

**Algorithm 2** *(Synthesis without lifting)*

1. Ignore the set $Q_m$ of marked states and convert the automaton $\mathcal{P} = (\Sigma \cup \{\epsilon\}, Q, \delta, q_0, Q_s)$ to a deterministic automaton $\hat{\mathcal{P}} = Acc(\Sigma, \hat{Q}, \hat{\delta}, \hat{q}_0, \hat{Q}_s)$, where

$$\hat{Q} := 2^Q;$$
$$\hat{\delta}(\hat{q}, \sigma) := \{q' \in Q : (\exists q \in \hat{q})q' \in \epsilon^*(\delta(q, \sigma))\};$$
$$\hat{q}_0 := \{q' \in Q : q' \in \epsilon^*(q_0)\};$$
$$\hat{Q}_s := \{\hat{q} \in \hat{Q} : (\forall u \in \Sigma_{uc}^*)\hat{\delta}(\hat{q}, u) \subseteq Q_s\};$$

2. If $\hat{Q}_s = \hat{Q}$, go to 7.

3. Set

$$\hat{\delta} := \hat{\delta}|_{\hat{Q}_s};$$
$$\hat{Q} := \{\hat{q} \in \hat{Q}_s : (\exists s \in \Sigma^*)\hat{q} = \hat{\delta}(\hat{q}_0, s)\},$$

   and form the product automaton

$$\mathcal{P}' := (\Sigma \cup \{\epsilon\}, Q \times \hat{Q}, \delta', (q_0, \hat{q}_0), Q_m \times \hat{Q}),$$

   where

$$\delta'((q, \hat{q}), \sigma) := \begin{cases} (\delta(q, \sigma), \hat{\delta}(\hat{q}, \sigma)) & \text{if both } \delta(q, \sigma) \text{ and } \hat{\delta}(\hat{q}, \sigma) \text{ are defined} \\ \text{undefined} & \text{otherwise.} \end{cases}$$

4. By trimming $\mathcal{P}'$ compute the set:

$$Q_t := \{q \in Q : (\exists \hat{q} \in \hat{Q})(q, \hat{q}) \text{ is accessible in } \mathcal{P}' \text{ from } (q_0, \hat{q}_0)$$
$$\text{and co} - \text{accessible in } \mathcal{P}' \text{ to } Q_m \times \hat{Q}\};$$

5. If $Q_t = Q_s$, go to 7. Otherwise, set

$$Q_s := Q_t;$$
$$\hat{Q}_s := \{\hat{q} \in \hat{Q} : (\forall u \in \Sigma_{uc}^*)\hat{\delta}(\hat{q}, u) \subseteq Q_s\};$$

6. Go to 3

7. Define the supervisor $\gamma$:

$$\gamma(s) = \{\sigma \in \Sigma_c : \hat{\delta}(\hat{q}_0, s\sigma) \text{ is not defined}\}.$$

∎

To prove that Algorithm 2 designs the correct supervisor in a finite number of steps (for finite automata), we first define, for languages $B$ and $M$ with $B \subseteq M = \overline{M} \subseteq (\Sigma \cup \Sigma')^*$,

$$sup\mathcal{N}(B) = \overline{B} - P^{-1}P(M - \overline{B})(\Sigma \cup \Sigma')^*$$
$$\Omega(B, M) = M \cap P^{-1}(P(supN(B)) - ((P(M) - P(supN(B)))/\Sigma_{uc}^*)\Sigma^*)$$
$$\Delta(B, M) = B \cap \Omega(B, M).$$

where $L/\Sigma_{uc}^* = \{s \in \Sigma^* : (\exists u \in \Sigma_{uc}^*)su \in L\}$. In the above, the operator $supN(B)$ calculates the supremal normal sublanguage of $\overline{B}$ (with respect to $M$) [1], the operator $\Omega(B, M)$ generates the supremal controllable and normal sublanguage of $\overline{B}$ (with respect to $M$) [1] and the operator $\Delta(B, M)$ intersects $\Omega(B, M)$ with $B$.

Suppose we apply these operators repeatedly with respect to the lifted automaton $\tilde{\mathcal{P}}$ and the corresponding legal language $E$ as follows.

$$M_0 = L(\tilde{\mathcal{P}}), \qquad B_0 = E$$
$$M_{i+1} = \Omega(B_i, M_i), \quad B_{i+1} = \Delta(B_i, M_i), \quad i = 0, 1, 2, \dots$$

Then we can show that $B_i$ converges to $supC\mathcal{N}(E)$ in the following

**Lemma 1** If there exists a positive integer $N$ such that $B_{N+1} = B_N$, then

$$B_N = supC\mathcal{N}(E).$$

**Proof**

Omitted.

∎

Using the above lemma, we can prove the following theorem, which states the correctness of Algorithm 2.

**Theorem 3** The supervisor synthesized using Algorithm 2 is nonblocking and satisfies

$$L_m(\gamma/\tilde{\mathcal{P}}) = supC\mathcal{N}(E).$$

**Outline of Proof**

We only give an outline of the proof because its details are tedious and provide no additional insight.

It is clear that in Algorithm 2, the first part of Step 1 is equivalent to calculating $sup\mathcal{N}(B_0)$ (without explicitly introducing $\Sigma'$) and the first part of Step 5 calculates $sup\mathcal{N}(B_i)$. The second parts (where $\hat{Q}_s$ is calculated) of Steps 1 and 5 calculate

$$P\Omega(B_i, M_i) = P(sup\mathcal{N}(B_i)) - ((P(M_i) - P(sup\mathcal{N}(B_i)))/\Sigma_{uc}^*)\Sigma^*$$

Steps 3 and 4 are equivalent to calculating

$$B_i \cap P^{-1}P(\Omega(B_i, M_i))$$
$$= B_i \cap P^{-1}P(\Omega(B_i, M_i)) \cap M_i$$
$$= B_i \cap \Omega(B_i, M_i)$$
$$= \Delta(B_i, M_i).$$

Therefore, Algorithm 2 implements the recursive computation of $B_i$, and calculates $supC\mathcal{N}(E)$.

$\blacksquare$

# 5 Control under partial observation

We now consider the situation when not all the events in $\Sigma$ are observable and the supervisor must be based on a subset $\Sigma_o \subseteq \Sigma$ of observable events. In this case, the set of unobservable events in the lifted process, is $(\Sigma \cup \Sigma') - \Sigma_o$, and if we denote by $T : \Sigma^* \rightarrow \Sigma_o^*$ the projection operator, then the projection from $\Sigma \cup \Sigma'$ to $\Sigma_o$ is obtained by the composition of $T$ and $P$.

In view of Theorem 1, the existence (and synthesis) of a supervisor under partial observation for $\mathcal{P}$ is equivalent to that of the corresponding supervisor for $\tilde{\mathcal{P}}$, because Theorem 1 hold for any supervisor, and a supervisor under partial observation is a special case. Therefore, we obtain the following corollary to Theorem 2.1 in [15].

**Corollary 1** There exists a nonblocking partial observation supervisor $\gamma : TPL(\tilde{\mathcal{P}}) \rightarrow 2^{\Sigma_c}$ such that $\gamma/\mathcal{P} = \mathcal{P}_s$ if and only if $E$ is controllable (with respect to $\Sigma_c$ and $L(\tilde{\mathcal{P}})$) and observable (with respect to $\Sigma_o$ and $L(\tilde{\mathcal{P}})$).

The supervisor can be synthesized with respect to $\tilde{\mathcal{P}}$. However, since it is no longer true that all the controllable events are also observable, observability can no longer be replaced by normality. Consequently, since the supremal observable sublanguage may not exist, a unique optimal supervisor may not exist either. To overcome this difficulty, two approaches can be employed: (1) to synthesize a sub-optimal supervisor based on the supremal controllable and normal sublanguage (with respect to $\Sigma_o$); and (2) to synthesize a maximal controllable and observable sublanguage, which may not be unique. Both approaches have been studied extensively in the literature and will not be repeated here.

If the specification is a language specification, then $E$ is normal [9]. In such a case, as we shall show in the following lemma, $E$ is observable with respect to $\Sigma_o$ and $L(\tilde{\mathcal{P}})$ if and only if $PE$ is observable with respect to $\Sigma_o$ and $PL(\tilde{\mathcal{P}})$.

**Lemma 2** Let $B$ be normal with respect to $\Sigma$ and $L(\tilde{\mathcal{P}})$. Then $B$ is observable with respect to $\Sigma_o$ and $L(\tilde{\mathcal{P}})$ if and only if $PB$ is observable with respect to $\Sigma_o$ and $PL(\tilde{\mathcal{P}}) = L(\mathcal{P})$.

**Proof**

Omitted.

$\blacksquare$

Using the lemma, we can immediately obtain the following

**Corollary 2** For a nondeterministic system $\mathcal{P}$ and a language specification $L(\hat{\mathcal{H}})$, there exists a nonblocking partial observation supervisor $\gamma$ such that $L(\gamma/\mathcal{P}) = L(\hat{\mathcal{H}})$ if and only if $L(\hat{\mathcal{H}})$ is controllable and observable with respect to $L(\mathcal{P})$.

This result was obtained in [14], where only language specifications were considered. The results in this section show that there is no need to treat the unobservable events $\Sigma_{uo} = \Sigma - \Sigma_o$ differently from the events $\Sigma'$, except that some events in $\Sigma_{uo}$ may be controllable. As a consequence, the supervisor synthesis may be more complex.

# 6 Decentralized control

The design of decentralized supervisors for nondeterministic systems can also be dealt with by using the deterministic theory and the lifting procedure. Since the methodology is quite analogous to what we have seen, we shall only outline the approach.

Without lose of generality, we may consider the case of two (decentralized) supervisors $\gamma_1$ and $\gamma_2$. For $i = 1, 2$, $\gamma_i$ can observe events in $\Sigma_{io} \subseteq \Sigma$ and control events in $\Sigma_{ic} \subseteq \Sigma$. Letting $T_i : \Sigma^* \to \Sigma_{io}^*$ denote the projections, we can can write the supervisors $\gamma_i$ as maps

$$\gamma_i : T_i L(\mathcal{P}) \to 2^{\Sigma_{ic}}.$$

As in [24], an event is enabled if it is enabled by both supervisors. The following existence result is then a corollary of Theorem 4.1 of [24].

**Corollary 3** There exists two nonblocking decentralized supervisors $\gamma_1$ and $\gamma_2$ such that $(\gamma_1 \wedge \gamma_2)/\mathcal{P} = \mathcal{P}_s$ if and only if $E$ is controllable (with respect to $\Sigma_{1c} \cup \Sigma_{2c}$) and co-observable.

Therefore, we conclude that both decentralized control and control under partial observation of nondeterministic systems can be synthesized by the existing methods for deterministic systems if we lift the corresponding processes.

# 7  Computational complexity

Since, in general, a supervisor synthesis problem under partial observation is of exponential complexity in terms of the number of transitions in the automata, it may be expected that the complexity of supervisor synthesis for nondeterministic systems also be exponential. Denote the number of states in an automaton $\mathcal{P}$ by $|\mathcal{P}|$ and the number of events by $|\Sigma|$. We outline the complexity analysis as follows.

Algorithm 1 involves two essential steps: (1) the procedure Extend that lifts $\mathcal{P}$ to a deterministic one, and (2) controller synthesis with respect to the lifted automaton. The procedure Extend adds at most $|\mathcal{P}| \times |\Sigma|$ states and $|\mathcal{P}|$ event labels to the process. The lifted automaton has, therefore, at most $|\mathcal{P}|(|\Sigma| + 1)$ states and $|\mathcal{P}| + |\Sigma|$ event labels. The complexity of executing Extend is of order $|\mathcal{P}|(|\Sigma| + 1)(|\mathcal{P}| + |\Sigma|)$. The synthesis of the optimal controller for the lifted process cannot be executed "on-line" because of the nonblocking requirement and, therefore, Algorithm 1 is of complexity

$$O((|\Sigma| + |\mathcal{P}|)2^{|\mathcal{P}|(|\Sigma|+1)}).$$

For Algorithm 2, the complexity of executing Steps 3-6 is at most $|\Sigma||\mathcal{P}|2^{|\mathcal{P}|}$. These steps will be repeated at most $|\mathcal{P}|$ times. Therefore, Algorithm 2 is of complexity

$$O(|\Sigma||\mathcal{P}|^2 2^{|\mathcal{P}|}).$$

# References


[1] R. D. Brandt, V. Garg, R. Kumar, F. Lin, S. I. Marcus and W. M. Wonham, 1990. Formulas for calculating supremal controllable and normal sublanguages. *Systems & Control Letters, 15*, pp. 111-117.

[2] S. L. Chung, S. Lafortune and F. Lin, 1992. Limited lookahead policies in supervisory control of discrete event systems. *IEEE Transactions on Automatic Control, 37(12)*, pp. 1921-1935.

[3] M. Fabian and B. Lennartson, 1994. Object oriented supervisory control with a class of nondeterministic specifications, Report No CTH/RT/I-94/007, Chalmers University of Technology, Goteborg, Sweden.

[4] M. Hennesy, *Algebraic Theory of Processes*, MIT Press, 1988.

[5] M. Heymann, 1990. Concurrency and discrete event control. *IEEE Control Systems Magazine, 10(4)*, pp. 103-112.

[6] M. Heymann and G. Meyer, 1991. An algebra of discrete event processes. *NASA Technical Memorandum 102848*.

[7] M. Heymann and F. Lin, 1994. On-line control of partially observed discrete event systems. *Discrete Event Dynamic Systems: Theory and Applications, 4(3)*, pp. 221-236.

[8] M. Heymann and F. Lin, 1995. On observability and nondeterminism in discrete event control, *Proceedings of the 33rd Allerton conference on Communication Control and Computing*, pp. 136-145.

[9] M. Heymann and F. Lin, 1996. Discrete event control of nondeterministic systems. *CIS Report 9601*, Technion, Israel.

[10] M. Heymann and F. Lin, 1996. Nonblocking supervisory control of nondeterministic systems. *CIS Report 9620*, Technion, Israel.

[11] C.A.R. Hoare, *Communicating Sequential Processes*, Prentice Hall, 1985.

[12] K. Inan, 1994. Nondeterministic supervision under partial observation. in G. Cohen and J.-P. Quadrat, Eds., *11th International Conference on Analysis and Optimization of Systems*, pp. 39-48, Springer Verlag.

[13] R. Kumar and M. A. Shayman, 1993. Non-blocking supervisory control of nondeterministic systems via prioritized synchronization. *Technical Research Report, T.R. 93-58*, Institute for Systems Research, University of Maryland.

[14] R. Kumar and M. A. Shayman, 1994. Supervisory control under partial observation of nondeterministic systems via prioritized synchronization. *Technical Report*, Department of Electrical Engineering, University of Kentucky.

[15] F. Lin and W. M. Wonham, 1988. On observability of discrete event systems. *Information Sciences, 44(3)*, pp. 173-198.

[16] F. Lin and W. M. Wonham, 1988. Decentralized supervisory control of discrete-event systems. *Information Sciences, 44(3)*, pp. 199-224.

[17] F. Lin and W. M. Wonham, 1990. Decentralized control and coordination of discrete event systems with partial observation. *IEEE Transactions on Automatic Control, 35(12)*, pp. 1330-1337.

[18] F. Lin and W. M. Wonham, 1994. Supervisory control of timed discrete event systems under partial observation, *IEEE Transactions on Automatic Control, 40(3)*, pp. 558-562.

[19] R. Milner, *A Calculus of Communicating Systems*, LNCS 94, Springer Verlag, 1980.

[20] A. Overkamp, 1994. Supervisory control for nondeterministic systems. *Proccedings of 11th International Conference on Analysis and Optimization of Systems*, pp. 59-65.

[21] A. Overkamp, 1994. Partial observation and partial specification in nondeterministic discrete-event systems, *preprint*.

[22] R. J. Ramadge and W. M. Wonham, 1987. Supervisory control of a class of discrete event processes. *SIAM J. Control and Optimization, 25(1)*, pp. 206-230.

[23] P. J. Ramadge and W. M. Wonham, 1989. The control of discrete event systems. *Proceedings of IEEE, 77(1)*, pp. 81-98.

[24] K. Rudie and W. M. Wonham, 1992. Think globally, act locally: decentralized supervisory control. *IEEE Transactions on Automatic Control, 37(11)*, pp. 1692-1708.

[25] M. Shayman and R. Kumar, 1995. Supervisory control of nondeterministic systems with driven events via prioritized synchronization and trajectory models. SIAM Journal of Control and Optimization, 33(2), pp. 469-497.

[26] J. N. Tsitsiklis, 1989. On the control of discrete-event dynamical systems. *Mathematics of Control, Signals, and Systems, 2(1)*, pp. 95-107.

Michael Heymann
Department of Computer Science
Technion, Israel Institute of Technology
Haifa 32000, Israel
e-mail: `heymann@cs.technion.ac.il`

Feng Lin
Department of Electrical and Computer Engineering
Wayne State University
Detroit, MI 48202, USA
e-mail: `flin@ece.eng.wayne.edu`

# Noncommutative Power Series and Formal Lie-algebraic Techniques in Nonlinear Control Theory

Matthias Kawski[1]
Arizona State University, Tempe, USA

Héctor J. Sussmann[2]
Rutgers University, New Brunswick, USA

**Abstract**: In nonlinear control, it is helpful to choose a formalism well suited to computations involving solutions of controlled differential equations, exponentials of vector fields, and Lie brackets. We show by means of an example —the computation of control variations that give rise to the Legendre-Clebsch condition— how a good choice of formalism, based on expanding diffeomorphisms as products of exponentials, can simplify the calculations. We then describe the algebraic structure underlying the formal part of these calculations, showing that it is based on the theory of formal power series, Lie series, the Chen series —introduced in control theory by M. Fliess— and the formula for the dual basis of a Poincaré-Birkhoff-Witt basis arising from a generalized Hall basis of a free Lie algebra.

## 1  Introduction

In the theory of nonlinear control systems, A. Agrachev and R. Gamkrelidze introduced a formalism for computing the flow maps arising from various controls, based on "chronological exponentials." It turns out that, underlying this formalism, there is a rich and interesting algebraic structure, involving algebras of formal power series (whose use in control theory was advocated by M. Fliess, cf., e.g, [7]), exponential Lie series, and the Chen-Fliess series. Since controls give rise to diffeomorphisms, it is natural to seek asymptotic expansions of flow maps as products of exponentials.

We will outline how such an expansion can be obtained, show how it can be applied to the derivation of high-order necessary conditions for optimality, and then describe the basic algebraic fact underlying such an expansion, namely, the formula for the dual basis of a Poincaré-Birkhoff-Witt basis of the universal enveloping algebra of a free Lie algebra.

---

[1]Supported in part by NSF-grant DMS 93-08289
[2]Supported in part by NSF Grant DMS95-00798 and AFOSR Grant 0923.

# 2 Manifolds, Vector Fields, and Control Systems

*Throughout this paper, "smooth" means "of class $C^\infty$," and "manifold" means "finite-dimensional, second-countable smooth manifold without boundary." We use $TM$, $T^*M$, $T_pM$, $T_p^*M$, to denote, respectively, the tangent and cotangent bundles of a manifold $M$, and the tangent and cotangent spaces at a point $p \in M$.*

If $M$ is a manifold, we use $\mathcal{E}(M)$ to denote the commutative algebra —over $\mathbb{R}$— of smooth real-valued functions on $M$, topologized in the usual way. (A sequence $\{\varphi_j\}$ converges to a limit $\varphi$ in $\mathcal{E}(M)$ iff for every $k \geq 0$ and every $k$-tuple $(X_1, \ldots, X_k)$ of smooth vector fields on $M$ the functions $X_1 X_2 \ldots X_k \varphi_j$ converge to $X_1 X_2 \ldots X_k \varphi$ uniformly on compact sets.) We let $\mathcal{E}'(M)$ denote the dual space of $\mathcal{E}(M)$, i.e. the space of compactly supported Schwartz distributions on $M$.

It is well known that $M$ is completely determined by $\mathcal{E}(M)$, in the following sense. A point $p$ of $M$ gives rise to a linear functional $\delta_p \in \mathcal{E}'(M)$ —the Dirac delta function at $p$— defined by $\delta_p(\varphi) = \varphi(p)$. This linear functional is *multiplicative*, i.e. is a homomorphism of $\mathbb{R}$-algebras. For any commutative algebra $A$ over a field $\mathbf{k}$, let us use $\sigma(A)$ to denote the *spectrum* of $A$, i.e. the set of all maps $\mu : A \to \mathbf{k}$ that are algebra homomorphisms and do not vanish identically. Then *the map $p \to \delta_p$ is a bijection from $M$ to $\sigma(\mathcal{E}(M))$.* So a manifold can be canonically identified with the spectrum of its algebra of smooth functions.

Via the map $p \to \delta_p$, we regard $M$ as embedded in $\mathcal{E}'(M)$. Moreover, many other objects related to $M$ can also be naturally regarded as members of $\mathcal{E}'(M)$. For example, $TM$ is embedded in $\mathcal{E}'(M)$ as follows: a tangent vector $v$ at a point is a linear functional on $\mathcal{E}(M)$, which maps $\varphi \in \mathcal{E}(M)$ to $v\varphi$. Since this functional is clearly continuous, $v$ belongs to $\mathcal{E}'(M)$.

If $X$ and $Y$ are smooth vector fields, then the product $XY$, evaluated at $p$, is a well defined member of $\mathcal{E}'(M)$, namely, the map $\varphi \to (XY\varphi)(p)$. So the product $XY$ —i.e. the map $p \to XY(p)$— is a well defined map from $M$ to $\mathcal{E}'(M)$. The difference $[X, Y] = XY - YX$ —the *Lie bracket* of $X$ and $Y$— is also a map from $M$ to $\mathcal{E}'(M)$, that happens to take values in $TM$, and is in fact a vector field.

On the other hand, $\mathcal{E}'(M)$ is a topological linear space, so linear operations and limiting processes that in principle appear not to make intrinsic sense on $M$ can be meaningful in $\mathcal{E}'(M)$. For example, if $\gamma : [0, \varepsilon] \to M$ is a curve, and $\gamma(0) = p$, then we would like to define the tangent vector $\dot{\gamma}(0)$ by just letting

$$\dot{\gamma}(0) \stackrel{\text{def}}{=} \lim_{h \to 0} h^{-1}(\gamma(h) - p). \tag{1}$$

This may look unacceptable, since $M$ is not a linear space, and $\gamma(h) - p$ does not make sense. However, Formula (1) is perfectly meaningful, and gives the right answer, if we regard it as an identity in $\mathcal{E}'(M)$, since $\lim_{h \to 0} h^{-1}(\gamma(h) - p)$ is the distribution that maps $\varphi \in \mathcal{E}(M)$ to the number $\lim_{h \to 0} h^{-1}(\varphi(\gamma(h)) - \varphi(p))$, i.e. directional differentiation at $p$ in the direction of $\dot{\gamma}(0)$.

To make the formalism work well, it is convenient to write $\pi\varphi$ —rather than $\pi(\varphi)$— for the value of the linear functional $\pi \in \mathcal{E}'(M)$ at the function $\varphi \in \mathcal{E}(M)$.

With this notation, when $p \in M$ and $\varphi \in \mathcal{E}(M)$ the value $\varphi(p)$ is now written $p\varphi$. If $f$ is a smooth vector field and $\varphi \in \mathcal{E}(M)$, then $f\varphi$ is in $\mathcal{E}(M)$, and the number usually written as $(f\varphi)(p)$ now becomes $p(f\varphi)$. On the other hand, if we also write $pf$ rather than the more common $f(p)$, the number $p(f\varphi)$ is also equal to $(pf)\varphi$, so we can just write $pf\varphi$ without parentheses.

More generally, maps $\Phi$ from $M$ to any manifold $N$ should act on points of $M$ on the right, so we will write $p\Phi$ rather than $\Phi(p)$. A smooth map $\Phi : M \to N$ gives rise to a map $\tilde{\Phi} : \mathcal{E}'(M) \to \mathcal{E}'(N)$, defined by letting $\tilde{\Phi}(D)$, for $D \in \mathcal{E}'(M)$, be the distribution on $N$ given by $\tilde{\Phi}(D)(\psi) = D(\psi \circ \Phi)$, i.e., in our formalism, $(D\tilde{\Phi})\psi \stackrel{\text{def}}{=} D(\Phi\psi)$, where, of course, $\Phi\psi$ is the function usually referred to as the "pullback" $\Phi^*(\psi)$ of $\psi$ to $M$ via $\Phi$. (Naturally, since maps act on the right, $\Phi\psi$ is the composite map obtained by applying $\Phi$ first, and then $\psi$.) When $D$ is actually a point $p \in M$, then $(D\tilde{\Phi})\psi$ is what is normally written as $\psi(\Phi(p))$, so $p\tilde{\Phi} = p\Phi$. So $\tilde{\Phi}$ is an extension to $\mathcal{E}'(M)$ of the map $\Phi$, originally defined on the subset $M$ of $\mathcal{E}'(M)$. It is convenient to just use $\Phi$ for this extension. With this notation, if $p \in M$ and $v$ is a tangent vector at $p$, then $v\Phi$ is the tangent vector at $p\Phi$ usually written as $\Phi_*(v)$, or $D\Phi(p).v$.

In particular, we will use $e^{tf}$ to denote the time $t$ flow map arising from a vector field $f$, and will write it as acting on the right. So $t \to pe^{tf}$ is the integral curve $\xi_{f,p}$ of $f$ that goes through $p$ at time 0. This curve satisfies the equation

$$(d/dt)(pe^{tf}) = pe^{tf}f, \tag{2}$$

which is a much more elegant way of writing the formula $(d/dt)(\xi_{f,p}(t)) = f(\xi_{f,p}(t))$, and amply justifies the use of the exponential notation.

With this formalism, several important formulas involving Lie brackets become completely trivial formally, and the formal calculations can be rigorously justified by working in $\mathcal{E}'(M)$. For example, let $f, g$ be vector fields, and let $p \in M$. Then

$$\begin{aligned}
&(d/dt)(pe^{tf}e^{tg}e^{-tf}e^{-tg}) \\
= {}& pe^{tf}fe^{tg}e^{-tf}e^{-tg} + pe^{tf}e^{tg}ge^{-tf}e^{-tg} - pe^{tf}e^{tg}e^{-tf}fe^{-tg} - pe^{tf}e^{tg}e^{-tf}e^{-tg}g,
\end{aligned}$$

whose value for $t = 0$ is $pf + pg - pf - pg$, i.e. 0.

The second derivative $\frac{d^2}{dt^2}(pe^{tf}e^{tg}e^{-tf}e^{-tg})$ is then equal to

$$\begin{aligned}
&pe^{tf}f^2e^{tg}e^{-tf}e^{-tg} + pe^{tf}fe^{tg}ge^{-tf}e^{-tg} - pe^{tf}fe^{tg}e^{-tf}fe^{-tg} - pe^{tf}fe^{tg}e^{-tf}e^{-tg}g \\
&+ pe^{tf}fe^{tg}ge^{-tf}e^{-tg} + pe^{tf}e^{tg}g^2e^{-tf}e^{-tg} - pe^{tf}e^{tg}ge^{-tf}fe^{-tg} - pe^{tf}e^{tg}ge^{-tf}e^{-tg}g \\
&- pe^{tf}fe^{tg}e^{-tf}fe^{-tg} - pe^{tf}e^{tg}ge^{-tf}fe^{-tg} + pe^{tf}e^{tg}e^{-tf}f^2e^{-tg} + pe^{tf}e^{tg}e^{-tf}fe^{-tg}g \\
&- pe^{tf}fe^{tg}e^{-tf}e^{-tg}g - pe^{tf}e^{tg}ge^{-tf}e^{-tg}g + pe^{tf}e^{tg}e^{-tf}fe^{-tg}g + pe^{tf}e^{tg}e^{-tf}e^{-tg}g^2,
\end{aligned}$$

whose value for $t = 0$ is $2pfg - 2pgf$, i.e. $2p[f, g]$. So we have shown that

$$pe^{tf}e^{tg}e^{-tf}e^{-tg} = p + t^2p[f, g] + O(t^2) \quad \text{as } t \to 0, \tag{3}$$

which is the familiar formula describing how the Lie bracket measures the failure to close of the "square" described by the left-hand side of (3).

As a second example, we write another familiar formula, namely,

$$(d/dt)(pe^{tf}ge^{-tf}) = pe^{tf}fge^{-tf} - pe^{tf}gfe^{-tf} = pe^{tf}[f,g]e^{-tf}, \qquad (4)$$

which says that, if we define a vector field $v_g$ along an integral curve of $f$ by letting $v_g(t) = pe^{tf}g$, and then move $v_g(t)$ back to $T_pM$ via the differential of the diffeomorphism $e^{-tf}$ —which sends $pe^{tf}$ to $p$— then the result is the derivative of the pullback of $v_{[f,g]}(t)$ via the differential of $e^{-tf}$.

Now consider a control system of the form

$$\dot{x} = x(f + ug), \quad |u| \le 1, \qquad (5)$$

where $f$ and $g$ are smooth vector fields on $M$, and we are using the previously described formalism in which vector fields act on the right.

For a control $\eta : [a, b] \to [-1, 1]$, let us use the expression $\boxed{xe^{\int_a^t (f+\eta(s)g)ds}}$ to denote the point $\xi(t)$, if $\xi$ is the trajectory of (5) corresponding to $\eta$ and the initial condition $\xi(a) = x$. (This "chronological exponential" formalism was introduced by Agrachev and Gamkrelidze. Notice that $xe^{\int_a^t (f+\eta(s)g)ds} = xe^{\int_a^t (f+\eta(s)g)ds}$ if $\eta$ is constant, and that the derivative of $xe^{\int_a^t (f+\eta(s)g)ds}$ is $xe^{\int_a^t (f+\eta(s)g)ds}(f+\eta(t)g)$ for a. e. $t$, justifying the use of an exponential notation.)

Given a measurable control $\eta : [a, b] \to [-1, 1]$, a *point variation of $\eta$ at a time* $t_0 \in [a, b]$ is a family $\boldsymbol{\eta} = \{\eta^\varepsilon\}_{0 \le \varepsilon \le \bar{\varepsilon}}$ of controls $\eta^\varepsilon : [a, b] \to [-1, 1]$ such that $\eta^0 = \eta$, having the property that for every $\delta > 0$ there exists $\bar{\varepsilon}(\delta) \in ]0, \bar{\varepsilon}]$ for which the set $\{t : \eta^\varepsilon(t) \ne \eta(t)\} \cap [t_0 - \delta, t_0 + \delta]$ is empty whenever $\varepsilon \in ]0, \bar{\varepsilon}(\delta)]$.

For a point variation $\boldsymbol{\eta}$ and an $x \in M$ we define the *endpoint curve* $\gamma_{\boldsymbol{\eta},x}$ by

$$\gamma_{\boldsymbol{\eta},x}(\varepsilon) = xe^{\int_a^b (f+\eta^\varepsilon(s)g)ds}. \qquad (6)$$

If $\gamma_{\boldsymbol{\eta},x}(\varepsilon)$ is continuous near $\varepsilon = 0$ and differentiable at $\varepsilon = 0$, then the vector

$$\tilde{v}_{\boldsymbol{\eta},x} \overset{\text{def}}{=} (d/d\varepsilon)\Big|_{\varepsilon=0}\Big(xe^{\int_a^b (f+\eta^\varepsilon(s)g)ds}\Big) \qquad (7)$$

is the *terminal variational vector* corresponding to the variation $\boldsymbol{\eta}$. The pullback of this vector to the point $xe^{\int_a^t (f+\eta(s)g)ds}$ via the diffeomorphism $\Big(e^{\int_t^b (f+\eta(s)g)ds}\Big)^{-1}$ is the *variational vector* $v_{\boldsymbol{\eta},x,t}$ generated at time $t$ by $\boldsymbol{\eta}$ and the initial condition $x$. So, if $\delta(\varepsilon) \ge 0$ is chosen for every $\varepsilon \in ]0, \bar{\varepsilon}]$ in arbitrary fashion, subject only to the condition that $\{t : \eta^\varepsilon(t) \ne \eta(t)\} \subseteq [a, t_0 + \delta(\varepsilon)]$, then

$$v_{\boldsymbol{\eta},x,t_0} = (d/d\varepsilon)\Big|_{\varepsilon=0}\Big(xe^{\int_a^{t_0+\delta(\varepsilon)} (f+\eta^\varepsilon(s)g)ds}\Big(e^{\int_{t_0}^{t_0+\delta(\varepsilon)} (f+\eta(s)g)ds}\Big)^{-1}\Big). \qquad (8)$$

Point variations and their corresponding variational vectors occur in the Pontryagin Maximum Principle and its "high-order" generalizations. For example, a necessary condition for a trajectory $\xi$ of (5) and corresponding control $\eta$ to have the property that the reachable set from $\xi(a)$ is not a neighborhood of $\xi(b)$ is that there exist a nontrivial "adjoint vector" —i.e. an absolutely continuous field of covectors $[a, b] \ni t \to p(t) \in T^*_{\xi(t)}M$ along $\xi$ that satisfies the *adjoint equation* $\boxed{\dot{p}(t) = -p(t).(\partial/\partial x)(f + \eta(t)g)(\xi(t))}$ for a.e. $t$— for which

$$\langle p(t), (f + \eta(t)g)(\xi(t))\rangle = \max\{\langle p(t), (f + ug)(\xi(t))\rangle : -1 \le u \le 1\} \qquad (9)$$

for almost every $t$. Condition (9) is equivalent to the statement that, for almost all $t$, $\langle p(t), v \rangle \leq 0$ whenever $v$ is a tangent vector at $\xi(t)$ such that $v$ is of the form $\xi(t)((u - \eta(t))g)$ for some $u \in [-1, 1]$. These vectors $v$ are precisely the variational vectors arising from "needle variations" $\boldsymbol{\eta}_{u,t}$ at time $t$.

In various "high-order" versions of the Maximum Principle, Condition (9) is replaced by the much stronger requirement that $\langle p(t), v \rangle \leq 0$ whenever $v$ is a tangent vector at $\xi(t)$ which is of the form $v_{\boldsymbol{\eta},\xi(a),t}$ for some point variation of $\eta$ at $t$. (The precise statement of the high-order maximum principle requires that the variational vectors satisfy an extra "compatibility" condition. We will ignore this complication, and point out that the compatibility condition is always satisfied when all the pairs $(t, v)$ under considerations satisfy *Knobloch's condition* (cf. Knobloch [10]): $v = \lim_{j \to \infty} v_j$, where the $v_j$ are variational vectors arising from point variations $\boldsymbol{\eta}_j$ at time $t_j$ and $t_j \to t$, $t_j \neq t$.)

**Theorem 2.1** *Let* $\xi : [a, b] \to M$ *be a trajectory of (5) corresponding to a measurable control* $\eta : [a, b] \to [-1, 1]$. *Let* $\bar{t} \in [a, b[$ *be such that* $|\eta(\bar{t})| < 1$ *and* $\bar{t}$ *is a Lebesgue point of* $\eta$. *Then there exists a point variation* $\boldsymbol{\eta}$ *of* $\eta$ *at* $\bar{t}$ *that gives rise to a variational vector* $v = v_{\boldsymbol{\eta},\xi(a),\bar{t}} = -[g, [f, g]](\xi(\bar{t}))$.

The vector $v$ given by the above theorem clearly satisfies Knobloch's condition, since the set of Lebesgue points of $\eta$ has full measure, and every interval $[\bar{t}, \bar{t} + \varepsilon]$ must contain, for small enough $\varepsilon$, a subset of positive measure where $|\eta| < 1$, since otherwise the integral $\int_{\bar{t}}^{\bar{t}+\varepsilon} |\eta(t) - \eta(\bar{t})| dt$ would be bounded below by a positive constant times $\varepsilon$, contradicting the hypothesis that $\bar{t}$ is a Lebesgue point of $\eta$.

Theorem 2.1 says that at almost all times $t$ such that $|\eta(t)| < 1$ the adjoint vector $p$ may be required to satisfy the inequality $\langle p(t), [g, [f, g]](\xi(t)) \rangle \geq 0$, in addition to the condition that $\langle p(t), g(\xi(t)) \rangle = 0$, which follows from (9). Since the Hamiltonian $H$ is given by $H(x, p, u) = \langle p, (f + ug)(x) \rangle$, we see that the new inequality says that $(\partial/\partial u)(d^2/dt^2)(\partial H/\partial u) \geq 0$, along our trajectory, which is the usual Legendre-Clebsch condition.

*Proof of Theorem 2.1.* Without loss of generality, we assume that $\bar{t} = 0$. We are going to construct our variation $\boldsymbol{\eta}$ by letting $\eta^\varepsilon$ be, for sufficiently small $\varepsilon$, a control of the form $\eta^\varepsilon = \eta + \theta^\rho$, with $\rho = k\varepsilon^{1/3}$, $k$ a constant to be chosen later, and $\theta^\rho$ a function that vanishes outside the interval $[0, \rho]$. Once we have explained how to choose $\theta^\rho$, we will want to study the effect of the resulting control $\eta + \theta^\rho$ on the interval $[0, \rho]$, and for this purpose we will first look at an arbitrary control $u : [0, \rho] \to [-1, 1]$ and study the map $e^{\int_0^\rho (f+u(t)g)dt}$. In particular, we will want to compute a product asymptotic expression for his map that will be exact modulo errors of order $o(\rho^3)$. Once this is done, we will try to choose $\theta^\rho$ so as to match all the factors of this expansion except one, and the one factor that is not matched will become the leading term of the variation we want.

So, let $u : [0, \rho] \to [-1, 1]$ be measurable, and write $x_1(t) = x \, e^{\int_0^t (f+u(s)g)ds}$, for a fixed $x \in M$. Let $x_1(t) = x_2(t)e^{tf}$. Then $\dot{x}_1(t) = \dot{x}_2(t)e^{tf} + x_2(t)e^{tf}f = \dot{x}_2(t)e^{tf} + x_1(t)f$.

Comparing this with $\dot{x}_1(t) = x_1(t)(f + u(t)g)$, we find $\dot{x}_2(t)e^{tf} = x_1(t)u(t)g$, so $\dot{x}_2(t) = x_2(t)e^{tf}u(t)ge^{-tf}$.

Now, if $X$, $Y$ are vector fields, then $(d/dt)e^{tY}Xe^{-tY} = e^{tY}[Y, X]e^{-tY}$ (cf. (4)), so $e^{tY}Xe^{-tY} = X + R_1(X, Y, t)$, where $R_1(X, Y, t) = \int_0^t e^{sY}[Y, X]e^{-sY}ds$.

Iterating this we get

$$e^{tY}Xe^{-tY} = X + t[Y, X] + R_2(X, Y, t)\,, \tag{10}$$

where
$$R_2(X, Y, t) = \int_0^t \int_0^{s_1} e^{s_2 Y}[Y, [Y, X]]e^{-s_2 Y}ds_2 ds_1\,. \tag{11}$$

Then
$$e^{tY}Xe^{-tY} = X + t[Y, X] + \frac{t^2}{2}[Y, [Y, X]] + R_3(X, Y, t)\,, \tag{12}$$

where
$$R_3(X, Y, t) = \int_0^t \int_0^{s_1} \int_0^{s_2} e^{s_3 Y}[Y, [Y, [Y, X]]]e^{-s_3 Y}ds_3 ds_2 ds_1\,. \tag{13}$$

Then
$$e^{tf}ge^{-tf} = g + t[f, g] + \frac{t^2}{2}[f, [f, g]] + R_3(g, f, t)\,, \tag{14}$$

so $\dot{x}_2(t) = x_2(t)(u(t)g + X_t)$, with $X_t = tu(t)[f, g] + \frac{t^2 u(t)}{2}[f, [f, g]] + u(t)R_3(g, f, t)$.

Now let $\boxed{U_1(t) = \int_0^t u(s)ds}$, and write $x_2(t) = x_3(t)e^{U_1(t)g}$. Then

$$\dot{x}_2(t) = \dot{x}_3(t)e^{U_1(t)g} + x_3(t)e^{U_1(t)g}u(t)g = \dot{x}_3(t)e^{U_1(t)g} + x_2(t)u(t)g\,. \tag{15}$$

So
$$\dot{x}_3(t) = x_3(t)e^{U_1(t)g}X_t e^{-U_1(t)g}\,. \tag{16}$$

We now use (10) and write $e^{U_1(t)g}X_t e^{-U_1(t)g} = X_t + U_1(t)[g, X_t] + R_2(X_t, g, U_1(t))$, so

$$\begin{aligned}
\dot{x}_3(t) &= x_3(t)\Big(X_t + U_1(t)[g, X_t] + R_2(X_t, g, U_1(t))\Big) \\
&= x_3(t)\Big(tu(t)[f, g] + \tfrac{t^2 u(t)}{2}[f, [f, g]] + tu(t)U_1(t)[g, [f, g]] + S_1(t)\Big),
\end{aligned}$$

where $S_1(t)$ is a complicated expression which is $O(t^3)$. Now define $U_2$ by letting $\boxed{U_2(t) = \int_0^t su(s)ds}$. Then write $x_3(t) = x_4(t)e^{U_2(t)[f,g]}$. We then get

$$\dot{x}_4(t) = x_4(t)\Big(\frac{t^2 u(t)}{2}[f, [f, g]] + tu(t)U_1(t)[g, [f, g]] + S_2(t)\Big)\,, \tag{17}$$

where $S_2(t)$ is also $O(t^3)$. Next, define $\boxed{U_3(t) = \tfrac{1}{2}\int_0^t s^2 u(s)ds}$, and then write $x_4(t) = x_5(t)e^{U_3(t)[f,[f,g]]}$. Then

$$\dot{x}_5(t) = x_5(t)\Big(tu(t)U_1(t)[g, [f, g]] + S_3(t)\Big)\,, \tag{18}$$

where $S_3(t)$ is also $O(t^3)$. Finally, if we let $\boxed{U_4(t) = \int_0^t su(s)U_1(s)ds}$, we can write $x_5(t) = x_6(t)e^{U_4(t)[g,[f,g]]}$, and conclude that $\dot{x}_6(t) = x_6(t)S_4(t)$, where $S_4(t)$ is $O(t^3)$. Since $x_6(0) = x$, we have $x_6(t) = x + O(t^4)$. Then

$$\boxed{x_1(t) = xe^{U_4(t)[g,[f,g]]}e^{U_3(t)[f,[f,g]]}e^{U_2(t)[f,g]}e^{U_1(t)g}e^{tf} + O(t^4)}\,, \tag{19}$$

i.e.
$$\boxed{e^{\int_0^t (f+u(s)g)ds} = e^{U_4(t)[g,[f,g]]}e^{U_3(t)[f,[f,g]]}e^{U_2(t)[f,g]}e^{U_1(t)g}e^{tf} + O(t^4)}\,. \tag{20}$$

In Formulas (19), (20), the product of five exponentials that appears in the right-hand side is the initial part —up to terms of degree 3— of the general *product expansion of the Chen series*, a purely formal Lie-algebraic result that gives, on a formal level, an expansion of a the diffeomorphism $e^{\int_0^t (f+u(s)g)ds}$ as a product of diffeomorphisms. The formulas themselves are special cases of the general proposition that *the Chen series associated to an expression such as* $e^{\int_0^t (f+u(s)g)ds}$ *gives an asymptotic expansion for this expression.*

We now use these asymptotic formulas to construct our variation. In our computations, we will use $U_i^\zeta$ to denote, for $i = 1, 2, 3, 4$, the functions $U_i$ arising from a particular control $\zeta$.

Fix once and for all a $c > 0$ such that $-1 \le \eta(0) - 2c$ and $\eta(0) + 2c \le 1$. Define the "good set" $G_\rho$ and the "bad set" $B_\rho$ by

$$G_\rho \stackrel{\text{def}}{=} \left\{ t \in [0, \rho] : |\eta(t) - \eta(0)| \le c \right\}, \quad B_\rho = [0, \rho] \backslash G_\rho. \tag{21}$$

Use $|E|$ to denote the Lebesgue measure of a measurable subset $E$ of $\mathbb{R}$. Then $c|B_\rho| \le \int_0^\rho |\eta(t) - \eta(0)|dt = o(\rho)$, so $|B_\rho| = o(\rho)$, and then $|G_\rho| = \rho - o(\rho)$.

Let $P^\rho(t) = p_0^\rho + p_1^\rho t + p_2^\rho t^2 + p_3^\rho t^3$ be the unique cubic polynomial such that

$$\int_{G_\rho} t^i P^\rho(t)\, dt = \rho^4 \delta_{i3} \quad \text{for} \quad i = 0, 1, 2, 3\,, \tag{22}$$

where "$\delta$" is Kronecker's delta. (Clearly, $P^\rho$ exists and is unique for small $\rho$.)

If $\hat{G}_\rho = \{s \in [0, 1] : \rho s \in G_\rho\}$, then $|\hat{G}_\rho| \to 1$ as $\rho \to 0$. Define $\hat{G}^0 = [0, 1]$. For $\rho \ge 0$, let $Q^\rho(s) = \sum_{i=0}^3 q_i^\rho t^i$ be such that $\int_{\hat{G}_\rho} Q^\rho(s)s^i ds = \delta_{i3}$ for $i = 0, 1, 2, 3$. Then

$$P^\rho(t) = Q^\rho(\rho^{-1}t) = \sum_{i=0}^3 q_i^\rho \rho^{-i}t^i \quad \text{if} \quad \rho > 0\,. \tag{23}$$

As $\rho \to 0$, the $q_i^\rho$ converge to $q_i^0$. (Explicitly, $Q^0(t) = 2800t^3 - 4200t^2 + 1680t - 140$.) Let $\kappa = 1 + \max\{|Q^0(s)| : 0 \le s \le 1\}$. Then $|Q^\rho(t)| \le \kappa$ for all $t \in [0, 1]$, if $\rho$ is small.

Take
$$\theta^\rho(t) = \frac{c}{\kappa} P^\rho(t) \text{ for } t \in G_\delta\,, \quad \theta^\rho(t) = 0 \text{ for } t \notin G_\delta\,. \tag{24}$$

Let $\eta^\varepsilon = \eta + \theta^\rho$. Then $\eta^\varepsilon$ is admissible if $\varepsilon$ is small, and $U_i^{\eta^\varepsilon}(\rho) = U_i^\eta(\rho)$ for $i = 1, 2, 3$. We now compute $U_4^{\eta^\varepsilon}(\rho) - U_4^\eta(\rho)$. For a general control $\zeta$, we have

$$U_4^\zeta(t) = \int_0^t s\zeta(s)U_1^\zeta(s)ds = \frac{1}{2}\int_0^t s\frac{d}{ds}(U_1^\zeta(s))^2 ds\,. \tag{25}$$

Integration by parts then yields $U_4^\zeta(\rho) = \frac{\rho}{2}(U_1^\zeta(\rho))^2 - \frac{1}{2}\int_0^\rho (U_1^\zeta(s))^2 ds$. Since $U_1^{\eta^\varepsilon}(\rho)$ is equal to $U_1^\eta(\rho)$, we have $U_4^{\eta^\varepsilon}(\rho) - U_4^\eta(\rho) = \frac{1}{2}\int_0^t \left( U_1^\eta(s)^2 - U_1^{\eta^\varepsilon}(s)^2 \right)ds$.

If we let $\Theta^\rho(t) = \int_0^t \theta^\rho(s)ds$, then $U_1^{\eta^\varepsilon}(t) = U_1^\eta(t) + \Theta^\rho(t)$, from which it follows that $U_1^{\eta^\varepsilon}(t)^2 = U_1^\eta(t)^2 + \Theta^\rho(t)^2 + 2U_1^\eta(t)\Theta^\rho(t)$, and then

$$U_4^{\eta^\varepsilon}(\rho) - U_4^\eta(\rho) = -\frac{1}{2}\int_0^\rho \Theta^\rho(s)^2 ds - \int_0^\rho U_1^\eta(s)\Theta^\rho(s)ds\,. \tag{26}$$

Now, if $0 \le s \le 1$, we have

$$\Theta^\rho(\rho s) = \int_0^{\rho s} \theta^\rho(w)dw = \frac{c}{\kappa}\int_0^{\rho s} P^\rho(w)dw = \frac{c\rho}{\kappa}\int_0^s P^\rho(\rho t)dt = \frac{c\rho}{\kappa}\int_0^s Q^\rho(t)dt\,. \tag{27}$$

Then

$$\int_0^\rho \Theta^\rho(t)^2 dt = \rho \int_0^1 \Theta^\rho(\rho s)^2 ds = \frac{c^2\rho^3}{\kappa^2} \int_0^1 \left( \int_0^s Q^\rho(t)dt \right)^2 ds = \frac{\nu c^2\rho^3}{\kappa^2} + o(\rho^3), \quad (28)$$

where $\nu = \int_0^1 \left( \int_0^s Q^0(t)dt \right)^2 ds$.

On the other hand, the integral $I = \int_0^\rho U_1^\eta(s)\Theta^\rho(s)ds$ can also be computed by parts, and is equal to $V^\eta(\rho)\Theta^\rho(\rho) - \int_0^\rho V^\eta(s)\theta^\rho(s)ds$, where $V^\eta(t) = \int_0^t U_1^\eta(s)ds$. Since $\Theta^\rho(\rho) = \int_0^\rho \theta^\rho(t)dt = 0$, $I$ is in fact equal to $- \int_0^\rho V^\eta(s)\theta^\rho(s)ds$. Since $U_1^\eta(t) = \int_0^t \eta(s)ds$, we have $U_1^\eta(t) = t\eta(0) + o(t)$. Then $V^\eta(t) = \frac{t^2}{2}\eta(0) + o(t^2)$. Since $\int_0^\rho t^2\theta^\rho(t)dt = 0$, we have $I = o(\rho^3)$.

Therefore

$$U_4^{\eta^\epsilon}(\rho) - U_4^\eta(\rho) = -\frac{\nu c^2\rho^3}{2\kappa^2} + o(\rho^3) = -\frac{\nu c^2 k^3 \varepsilon}{2\kappa^2} + o(\varepsilon).$$

Choosing $k$ such that $\nu c^2 k^3 = 2\kappa^2$, we have $U_4^{\eta^\epsilon}(\rho) - U_4^\eta(\rho) = -\varepsilon + o(\rho^3)$. Then, if $x \in M$, we have

$$x e^{\int_0^\rho (f+\eta^\epsilon(s)g)ds} \left( e^{\int_0^\rho (f+\eta(s)g)ds} \right)^{-1} = -\varepsilon\, x\, [g,[f,g]] + o(\varepsilon),$$

and our proof is complete.

# 3   Basic algebraic structures

*In this and the following sections, we work with a fixed field* **k** *of scalars, assumed to be of characteristic zero. We use LS, AA, LA, CA, CP as abbreviations for "linear space," "associative algebra," "Lie algebra," "chronological algebra," and "chronological product," respectively. All LS's, AA's, LA's, and CA's assumed to be over* **k**. *Every AA* $\mathcal{A}$ *is automatically regarded as a LA, with the Lie bracket* $[x,y]$ *of* $x,y \in \mathcal{A}$ *defined by* $[x,y] = xy - yx$.

The calculations of the preceding sections were carried out for vector fields, but it is clear that a large part of what was done was purely formal. The purpose of this section is to exhibit the basic algebraic structures underlying the formal part of our arguments, to review some of their properties, and to fix terminology and notation. For a detailed description the reader is referred to [16].

We recall that, if $L$ is a LA, then a *universal enveloping algebra* (abbr. UAE) of $L$ is a pair $(\mu, U)$ such that $U$ is an AA and $\mu : L \to U$ is a LA-homomorphism, having the property that, if $\mu' : L \to U'$ is any other LA-homomorphism from $L$ to an AA $U'$, then there exists a unique AA-homomorphism $\nu : U \to U'$ such that $\nu\mu = \mu'$. The existence of $(\mu, U)$ is trivial, since one can let $U$ be the quotient of the tensor algebra $T(L)$ over $L$ by the two-sided ideal generated by the elements $x \otimes y - y \otimes x - [x,y]$, for $x, y \in L$. The *Poincaré-Birkhoff-Witt Theorem* (abbr. PBWT) says that, if $(\mu, U)$ is a UEA of $L$, $\mathcal{B}$ is any basis of $L$, endowed with a total ordering $\preceq$, and we use $PWB(\mathcal{B}, \preceq)$ to denote the set of all products $\mu(B_1) \cdot \mu(B_2) \cdot \ldots \cdot \mu(B_m)$, for all finite sequences $(B_1, B_2, \ldots, B_m)$ of members of $\mathcal{B}$ such that $B_1 \preceq B_2 \preceq \ldots \preceq B_m$, then $PWB(\mathcal{B}, \preceq)$ is a basis of $U$. It follows from the PBWT that $\mu$ *is necessarily injective.* In view of this, we will only consider from now on UEA's $(\mu, U)$ such that $L \subseteq U$ and $\mu : L \to U$ is the inclusion map.

Let $\mathbf{X}$ be a set (also called an *alphabet*) of noncommuting indeterminates (or *letters*). Use $W(\mathbf{X})$ to denote the set $\cup_{n=0}^{\infty} \mathbf{X}^n$, i.e. the union of the Cartesian products $\mathbf{X}^n$ of $n$ copies of $\mathbf{X}$. We refer to $W(\mathbf{X})$ as the *free monoid* generated by $\mathbf{X}$. The members of $W(\mathbf{X})$ —called *monomials*, or *words*— are the finite sequences $w = (x_1, x_2, \dots, x_n)$ of letters, for which we will use the notation $w = x_1 x_2 \dots x_n$. If $w \in \mathbf{X}^n$ then $w$ is a *word of length* $n$, or a *monomial of degree* $n$, and we write $n = |w|$. The *product* $ww'$ of $w = x_1 x_2 \dots x_n$, $w' = x'_1 x'_2 \dots x'_\nu$ is just the concatenation $x_1 x_2 \dots x_n x'_1 x'_2 \dots x'_\nu$. Clearly, $|ww'| = |w| + |w'|$. Also, we write $lett(w)$ to denote the set of all letters $x \in \mathbf{X}$ that occur in $w$. We write 1 —or $\emptyset$— to denote the only word of length zero, i.e. the empty word. Clearly, $1w = w1 = w$ for all $w \in W(\mathbf{X})$. We use $W^+(\mathbf{X})$ to denote the set $\{w \in W(\mathbf{X}) : w \neq 1\}$.

We let $A_{\mathbf{k}}(\mathbf{X})$ denote the free AA generated by $\mathbf{X}$ with coefficients in $\mathbf{k}$. Then $A_{\mathbf{k}}(\mathbf{X})$ is the set of all formal linear combinations $p = \sum_{w \in W(\mathbf{X})} p_w w$ of monomials with coefficients $p_w \in \mathbf{k}$ such that the set $\{w : p_w \neq 0\}$ is finite. The members of $A_{\mathbf{k}}(\mathbf{X})$ are the *noncommuting polynomials* in the letters of $\mathbf{X}$. We let $A_{\mathbf{k}}^+(\mathbf{X}) \overset{\text{def}}{=} \{P = \sum_w p_w w \in A_{\mathbf{k}}(\mathbf{X}) : p_1 = 0\}$, so $A_{\mathbf{k}}^+(\mathbf{X})$ is a two-sided ideal of $A_{\mathbf{k}}(\mathbf{X})$.

We use $\hat{A}_{\mathbf{k}}(\mathbf{X})$ to denote the completion of $A_{\mathbf{k}}(\mathbf{X})$ with respect to the uniform structure in which a basis of "entourages" of the diagonal is given by the sets $\mathcal{U}_F = \{(P, Q) \in A_{\mathbf{k}}(\mathbf{X}) \times A_{\mathbf{k}}(\mathbf{X}) : P - Q \in U_F\}$, for all finite subsets $F$ of $W(\mathbf{X})$, where $U_F = \{P = \sum_w p_w w \in A_{\mathbf{k}}(\mathbf{X}) : p_w = 0 \text{ whenever } w \in F\}$. Then $\hat{A}_{\mathbf{k}}(\mathbf{X})$ is the set of all sums $S = \sum_w s_w w$, with $\{s_w\}_{w \in W(\mathbf{X})}$ an arbitrary family of members of $\mathbf{k}$ (so in particular $\{w : s_w \neq 0\}$ is not required to be finite). The members of $\hat{A}_{\mathbf{k}}(\mathbf{X})$ are then the *formal power series in the letters of* $\mathbf{X}$ *with coefficients in* $\mathbf{k}$. The multiplication map $(P, Q) \to PQ$ from $A_{\mathbf{k}}(\mathbf{X}) \times A_{\mathbf{k}}(\mathbf{X})$ to $A_{\mathbf{k}}(\mathbf{X})$ is uniformly continuous, so it extends to a continuous map —also denoted by $(S, T) \to ST$— from $\hat{A}_{\mathbf{k}}(\mathbf{X}) \times \hat{A}_{\mathbf{k}}(\mathbf{X})$ to $\hat{A}_{\mathbf{k}}(\mathbf{X})$, with respect to which $\hat{A}_{\mathbf{k}}(\mathbf{X})$ is an AA.

Associated to the basis $W(\mathbf{X})$ of $A_{\mathbf{k}}(\mathbf{X})$ is a pairing $<\cdot, \cdot> : \hat{A}_{\mathbf{k}}(\mathbf{X}) \times A_{\mathbf{k}}(\mathbf{X}) \to \mathbf{k}$, that sends $(S, T) \in \hat{A}_{\mathbf{k}}(\mathbf{X}) \times A_{\mathbf{k}}(\mathbf{X})$ to $<S, T> = \sum_w S_w T_w$, if $S = \sum_w S_w w$ and $T = \sum_w T_w w$. Then $S = \sum_w <S, w> w$ if $S \in \hat{A}_{\mathbf{k}}(\mathbf{X})$, so $<S, T> = \sum_w <S, w> <T, w>$. With the pairing $<\cdot, \cdot>$, $\hat{A}_{\mathbf{k}}(\mathbf{X})$ is the algebraic dual of $A_{\mathbf{k}}(\mathbf{X})$. Moreover, the continuous linear functionals on $\hat{A}_{\mathbf{k}}(\mathbf{X})$ are exactly the maps $S \to <S, T>$, for $T \in A_{\mathbf{k}}(\mathbf{X})$, so $A_{\mathbf{k}}(\mathbf{X})$ is the topological dual of $\hat{A}_{\mathbf{k}}(\mathbf{X})$.

We use $L_{\mathbf{k}}(\mathbf{X})$ to denote the Lie subalgebra generated by $\mathbf{X}$ of the LA $A_{\mathbf{k}}(\mathbf{X})$. It is well known —and follows easily from the PBWT— that $L_{\mathbf{k}}(\mathbf{X})$ is a realization of the *free LA over* $\mathbf{X}$. (That is, if $\Lambda$ is any LA over $\mathbf{k}$ and $\mu : \mathbf{X} \to \Lambda$ is a mapping, then $\mu$ can be extended in a unique way to a LA-homomorphism $\tilde{\mu} : L_{\mathbf{k}}(\mathbf{X}) \to \Lambda$.) It is also easy to show that $A_{\mathbf{k}}(\mathbf{X})$ is a UAE of $L_{\mathbf{k}}(\mathbf{X})$.

A formal series $S \in \hat{A}_{\mathbf{k}}(\mathbf{X})$ is called a *Lie series* if for every $n$ the homogeneous component $S^{hom,n} \overset{\text{def}}{=} \sum_{|w|=n} <S, w> w$ is in $L_{\mathbf{k}}(\mathbf{X})$. We use $\hat{L}_{\mathbf{k}}(\mathbf{X})$ to denote the set of all Lie series. It is easy to see that $\hat{L}_{\mathbf{k}}(\mathbf{X})$ is a Lie subalgebra of $\hat{A}_{\mathbf{k}}(\mathbf{X})$.

The family $\{v \otimes w\}_{v,w \in W(\mathbf{X})}$ is a basis of the tensor product $A_\mathbf{k}(\mathbf{X}) \otimes A_\mathbf{k}(\mathbf{X})$, so every $P \in A_\mathbf{k}(\mathbf{X}) \otimes A_\mathbf{k}(\mathbf{X})$ is expressible in a unique way as a sum $P = \sum_{v,w} p_{v,w} v \otimes w$ such that $\{(v,w) : p_{v,w} \neq 0\}$ is finite. We use $A_\mathbf{k}(\mathbf{X}) \hat{\otimes} A_\mathbf{k}(\mathbf{X})$ to denote the completion of $A_\mathbf{k}(\mathbf{X}) \otimes A_\mathbf{k}(\mathbf{X})$ with respect to the uniform structure in which a basis of "entourages" of the diagonal is given by the sets

$$\mathcal{U}_F^2 = \{(P,Q) \in (A_\mathbf{k}(\mathbf{X}) \otimes A_\mathbf{k}(\mathbf{X})) \times (A_\mathbf{k}(\mathbf{X}) \otimes A_\mathbf{k}(\mathbf{X})) : P - Q \in U_F^2\},$$

for all finite subsets $F$ of $W(\mathbf{X}) \times W(\mathbf{X})$, where

$$U_F^2 = \{P = \sum_{v,w} p_{v,w} v \otimes w \in A_\mathbf{k}(\mathbf{X}) \otimes A_\mathbf{k}(\mathbf{X}) : p_{v,w} = 0 \text{ whenever } (v,w) \in F\}.$$

Then $A_\mathbf{k}(\mathbf{X}) \hat{\otimes} A_\mathbf{k}(\mathbf{X})$ can be thought of as the set of all sums $S = \sum_{v,w} s_{v,w} v \otimes w$, with $\{s_{v,w}\}_{v,w \in W(\mathbf{X})}$ a completely arbitrary family of members of $\mathbf{k}$ (so in particular $\{(v,w) : s_{v,w} \neq 0\}$ is not required to be finite). The members of $A_\mathbf{k}(\mathbf{X}) \hat{\otimes} A_\mathbf{k}(\mathbf{X})$ are exactly the formal sums $S = \sum_{v \in W(\mathbf{X})} v \otimes S_v$, with all the $S_v$ in $\hat{A}_\mathbf{k}(\mathbf{X})$, so $A_\mathbf{k}(\mathbf{X}) \hat{\otimes} A_\mathbf{k}(\mathbf{X})$ is the space $\hat{A}_{\hat{A}_\mathbf{k}(\mathbf{X})}(\mathbf{X})$ of formal power series in $\mathbf{X}$ with co-efficients in $\hat{A}_\mathbf{k}(\mathbf{X})$. There is a natural pairing $(S,P) \to \, <S,P>$ that sends $S \in A_\mathbf{k}(\mathbf{X}) \hat{\otimes} A_\mathbf{k}(\mathbf{X})$, $P \in A_\mathbf{k}(\mathbf{X}) \otimes A_\mathbf{k}(\mathbf{X})$ to $<S,P> = \sum_{v,w} s_{v,w} p_{v,w} \in \mathbf{k}$, if $S = \sum_{v,w} s_{v,w} v \otimes w \in A_\mathbf{k}(\mathbf{X}) \hat{\otimes} A_\mathbf{k}(\mathbf{X})$, $P = \sum_{v,w} p_{v,w} v \otimes w \in A_\mathbf{k}(\mathbf{X}) \otimes A_\mathbf{k}(\mathbf{X})$. With respect to this pairing, $A_\mathbf{k}(\mathbf{X}) \hat{\otimes} A_\mathbf{k}(\mathbf{X})$ is the algebraic dual of $A_\mathbf{k}(\mathbf{X}) \otimes A_\mathbf{k}(\mathbf{X})$, and $A_\mathbf{k}(\mathbf{X}) \otimes A_\mathbf{k}(\mathbf{X})$ is the topological dual of $A_\mathbf{k}(\mathbf{X}) \hat{\otimes} A_\mathbf{k}(\mathbf{X})$.

An important subspace of $A_\mathbf{k}(\mathbf{X}) \hat{\otimes} A_\mathbf{k}(\mathbf{X})$ is $A_\mathbf{k}(\mathbf{X}) \tilde{\otimes} A_\mathbf{k}(\mathbf{X})$, the set of all sums $S = \sum_{v,w} s_{v,w} v \otimes w$ such that for each $v$ the set $\{w : s_{v,w} \neq 0\}$ is finite. The members of $A_\mathbf{k}(\mathbf{X}) \tilde{\otimes} A_\mathbf{k}(\mathbf{X})$ are the sums $S = \sum_{v \in W(\mathbf{X})} v \otimes S_v$ with all the $S_v$ in $A_\mathbf{k}(\mathbf{X})$, so $A_\mathbf{k}(\mathbf{X}) \tilde{\otimes} A_\mathbf{k}(\mathbf{X})$ is the space $\hat{A}_{A_\mathbf{k}(\mathbf{X})}(\mathbf{X})$ of formal power series in $\mathbf{X}$ with coefficients in $A_\mathbf{k}(\mathbf{X})$. Finally, the tensor product $\hat{A}_\mathbf{k}(\mathbf{X}) \otimes \hat{A}_\mathbf{k}(\mathbf{X})$ is naturally identified with the set of all sums $S = \sum_{v,w} s_{v,w} v \otimes w$ such that the matrix $(s_{v,w})_{v,w \in W(\mathbf{X})}$ has finite rank.

Clearly, $A_\mathbf{k}(\mathbf{X}) \hat{\otimes} A_\mathbf{k}(\mathbf{X})$ is an AA, in which the product $PQ$ is given by $PQ = \sum_{v,w} (PQ)_{v,w} v \otimes w$, where

$$(PQ)_{v,w} = \sum_{v',v'',w',w'' \in W(\mathbf{X}), \, v'v''=v, \, w'w''=w} \, <v' \otimes w', P> \, <v'' \otimes w'', W>.$$

(The summations defining the $(PQ)_{v,w}$ are clearly finite.) Then $A_\mathbf{k}(\mathbf{X}) \otimes A_\mathbf{k}(\mathbf{X})$, $A_\mathbf{k}(\mathbf{X}) \tilde{\otimes} A_\mathbf{k}(\mathbf{X})$, and $\hat{A}_\mathbf{k}(\mathbf{X}) \otimes \hat{A}_\mathbf{k}(\mathbf{X})$ are subalgebras of $A_\mathbf{k}(\mathbf{X}) \hat{\otimes} A_\mathbf{k}(\mathbf{X})$.

The *diagonal map* $\Delta : A_\mathbf{k}(\mathbf{X}) \to A_\mathbf{k}(\mathbf{X}) \otimes A_\mathbf{k}(\mathbf{X})$ is the k-algebra homomorphism defined on generators $x \in \mathbf{X}$ by $\Delta(x) = x \otimes 1 + 1 \otimes x$. It is easy to see that $\Delta$ is uniformly continuous. So $\Delta$ extends uniquely to a continuous algebra homomorphism —also called $\Delta$— from $\hat{A}_\mathbf{k}(\mathbf{X})$ to $A_\mathbf{k}(\mathbf{X}) \hat{\otimes} A_\mathbf{k}(\mathbf{X})$. Clearly, $\Delta(S) = \sum_n \Delta(S^{hom,n})$ for $S \in \hat{A}_\mathbf{k}(\mathbf{X})$.

The well known *Friedrichs' criterion* —which is a fairly easy consequence of the PBWT— says that, if $S \in \hat{A}_\mathbf{k}(\mathbf{X})$, then $S \in \hat{L}_\mathbf{k}(\mathbf{X})$ iff $\boxed{\Delta(S) = S \otimes 1 + 1 \otimes S}$.

The *shuffle product* is the bilinear map $\text{ш} : A_\mathbf{k}(\mathbf{X}) \times A_\mathbf{k}(\mathbf{X}) \to A_\mathbf{k}(\mathbf{X})$ such that

$$<S, v \, \text{ш} \, w> = <\Delta(S), v \otimes w> \qquad \text{for } S \in \hat{A}_\mathbf{k}(\mathbf{X}), \, v, w \in A_\mathbf{k}(\mathbf{X}). \tag{29}$$

So $ш$, regarded as a linear map $A_{\mathbf{k}}(\mathbf{X}) \otimes A_{\mathbf{k}}(\mathbf{X}) \to A_{\mathbf{k}}(\mathbf{X})$, is the transpose of $\Delta$.

It is also possible to characterize the shuffle product recursively, by first letting $w \, ш \, 1 = 1 \, ш \, w = w$, and then defining $(xv) \, ш \, (yw) = x(v \, ш \, (yw)) + y((xv) \, ш \, w)$ for $x, y \in \mathbf{X}$, $v, w \in W(\mathbf{X})$. It is easy to show that $A_{\mathbf{k}}(\mathbf{X})$, endowed with $ш$, is an associative and commutative algebra.

Friedrichs' criterion is equivalent to the statement that *an element $S \in \hat{A}_{\mathbf{k}}(\mathbf{X})$ is a Lie series if and only if (i) $< S, 1 > = 0$, and (ii) $S$ is orthogonal to all nontrivial shuffles, i.e. $< S, v \, ш \, w > = 0$ for all $v, w \in W^+(\mathbf{X})$.*

Next, we let $\hat{G}_{\mathbf{k}}(\mathbf{X}) \subseteq \hat{A}_{\mathbf{k}}(\mathbf{X})$ denote the set of all *exponential Lie series*, that is the set of all formal power series $S \in \hat{A}_{\mathbf{k}}(\mathbf{X})$ such that there is a $Z \in \hat{L}_{\mathbf{k}}(\mathbf{X})$ for which $S = \exp(Z) \stackrel{\text{def}}{=} \sum_{k=0}^{\infty} \frac{1}{k!} Z^k$. Friedrichs' criterion easily implies that a series $S \in \hat{A}_{\mathbf{k}}(\mathbf{X})$ is in $\hat{G}_{\mathbf{k}}(\mathbf{X})$ if and only if $S \neq 0$ and $\Delta(S) = S \otimes S$. From this it follows in particular that if $S \in \hat{G}_{\mathbf{k}}(\mathbf{X})$ then $< S, 1 > = 1$, so $S^{-1} = \sum_{k=0}^{\infty} (1 - S)^k$ exists. It also follows that $\hat{G}_{\mathbf{k}}(\mathbf{X})$ is a group under multiplication. (The fact that $S_1, S_2 \in \hat{G}_{\mathbf{k}}(\mathbf{X})$ implies $S_1 S_2 \in \hat{G}_{\mathbf{k}}(\mathbf{X})$ is the well known *Campbell-Hausdorff formula*. If $S_i = \exp(Z_i)$ for $i = 1, 2$, then $S_1 S_2 = \exp(\mathbf{P}(Z_1, Z_2))$, where

$$\boxed{\mathbf{P}(x, y) = x + y + \tfrac{1}{2}[x, y] + \tfrac{1}{12}[x, [x, y]] + \tfrac{1}{12}[y, [x, y]] + \ldots}$$

.)

Since $< \Delta(S), v \otimes w > = < S, v \, ш \, w >$, the condition that $\Delta(S) = S \otimes S$ is equivalent to the property that $< S, v \, ш \, w > = < S, v > < S, w >$ for all words $v, w$. i.e. that "the coefficients $< S, w >$ of $S$ satisfy the shuffle relations." This observation, known as *Ree's theorem*, says that $S \in \hat{G}_{\mathbf{k}}(\mathbf{X})$ if and only if the linear map $T \to < S, T >$, from $A_{\mathbf{k}}(\mathbf{X})$ to $\mathbf{k}$, is nonzero and multiplicative (with respect to $ш$), i.e. is an algebra homomorphism from $(A_{\mathbf{k}}(\mathbf{X}), ш)$ to $\mathbf{k}$ that sends 1 to 1.

There is a clear analogy between the facts of the previous paragraphs and our discussion in §2. The commutative algebra $A_{\mathbf{k}}(\mathbf{X})$ can be realized as an algebra of functions on $\hat{G}_{\mathbf{k}}(\mathbf{X})$, by mapping each $P \in A_{\mathbf{k}}(\mathbf{X})$ to the function $\tilde{P}$ given by $\tilde{P}(S) = < S, P >$ for $S \in \hat{G}_{\mathbf{k}}(\mathbf{X})$. Since $< S, P_1 > < S, P_2 > = < S, P_1 \, ш \, P_2 >$ for $S \in \hat{G}_{\mathbf{k}}(\mathbf{X})$, we see that under the map $P \to \tilde{P}$ the shuffle product in $A_{\mathbf{k}}(\mathbf{X})$ corresponds to ordinary pointwise multiplication of functions on $\hat{G}_{\mathbf{k}}(\mathbf{X})$. Moreover, $\hat{G}_{\mathbf{k}}(\mathbf{X})$ is embedded in $\hat{A}_{\mathbf{k}}(\mathbf{X})$, the dual of $A_{\mathbf{k}}(\mathbf{X})$, and Ree's theorem tells us that $\hat{G}_{\mathbf{k}}(\mathbf{X})$ is exactly the spectrum of $A_{\mathbf{k}}(\mathbf{X})$, i.e. that the nonzero linear functionals $S \in \hat{A}_{\mathbf{k}}(\mathbf{X})$ that are multiplicative are exactly those that belong to $\hat{G}_{\mathbf{k}}(\mathbf{X})$.

So $\hat{G}_{\mathbf{k}}(\mathbf{X})$ may be regarded as a formal analogue of the manifold $M$ of §2, with $A_{\mathbf{k}}(\mathbf{X})$ playing the role of $\mathcal{E}(M)$ and $\hat{A}_{\mathbf{k}}(\mathbf{X})$ that of $\mathcal{E}'(M)$. So it is natural to call the elements of $\hat{G}_{\mathbf{k}}(\mathbf{X})$ *formal points*. Clearly, $\hat{G}_{\mathbf{k}}(\mathbf{X})$ is a "formal Lie group," and $\hat{L}_{\mathbf{k}}(\mathbf{X})$ is its "Lie algebra." Pursuing our analogy, we define a *formal tangent vector* to $\hat{G}_{\mathbf{k}}(\mathbf{X})$ at a point $S \in \hat{G}_{\mathbf{k}}(\mathbf{X})$ to be a linear functional $V : A_{\mathbf{k}}(\mathbf{X}) \to \mathbf{k}$ such that $V(P \, ш \, Q) = V(P)Q(S) + V(Q)P(S)$ for all $P, Q \in A_{\mathbf{k}}(\mathbf{X})$. Using the identification of $\hat{A}_{\mathbf{k}}(\mathbf{X})$ with the dual of $A_{\mathbf{k}}(\mathbf{X})$, the linear functional $V$ is of the form $P \to < W, P >$ for some $W \in \hat{A}_{\mathbf{k}}(\mathbf{X})$, and we can write $W = SZ$, for $Z \in \hat{A}_{\mathbf{k}}(\mathbf{X})$, since $S$ is invertible. Then the functional $P \to < SZ, P >$ is a formal tangent vector at $S$ if and only if $< SZ, P \, ш \, Q > = < SZ, P > < S, Q > + < S, P > < SZ, Q >$ for

$P, Q \in A_\mathbf{k}(\mathbf{X})$, i.e. iff $<\Delta(SZ), P \otimes Q> = <SZ \otimes S, P \otimes Q> + <S \otimes SZ, P \otimes Q>$ for all $P, Q \in A_\mathbf{k}(\mathbf{X})$. This happens iff $\Delta(SZ) = SZ \otimes S + S \otimes SZ$, i.e. —since $\Delta(SZ) = \Delta(S)\Delta(Z) = (S \otimes S)\Delta(Z)$, and $SZ \otimes S + S \otimes SZ = (S \otimes S)(Z \otimes 1 + 1 \otimes Z)$— iff $\Delta(Z) = Z \otimes 1 + 1 \otimes Z$. So *the formal tangent vectors to $\hat{G}_\mathbf{k}(\mathbf{X})$ at $S$ are exactly the functionals $A_\mathbf{k}(\mathbf{X}) \ni P \to <SZ, P> \in \mathbf{k}$, for $Z \in \hat{L}_\mathbf{k}(\mathbf{X})$.* In particular, the members $Z$ of $\hat{L}_\mathbf{k}(\mathbf{X})$ must be thought of as tangent vectors to $\hat{G}_\mathbf{k}(\mathbf{X})$ at 1. If $Z \in \hat{L}_\mathbf{k}(\mathbf{X})$ then the map $S \to SZ$ is a *formal left-invariant vector field* on $\hat{G}_\mathbf{k}(\mathbf{X})$.

In agreement with the notation $pV\varphi$ introduced in §2, for a point $p$, a vector field $V$ and a function $\varphi$, the expression $<SZ, P>$ can be thought of as the result of applying $SZ$ —regarded as a tangent vector at $S$— to the function $P \in A_\mathbf{k}(\mathbf{X})$. Naturally, then, we define $L_Z$ —the operator of "formal Lie differentiation in the direction of $Z$"— to be the map that assigns to every $P \in A_\mathbf{k}(\mathbf{X})$ the function $Q = L_Z P \in A_\mathbf{k}(\mathbf{X})$ such that $<SZ, P> = <S, Q>$ for all $S \in \hat{G}_\mathbf{k}(\mathbf{X})$. Then $<SZ, P> = <S, Q>$ for all $S \in \hat{A}_\mathbf{k}(\mathbf{X})$, so $L_Z : A_\mathbf{k}(\mathbf{X}) \to A_\mathbf{k}(\mathbf{X})$ is just the transpose of the map $\hat{A}_\mathbf{k}(\mathbf{X}) \ni S \to SZ \in \hat{A}_\mathbf{k}(\mathbf{X})$.

For $S \in \hat{A}_\mathbf{k}(\mathbf{X})$, $P, Q \in A_\mathbf{k}(\mathbf{X})$, we have

$$<S, L_Z(P \, \text{ш} \, Q)> = <SZ, P \, \text{ш} \, Q> = <SZ, P><S, Q> + <S, P><SZ, Q>$$
$$= <S, L_Z P><S, Q> + <S, P><S, L_Z Q> = <S \otimes S, L_Z P \otimes Q + P \otimes L_Z Q>$$
$$= <\Delta(S), L_Z P \otimes Q + P \otimes L_Z Q> = <S, (L_Z P) \, \text{ш} \, Q + P \, \text{ш} \, (L_Z Q)> .$$

So $L_Z(P \, \text{ш} \, Q) = (L_Z P) \, \text{ш} \, Q + P \, \text{ш} \, (L_Z Q)$ for $P, Q \in A_\mathbf{k}(\mathbf{X})$, showing that $L_Z$ is a derivation on the algebra $A_\mathbf{k}(\mathbf{X})$ equipped with the shuffle product.

If $Z = x \in \mathbf{X}$, then $L_x$ is easily seen to be the map characterized by $L_x(wy) = 0$ if $y \neq x$, $w \in W(\mathbf{X})$, $L_x(wx) = w$ if $w \in W(\mathbf{X})$, and $L_x(1) = 0$.

This characterization implies, in particular, that *for every family $\{P^x\}_{x \in \mathbf{X}}$ of members of $A_\mathbf{k}(\mathbf{X})$ indexed by $\mathbf{X}$ there exists a unique $Q \in A_\mathbf{k}(\mathbf{X})$ such that $L_x Q = P^x$ for all $x \in \mathbf{X}$ and $<1, Q> = 0$.* (Indeed, letting $P^x = \sum_{w \in W(\mathbf{X})} p^x_w w$, $Q = \sum_{w \in W(\mathbf{X})} q_w w$, we have $L_x Q = \sum_{w \in W(\mathbf{X})} q_{wx} w$, so $Q$ satisfies our conditions iff $q_1 = 0$ and $q_{wx} = p^x_w$ for all $w \in W(\mathbf{X})$, $x \in \mathbf{X}$, from which the existence and uniqueness of $Q$ follows trivially.)

# 4 Chronological algebras, iterated integrals, and Chen series

Chronological algebras play a fundamental role in control (cf. [2, 3, 18, 9]), simplify formulas in combinatorics (cf. [8]), and are closely related to the *Leibniz algebras* that have recently been investigated in the algebraic literature (cf. [11, 12]).

A *(right) chronological algebra* is a linear space $A$ over a field $\mathbf{k}$ endowed with a bilinear operation $* : A \times A \to A$ that satisfies the *right chronological identity*:

$$x * (y * z) = (x * y) * z + (y * x) * z \qquad \text{for all } x, y, z \in A . \tag{30}$$

(One can also define *left CA's*, in which the identity $(x * y) * z = x * (y * z) + x * (z * y)$ holds. In this note only right CA's will be used, so we will omit the word "right." The notion of CA introduced here is similar to that of Agrachev and Gamkrelidze [3], which also has been studied under the name of *Leibniz algebra* by Loday [11, 12] and others. The key identity for that other notion is the formula $(x\sharp y)\sharp z - (y\sharp x)\sharp z = x\sharp(y\sharp z) - y\sharp(x\sharp z)$, which says that $L_{x\sharp y - y\sharp x} = [L_x, L_y]$, where $L_u$ is the map $z \to u\sharp z$. A typical example of a CA in this other sense is obtained by tensoring a Lie algebra with a CA in our sense.)

If $(A, *)$ is a CA, and we define $P \shuffle Q \overset{\text{def}}{=} P * Q + Q * P$ for $P, Q \in A$, then it is easy to see that $\shuffle$ is commutative and associative.

As a first example of a CP, we define $P * Q$, for $P, Q \in A_\mathbf{k}(\mathbf{X})$, by letting $P * Q$ be the unique $R \in A_\mathbf{k}(\mathbf{X})$ such that $L_x(R) = P \shuffle L_x Q$ for all $x \in \mathbf{X}$ and $<1, R> = 0$. (The existence and uniqueness of $R$ follows from the last remark of the previous section.) The CP then satisfies $L_x(P * Q) = P \shuffle L_x Q$ for all $x \in \mathbf{X}$. It then follows easily that $P * Q + Q * P = P \shuffle Q$ whenever $<1, P> \, <1, Q> = 0$, since $L_x(P * Q + Q * P) = L_x(P * Q) + L_x(Q * P) = P \shuffle L_x Q + Q \shuffle L_x P = L_x(P \shuffle Q)$, $<1, P * Q + Q * P> = 0$, and $<1, P \shuffle Q> = <1, P> \, <1, Q>$. This implies that, on the algebra $A_\mathbf{k}^+(\mathbf{X})$, the map $*$ is a CP, since

$$L_x(P * (Q * R)) = P \shuffle L_x(Q * R) = P \shuffle (Q \shuffle L_x R) = (P \shuffle Q) \shuffle L_x R$$
$$= L_x((P \shuffle Q) * R) = L_x((P * Q + Q * P) * R) = L_x((P * Q) * R + (Q * P) * R),$$

so $P * (Q * R) = (P * Q) * R + (Q * P) * R$.

We refer to $(A_\mathbf{k}^+(\mathbf{X}), *)$ as the *free CA over* $\mathbf{k}$ *in the indeterminates* $\mathbf{X}$, because $(A_\mathbf{k}^+(\mathbf{X}), *)$ is a "free CA" in the usual sense: if $B$ is any CA over $\mathbf{k}$ and $\mu : \mathbf{X} \to B$ is a map, then $\mu$ can be extended to a unique CA-homomorphism $\tilde{\mu}$ from $A_\mathbf{k}^+(\mathbf{X})$ to $B$. (We construct $\tilde{\mu}$ recursively by letting $\tilde{\mu}(x) = \mu(x)$ for $x \in \mathbf{X}$, and $\tilde{\mu}(wx) = \tilde{\mu}(w) * \mu(x)$ for $x \in \mathbf{X}$, $w \in W(X)$. It is then not hard to verify that $\tilde{\mu}$ has the desired property, using the fact that $wx = w * x$. Uniqueness is trivial.)

There are numerous other examples of CA's. For example, if $\mathbf{k} = \mathbb{R}$, and $\mathcal{AC}$ is the space of locally absolutely continuous functions $f : [0, +\infty[ \to \mathbb{R}$ such that $f(0) = 0$, then we can define $f * g$, for $f, g$ in $\mathcal{AC}$, by $(f * g)(t) = \int_0^t f(s) g'(s)\, ds$, where $g'$ is the derivative of $g$. Then $f * g + g * f = fg$, and the CP identity says that $\int f (\int gh')' = \int (fg) h'$, which is trivially true. So $(\mathcal{AC}, *)$ is a CA over $\mathbb{R}$, and the commutative product associated to $*$ is just ordinary multiplication.

There are several natural chronological subalgebras of $(\mathcal{AC}, *)$. For example, the set of $f \in \mathcal{AC}$ such that $f \in C^\infty$, or the set $f \in \mathcal{AC}$ that are polynomial functions. In the latter example, a basis of the algebra is given by the monomials $x^m$, $m = 1, 2, \ldots$, and $x^n * x^m = \frac{1}{n} x^{n+m}$.

Finally, if $\mathbf{k}$ is any field of characteristic zero, then in the algebra $\mathbf{k}^+[X]$ of polynomials of a single variable over $\mathbf{k}$ with zero constant terms, we can define $x^n * x^m = \frac{m}{m+n} x^{n+m}$. Then $(\mathbf{k}^+[X], *)$ is a chronological algebra. Naturally, if $P, Q \in \mathbf{k}^+[X]$, then $P * Q = \int(PQ')$ where, for $S \in \mathbf{k}^+[X]$, $\int S$ is the unique polynomial $T \in \mathbf{k}^+[X]$ such that $T' = S$.

The $\mathbf{k}$-algebra $\mathbf{k}[X_1, \ldots X_m]$ of polynomials in $m$ variables with coefficients in $\mathbf{k}$ can be represented as an algebra of functions on $\mathbf{k}^m$, namely, the algebra

$\mathbf{k}[x_1, \ldots x_n]$ of polynomial functions in $m$ k-valued *variables,* i.e. the subalgebra of $\mathrm{Map}(\mathbf{k}^m, \mathbf{k})$ generated by the projection maps $\mathbf{k}^m \ni (p_1, \ldots, p_m) \to x_i(p) \overset{\mathrm{def}}{=} p_i \in \mathbf{k}$, where $\mathrm{Map}(\mathbf{k}^m, \mathbf{k})$ is the set of all maps from $\mathbf{k}^m$ to $\mathbf{k}$, regarded as an algebra with pointwise multiplication.

Similarly, we will represent the free CA $(A_\mathbf{k}(\mathbf{X}), *)$ as an algebra of "dynamic functionals" $\mathcal{U}_\mathbf{k}^\mathbf{X} \to \mathcal{U}_\mathbf{k}$, where $\mathcal{U}_\mathbf{k}$ is a CA of "time-varying scalars," i.e. of k-valued functions $t \to f(t)$ of "time." One has substantial freedom in choosing the basic CA $\mathcal{U}_\mathbf{k}$. Here, for the sake of clarity, we specialize to a familiar setting. We work with $\mathbf{k} = \mathbb{R}$, and choose $\mathcal{U}_\mathbf{k} = \mathcal{AC}$. Then $\mathcal{U}_\mathbf{k}^\mathbf{X}$ is the set of all families $\{U_x : x \in \mathbf{X}\}$ of locally absolutely continuous real-valued functions on $[0, \infty[$ that vanish at 0. We use $\pi_y$ to denote, for each $y \in \mathbf{X}$, the canonical projection $\mathcal{U}_\mathbf{k}^\mathbf{X} \ni \{U_x : x \in \mathbf{X}\} \to U_y \in \mathcal{U}_\mathbf{k}$.

We use $\mathrm{Map}(\mathcal{U}_\mathbf{k}^\mathbf{X}, \mathcal{U}_\mathbf{k})$ to denote the set of all maps from $\mathcal{U}_\mathbf{k}^\mathbf{X}$ to $\mathcal{U}_\mathbf{k}$. Then $\mathrm{Map}(\mathcal{U}_\mathbf{k}^\mathbf{X}, \mathcal{U}_\mathbf{k})$ is a CA under pointwise chronological multiplication: for $\Phi, \Psi$ in $\mathrm{Map}(\mathcal{U}_\mathbf{k}^\mathbf{X}, \mathcal{U}_\mathbf{k})$ and $U \in \mathcal{U}_\mathbf{k}^\mathbf{X}$, we define $(\Phi * \Psi)(U) = \Phi(U) * \Psi(U)$.

The CA $\mathcal{IIF}_\mathbf{k}(\mathbf{X})$ of *iterated integral functionals on* $\mathcal{U}_\mathbf{k}^\mathbf{X}$ is the chronological subalgebra of $\mathrm{Map}(\mathcal{U}_\mathbf{k}^\mathbf{X}, \mathcal{U}_\mathbf{k})$ generated by the set $\{\pi_x : x \in \mathbf{X}\}$. So, if we use $\mathcal{I}_\mathbf{k}^\mathbf{X}$ to denote the unique CA-homomorphism from $A_\mathbf{k}(\mathbf{X}) \to \mathrm{Map}(\mathcal{U}_\mathbf{k}^\mathbf{X}, \mathcal{U}_\mathbf{k})$ that sends $x$ to $\pi_x$ for each $x \in \mathbf{X}$, then $\mathcal{IIF}_\mathbf{k}(\mathbf{X}) = \mathcal{I}_\mathbf{k}^\mathbf{X}\big(A_\mathbf{k}(\mathbf{X})\big)$. Clearly, $\mathcal{I}_\mathbf{k}^\mathbf{X} : A_\mathbf{k}(\mathbf{X}) \to \mathcal{IIF}_\mathbf{k}(\mathbf{X})$ is a surjective CA-homomorphism. We will see later that $\mathcal{I}_\mathbf{k}^\mathbf{X}$ is also injective.

We now define the *Chen-Fliess series* $S_U$ of an input $U \in \mathcal{U}_\mathbf{k}^\mathbf{X}$. For this purpose, we first extend the CP notation and define $(F * G)(t) = \int_0^t F(s)G'(s)ds$ for functions $F, G$ on $[0, \infty[$ with values in any, not necessarily commutative, $\mathbb{R}$-algebra. The *universal control system with inputs in* $\mathcal{U}_\mathbf{k}^\mathbf{X}$ is the system

$$\Sigma(\mathbf{X}) : \qquad (d/dt)S(t) = S(t)\left( \sum_{x \in \mathbf{X}} x \dot{U}_x(t) \right), \quad S(0) = 1, \qquad (31)$$

evolving in $\hat{A}_\mathbf{k}(\mathbf{X})$. For any family $U = \{U_x\}_{x \in \mathbf{X}}$ of locally integrable functions on $[0, \infty[$, (31) has a unique solution $[0, \infty[ \ni t \to S_U(t) \in \hat{A}_\mathbf{k}(\mathbf{X})$, known as *the Chen-Fliess series for the input $U$.* Moreover, the Friedrichs criterion easily implies that *(31) actually evolves in* $\hat{G}_\mathbf{k}(\mathbf{X})$, i.e. that $S_U(t) \in \hat{G}_\mathbf{k}(\mathbf{X})$ for all $U, t$.

If we let $Z_U(t) = \sum_{x \in \mathbf{X}} x U_x(t)$, we see that (31) says that $\dot{S}_U = S_U \dot{Z}_U$ and $S_U(0) = 1$, i.e. that $S_U(t) = 1 + \int_0^t S_U(s) \dot{Z}_U(s)ds$ or, equivalently, $S_U = 1 + S_U * Z_U$. Using $1 * V = V$, this implies that $S_U = 1 + Z_U + (S * Z_U) * Z_U$. It is then easy to show, by successive iterations, that

$$S_U = 1 + Z_U + Z_U * Z_U + ((S_U * Z_U) * Z_U) * Z_U,$$
$$S_U = 1 + Z_U + Z_U * Z_U + (Z_U * Z_U) * Z_U + (((S_U * Z_U) * Z_U) * Z_U) * Z_U,$$

and so on, so that, finally, $S_U = \sum_{k=0}^\infty Z_U^{(*k)}$, where $Z_U^{(*k)}$ is defined recursively by $Z_U^{(*0)} = 1$, $Z_U^{(*(k+1))} = Z_U^{(*k)} * Z_U$.

Clearly, $Z_U^{(*k)} = \sum_{w \in W(\mathbf{X}):|w|=k} w U_w$, where $U_w$ is defined recursively by $U_\emptyset = 1$, $U_{wx} = U_w * U_x$ for $w \in W(\mathbf{X})$, $x \in \mathbf{X}$. So $\boxed{S_U(t) = \sum_{w \in W(\mathbf{X})} w\, U_w(t)}$. If $w \in W(\mathbf{X})$,

125

then $\mathcal{I}_{\mathbf{k}}^{\mathbf{X}}(w)$ is the functional that assigns to $U \in \mathcal{U}_{\mathbf{k}}^{\mathbf{X}}$ the function $U_w$. Therefore $\boxed{S_U(t) = \sum_{w \in W(\mathbf{X})} w\, \mathcal{I}_{\mathbf{k}}^{\mathbf{X}}(w)(U)(t)}$. We now define $\mathbf{CH}^{\mathbf{X}}$ to be the series

$$\mathbf{CH}^{\mathbf{X}} \overset{\text{def}}{=} \sum_{w \in W(\mathbf{X})} w \otimes \mathcal{I}_{\mathbf{k}}^{\mathbf{X}}(w)\,, \tag{32}$$

so $\mathbf{CH}^{\mathbf{X}} \in \hat{A}_{\mathcal{I}\mathcal{I}\mathcal{F}_{\mathbf{k}}(\mathbf{X})}(\mathbf{X})$. The natural evaluation pairing from the Cartesian product $\mathcal{I}\mathcal{I}\mathcal{F}_{\mathbf{k}}(\mathbf{X}) \times \mathcal{U}_{\mathbf{k}}^{\mathbf{X}}$ to $\mathcal{U}_{\mathbf{k}}$ that sends $(\Phi, U)$ to $\Phi(U)$ induces an "evaluation map" $\mathbf{E}^{\mathbf{X}} : \hat{A}_{\mathcal{I}\mathcal{I}\mathcal{F}_{\mathbf{k}}(\mathbf{X})}(\mathbf{X}) \times \mathcal{U}_{\mathbf{k}}^{\mathbf{X}} \to \hat{A}_{\mathcal{U}_{\mathbf{k}}}(\mathbf{X})$. It is then clear that, if $U \in \mathcal{U}_{\mathbf{k}}^{\mathbf{X}}$, then $\mathbf{E}^{\mathbf{X}}(\mathbf{CH}^{\mathbf{X}}, U) = S_U$. If we let

$$\overline{\mathbf{CH}}^{\mathbf{X}} \overset{\text{def}}{=} \sum_{w \in W(\mathbf{X})} w \otimes w\,, \tag{33}$$

then $\overline{\mathbf{CH}}^{\mathbf{X}} \in \hat{A}_{\mathbf{k}}(\mathbf{X}) \tilde{\otimes} A_{\mathbf{k}}(\mathbf{X})$, and $\mathbf{CH}^{\mathbf{X}} = \left(\mathrm{id} \otimes \mathcal{I}_{\mathbf{k}}^{\mathbf{X}}\right)(\overline{\mathbf{CH}}^{\mathbf{X}})$.

If $P \in \hat{A}_{\mathbf{k}}(\mathbf{X})$, then $\mathcal{I}_{\mathbf{k}}^{\mathbf{X}}(P)$ is an iterated integral functional, which can be evaluated at any input $U \in \mathcal{U}_{\mathbf{k}}^{\mathbf{X}}$, yielding a function $\varphi_{P,U} : [0, \infty[ \to \mathbb{R}$ given by $\varphi_{P,U}(t) = \mathcal{I}_{\mathbf{k}}^{\mathbf{X}}(P)(U)(t)$. If $P$ is a word $v \in W(\mathbf{X})$, then

$$<S_U(t), P> \; = \; <\textstyle\sum_w w\, \mathcal{I}_{\mathbf{k}}^{\mathbf{X}}(w)(U)(t), P> \; = \; \mathcal{I}_{\mathbf{k}}^{\mathbf{X}}(P)(U)(t) = \varphi_{P,U}(t)\,.$$

It follows by linearity that $\varphi_{P,U}(t) \; = \; < S_U(t), P >$ for all $P, U$. If $\mathcal{I}_{\mathbf{k}}^{\mathbf{X}}(P) = 0$ as a member of $\mathcal{I}\mathcal{I}\mathcal{F}_{\mathbf{k}}(\mathbf{X})$, then $\varphi_{P,U}(t) = 0$ for all $U, t$, so $< S_U(t), P > \; = 0$ for all $U, t$. It follows in particular that $< S, P > \; = 0$ for every member $S \in \hat{G}_{\mathbf{k}}(\mathbf{X})$ which is of the form $Q = e^{t_1 x_1} e^{t_2 x_2} \ldots e^{t_k x_k}$ for some $x_1, \ldots, x_k$ in $\mathbf{X}$, $t_1, \ldots, t_k$ in $\mathbb{R}$. Successive differentiations of these identities with respect to $t_1, \ldots, t_k$ yield $< x_1 x_2 \ldots x_k, P > \; = 0$. So $< w, P > \; = 0$ for every $w \in W(\mathbf{X})$, and then $P = 0$. This proves that *the map $\mathcal{I}_{\mathbf{k}}^{\mathbf{X}}$ is an isomorphism from $\hat{A}_{\mathbf{k}}(\mathbf{X})$ onto $\mathcal{I}\mathcal{I}\mathcal{F}_{\mathbf{k}}(\mathbf{X})$.*

Now that we know that the map $\mathcal{I}_{\mathbf{k}}^{\mathbf{X}}$ is an isomorphism, we can conclude that $\mathrm{id} \otimes \mathcal{I}_{\mathbf{k}}^{\mathbf{X}}$ is an isomorphism as well, so we can identify the spaces $\hat{A}_{\mathbf{k}}(\mathbf{X}) \tilde{\otimes} A_{\mathbf{k}}(\mathbf{X})$ and $\hat{A}_{\mathbf{k}}(\mathbf{X}) \tilde{\otimes} \mathcal{I}\mathcal{I}\mathcal{F}_{\mathbf{k}}(\mathbf{X})$. In particular, any expansion we obtain for $\mathbf{CH}^{\mathbf{X}}$ will yield a similar expansion for $\overline{\mathbf{CH}}^{\mathbf{X}}$.

We remark that the space $\hat{A}_{\mathbf{k}}(\mathbf{X}) \tilde{\otimes} A_{\mathbf{k}}(\mathbf{X})$ is naturally identified with the space $\mathrm{Hom}_{\mathbf{k}}(A_{\mathbf{k}}(\mathbf{X}), A_{\mathbf{k}}(\mathbf{X}))$ of linear endomorphisms of $A_{\mathbf{k}}(\mathbf{X})$, by assigning to each map $\Lambda \in \mathrm{Hom}_{\mathbf{k}}(A_{\mathbf{k}}(\mathbf{X}), A_{\mathbf{k}}(\mathbf{X}))$ the series $\sum_{w \in W(\mathbf{X})} w \otimes \Lambda(w)$. Under this identification, $\overline{\mathbf{CH}}^{\mathbf{X}}$ corresponds to the identity map of $A_{\mathbf{k}}(\mathbf{X})$. So the Chen series is none other than the identity map of $A_{\mathbf{k}}(\mathbf{X})$, modulo several natural identifications, showing that $\mathbf{CH}^{\mathbf{X}}$ is a natural object.

# 5  Exponential product expansions and dual PBW-bases

In §2 we showed how to compute the first few factors of an expansion as a product of exponentials of flow maps $e^{\int_0^t (\sum_{i=1}^m u_i(s) f_i)\, ds}$ determined by $m$ smooth vector fields $f$ and $g$, by means of successive applications of the method of variations of constants. In the situation discussed in §2 we had $m = 2$, and $u_1(t) \equiv 1$, and we just computed the first five factors. It turns out that the formal calculation can be pursued for any number of factors, for a general $m$, and for general inputs $u = \dot{U} \in \mathcal{AC}^m$.

Remarkably, the algebra works out in such a way that one obtains a formula expressing the Chen series $\mathbf{CH}^{\mathbf{X}}$ as an infinite product of exponentials

$$\mathbf{CH}^{\mathbf{X}} = \overleftarrow{\prod}_{B \in \mathcal{B}} e^{B \otimes \Phi_B} , \tag{34}$$

where $\mathcal{B}$ is any "generalized Hall basis" (abbr. GHB) of $\hat{A}_{\mathbf{k}}(\mathbf{X})$, the coefficients $\Phi_B \in \mathcal{IIF}_{\mathbf{k}}(\mathbf{X})$ are iterated integral functionals given by simple formulas, as explained below, and the symbol $\overleftarrow{\prod}$ indicates that the factors are ordered from right to left, following the ordering of $\mathcal{B}$.

GHB's arise when one seeks to spell out explicit combinatorial rules to write bases of $L_{\mathbf{k}}(\mathbf{X})$. Several such schemes have been proposed, but all were shown by Viennot [19] (see also [14] for a modern discussion) to arise from the same underlying principle, resulting in what is now known as GHB's, which is a special type of basis $\mathcal{B}$ of $L_{\mathbf{k}}(\mathbf{X})$, endowed with a total ordering $\preceq$. (We refer the reader to [19, 14] for the precise definition of a GHB.)

Applying the method of variation of constants, Formula (34) was derived in Sussmann [18] in 1986, together with an explicit recursive formula for the functionals $\Phi_B$. If $B = x \in \mathbf{X}$, then $\Phi_x = \pi_x$. If $B \in \mathcal{B}$ but $B \notin \mathbf{X}$, then write $B = \mathrm{ad}_{B_1}^{m_1} \mathrm{ad}_{B_2}^{m_2} \ldots \mathrm{ad}_{B_k}^{m_k}(B_{k+1})$, with the $B_i$ in $\mathcal{B}$, $B_1 \succ B_2 \succ \ldots \succ B_k$, $B_k \prec B_{k+1}$, and $m_1, \ldots, m_k$ positive integers —it is a fact that every $B \in \mathcal{B}$ can be so expressed— and then $\Phi_B$ is given by

$$\Phi_B = \Big( \prod_{i=1}^{k} \frac{1}{m_i!} \Phi_{B_i}^{m_i} \Big) * \Phi_{B_{k+1}} . \tag{35}$$

(The derivation given in [18] was for classical P. Hall bases, but the proof applies without change to any generalized Hall basis.)

Expanding the exponentials of (34), one gets the formula

$$\mathbf{CH}^{\mathbf{X}} = \sum_{k=0}^{\infty} \sum_{B_1 \succ B_2 \succ \ldots \succ B_k} \sum_{\mu_1, \mu_2, \ldots, \mu_k} B_1^{\mu_1} B_2^{\mu_2} \ldots B_k^{\mu_k} \otimes \frac{\Phi_{B_1}^{\mu_1} \Phi_{B_2}^{\mu_2} \ldots \Phi_{B_k}^{\mu_k}}{\mu_1! \mu_2! \ldots \mu_k!} . \tag{36}$$

Via the inverse of the isomorphism $\mathrm{id} \otimes \mathcal{I}_{\mathbf{k}}^{\mathbf{X}}$, and recalling that $\mathcal{I}_{\mathbf{k}}^{\mathbf{X}}$ is an isomorphism from $A_{\mathbf{k}}(\mathbf{X})$ with the shuffle product to $\mathcal{IIF}_{\mathbf{k}}(\mathbf{X})$ with ordinary multiplication, we can transform (36) into an expansion

$$\overline{\mathbf{CH}}^{\mathbf{X}} = \sum_{k=0}^{\infty} \sum_{B_1 \succ B_2 \succ \ldots \succ B_k} \sum_{\mu_1, \mu_2, \ldots, \mu_k} B_1^{\mu_1} B_2^{\mu_2} \ldots B_k^{\mu_k} \otimes \frac{\Psi_{B_1}^{\unicode{x29F5}, \mu_1} \unicode{x29F5} \Psi_{B_2}^{\unicode{x29F5}, \mu_2} \unicode{x29F5} \ldots \unicode{x29F5} \Psi_{B_k}^{\unicode{x29F5}, \mu_k}}{\mu_1! \mu_2! \ldots \mu_k!} , \tag{37}$$

where $\Psi_x = x$ for $x \in \mathbf{X}$ and, if $B \in \mathcal{B} \backslash \mathbf{X}$, then

$$\Psi_B = \Big( \frac{\Psi_{B_1}^{\unicode{x29F5}, m_1}}{m_1!} \unicode{x29F5} \frac{\Psi_{B_2}^{\unicode{x29F5}, m_2}}{m_2!} \unicode{x29F5} \ldots \unicode{x29F5} \frac{\Psi_{B_k}^{\unicode{x29F5}, m_k}}{m_k!} \Big) * \Psi_{B_{k+1}} , \tag{38}$$

if $B = \mathrm{ad}_{B_1}^{m_1}\mathrm{ad}_{B_2}^{m_2}\ldots\mathrm{ad}_{B_k}^{m_k}(B_{k+1})$, with the $B_i$ in $\mathcal{B}$, $B_1 \succ B_2 \succ \ldots \succ B_k$, $B_k \prec B_{k+1}$, and $m_1,\ldots,m_k$ positive integers. (The ш symbols accompanying the exponents are there as a reminder that all the powers are taken in the sense of the shuffle product.)

Formulas (37) and (38) give the expansion of $\overline{\mathrm{CH}}^{\mathbf{X}}$ —i.e. the identity element of $\mathrm{Hom}_{\mathbf{k}}(A_{\mathbf{k}}(\mathbf{X}), A_{\mathbf{k}}(\mathbf{X}))$, regarded as a member of $\hat{A}_{\mathbf{k}}(\mathbf{X})\tilde{\otimes}A_{\mathbf{k}}(\mathbf{X})$— in terms of the Poincaré-Birkhoff-Witt basis of $A_{\mathbf{k}}(\mathbf{X})$ associated to $\mathcal{B}$. (Recall that $A_{\mathbf{k}}(\mathbf{X})$ is a UEA of $L_{\mathbf{k}}(\mathbf{X})$.) So the coefficients $\dfrac{\Psi_{B_1}^{\text{ш},\mu_1}\,\text{ш}\,\Psi_{B_2}^{\text{ш},\mu_2}\,\text{ш}\ldots\text{ш}\,\Psi_{B_k}^{\text{ш},\mu_k}}{\mu_1!\mu_2!\ldots\mu_k!}$ appearing in (37) are the members of the *dual basis of the PBW basis arising from* $\mathcal{B}$.

The derivation outlined here was given for $\mathbf{k} = \mathbb{R}$, but one can easily see in various ways that the formula is valid for any field $\mathbf{k}$ of characteristic zero. (For example, (37) and (38) are identities between formal power series with rational coefficients, so they are valid over any field of characteristic zero.)

A similar formula was derived by Melançon and Reutenauer in 1989 in [13] by combinatorial means using rewriting systems. Using different notations, one can also find this formula in Schützenberger's 1958 notes [17].


# References


[1] Agrachev, A. and A. Sarychev, Abnormal sub-Riemannian geodesiscs: Morse Index and Rigidity, Ann. Inst. H. Poincaré **13** (1996), pp. 635-690.

[2] Agrachev, A. and R. Gamkrelidze, Exponential representation of flows and chronological calculus, Math. USSR Sbornik (Russian) **107**, N4 (1978), pp. 487-532.

[3] Agrachev, A. and R. Gamkrelidze, Chronological algebras and nonstationary vector fields, Journal Soviet Math. **17**, no.1 (1979), pp. 1650-1675.

[4] Bourbaki, N., Lie Groups and Lie algebras, Hermann, Paris, 1975.

[5] Chen, K. T., Integration of paths, geometric invariants, and a generalized Baker-Hausdorff formula, Annals of Math. **67** (1957), pp. 164-178.

[6] Crouch, P. and R. Grossman, Numerical Integration of Ordinary Differential Equations on Manifolds, J. Nonlinear Science **3** (1993), pp. 1-33.

[7] Fliess, M., Fonctionnelles causales non linéaires et indeterminées non commutatives, Bull. Soc. Math. France **109** (1981), pp.3-40.

[8] Kawski, M., Chronological algebras and nonlinear control, Proc. Asian Conf. Control, Tokyo, 1994.

[9] Kawski, M., Nonlinear control and combinatorics of words, in: Nonlinear Feedback and Optimal control, B. Jakubczyk and W. Respondek, eds., Dekker, 1997 (to appear).

[10] Knobloch, H. W., High Order Necessary Conditions in Optimal Control, Springer-Verlag, 1975.

[11] Loday, J.-L, Une version non commutative des algèbres de Lie: les algèbres de Leibniz, L'Enseignement Mathématique **39** (1993), pp. 268-293.

[12] Loday, J.-L., and T. Pirashvili, Universal enveloping algebras of Leibniz algebras and (co)homology, Math. Annalen **196** (1993), pp. 139-158.

[13] Melançon, G. and C. Reutenauer C., Lyndon words, free algebras and shuffles, Canadian J. Math. **XLI** (1989), pp. 577–591.

[14] Melançon, G. and Reutenauer C., Combinatorics of Hall trees and Hall words, J. Comb. Th. Ser. A **59** (1992), pp. 285-299.

[15] Ree, R. Lie elements and an algebra associated with shuffles, Annals of Math., **68** (1958), pp. 210–220.

[16] Reutenauer, C., Free Lie Algebras, London Math. Soc. monographs, new series **7**, Oxford, 1993.

[17] Schützenberger, M., Sèminaire Dubreil, Facultè de Sciences de Paris, 1958.

[18] Sussmann, H. A product expansion of the Chen series, in "Theory and Applications of Nonlinear Control Systems," C. I. Byrnes and A. Lindquist eds., Elsevier, North-Holland (1986), pp. 323–335.

[19] Viennot, G. Algèbres de Lie Libres et Monoïdes Libres, Lect. Notes, Math., 692, Springer, Berlin, 1978.

Matthias Kawski
Department of Mathematics
Arizona State University
Tempe, Arizona 85287, USA
Phone: (602) 965 3376
Fax: (602) 965 0461
kawski@asu.edu
http://math.la.asu.edu/~kawski

Héctor J. Sussmann
Department of Mathematics
Rutgers University
Piscataway, NJ 08855, USA
Phone: (732)445-5407
Fax: (732)445-5530
sussmann@hamilton.rutgers.edu
http://math.rutgers.edu/~sussmann

# Recent Progress on the Partial Stochastic Realization Problem

Anders Lindquist

Royal Institute of Technology, Stockholm, Sweden

To Paul Fuhrmann on the occasion of his 60th birthday

In view of Paul Fuhrmann's many important contributions to realization theory, it seems quite appropriate to devote this lecture to the stochastic partial realization problem, when today we are honoring him on his 60th birthday. Some ten years ago Christopher I. Byrnes and I launched a joint research program on this topic, and by now we have some results which I think might interest this audience [3, 4, 5, 6, 7, 8, 9, 10, 11, 12]. Some of these results we have been obtain in collaboration with S. V. Gusev in particular, but also A. S. Matveev and H. J. Landau. This short write-up is not a paper in itself but is merely intended to interest the audience in reading the papers [8, 7, 10, 11, 12] and also [32]. The stochastic partial realization problem has important applications in speech synthesis [17], spectral estimation [22, 33], stochastic systems theory [23], systems identification [32], and several other areas of systems and control.

## 1    The deterministic partial realization problem

Before turning to the main topic, let us consider for a moment the simpler deterministic partial realization problem; see, e.g., [24, 25, 21, 18]: Given a sequence of real numbers

$$c_0, c_1, c_2, \ldots, c_n \tag{1}$$

find a *partial realization* of (1), i.e., a matrix $F$, a column vector $g$ and a row vector $h$ such that

$$c_k = hF^{k-1}g \quad \text{for } k = 1, 2, \ldots, n. \tag{2}$$

This then defines an infinite extension

$$c_k = hF^{k-1}g \quad \text{for } k = n+1, n+2, \ldots \tag{3}$$

of (1) such that the rational function

$$v(z) = h(zI - F)^{-1}g + \tfrac{1}{2}c_0 \tag{4}$$

has the Laurent expansion

$$v(z) = \tfrac{1}{2}c_0 + c_1 z^{-1} + c_2 z^{-2} + \dots$$

in the neighborhood of infinity. If the dimension of the matrix is as small as possible, we say that the partial realization is *minimal*.

This problem can be motivated in the following way. Let us assume that we have a linear system

$$\xrightarrow{\;u\;} \boxed{w(z)} \xrightarrow{\;y\;}$$

and that we would like to find matrices $A, B, C, D$ so that

$$\begin{cases} x(t+1) = Ax(t) + Bu(t) \\ \quad\;\; y(t) = Cx(t) + Du(t) \end{cases} \tag{5}$$

models its input/output behavior. If we apply a impulse signal

$$u(t) = \begin{cases} 1 & \text{for } t = 0 \\ 0 & \text{for } t = 1, 2, 3, \dots \end{cases}$$

to the input, the output is

$$y(t) = \begin{cases} \tfrac{1}{2}c_0 & \text{for } t = 0 \\ CA^{t-1}B & \text{for } t = 1, 2, 3, \dots. \end{cases}$$

Thus, given the finite impulse response

$$y(0), y(1), \dots, y(n),$$

determining the matrices $A, B, C, D$ obviously amounts to solving the partial realization problem.

We shall call the dimension of a minimal partial realization the *algebraic degree* of (1). The algebraic degree of (1) could be anything between zero and $n$, but it has a generic value $\left[\frac{n}{2}\right]$, and all other values are rare.

The deterministic partial realization problem is equivalent to Padé approximation and, as shown in [21], the solution can be determined recursively via Lanczos' algorithm [27]. There is no guarantee that $F$ is Schur stable (all eigenvalues in the open unit disc). In fact, the property that $F$ is Schur stable is not even generic [2].

From (2) it should be clear that a minimal partial realization (4) satisfies

$$H_{ij} := \begin{bmatrix} c_1 & c_2 & \cdots & c_j \\ c_2 & c_3 & \cdots & c_{j+1} \\ \vdots & \ddots & & \vdots \\ c_i & c_{i+1} & \cdots & c_{i+j} \end{bmatrix} = \begin{bmatrix} h \\ hF \\ \vdots \\ hF^{i-1} \end{bmatrix} \begin{bmatrix} g & Fg & \cdots & F^{j-1}g \end{bmatrix} \tag{6}$$

for each $i, j$ such that $i + j = n$. Hence the maximum rank of these Hankel matrices is a lower bound for the algebraic degree of (1).

# 2     The stochastic partial realization problem

Now suppose $A$ is a stability matrix. Then, passing (normalized) white noise $\{u(t)\}$ through the filter

$$\text{white noise} \xrightarrow{u} \boxed{w(z)} \xrightarrow{y}$$

and letting it come to statistical steady state, we obtain a stationary output process $\{y(t)\}$ with spectral density

$$\Phi(z) = w(z)w(z^{-1}) = \sum_{k=-\infty}^{\infty} c_k z^{-k}, \tag{7}$$

where

$$c_k = \mathrm{E}\{y(t+k)y(t)\}. \tag{8}$$

In view of the ergodic property of the process $\{y(t)\}$, the covariance lags (8) may also be represented as

$$c_k = \lim_{T \to \infty} \frac{1}{T+1} \sum_{t=0}^{T} y_{t+k} y_t \tag{9}$$

almost surely, where

$$y_0, y_1, y_2, y_3, \cdots$$

is a realization of the process $\{y(t)\}$. In practice, however, we only have access to finite string of output data

$$y_0, y_1, y_2, \ldots, y_N. \tag{10}$$

In general $N$ will be large, and therefore

$$c_k = \frac{1}{N+1-k} \sum_{t=0}^{N-k} y_{t+k} y_t \tag{11}$$

will be a reasonably good estimate. However, we can only estimate a finite number of covariance lags

$$c_0, c_1, c_2, \ldots, c_n, \tag{12}$$

where $n << N$. For simplicity we assume that (12) is a *bona fide* partial covariance sequence in the sense that

$$T_n = \begin{bmatrix} c_0 & c_1 & c_2 & \cdots & c_n \\ c_1 & c_0 & c_1 & \cdots & c_{n-1} \\ \vdots & \vdots & \vdots & \ddots & \vdots \\ c_n & c_{n-1} & c_{n-2} & \cdots & c_0 \end{bmatrix}.$$

Of course, without loss of generality, we may normalize and take $c_0 = 1$. From now on we shall do this.

Now, reconstructing the filter $w(z)$ from the partial covariance sequence (12) is precisely the stochastic partial realization problem, first formulated in this context by Kalman [23]. Just as in the deterministic partial realization problem, we are faced with the problem of finding a triplet $(F, g, h)$ such that

$$hF^{k-1}g = c_k \quad \text{for } k = 1, 2, \ldots, n. \tag{13}$$

Then defining the infinite extension

$$c_k = hF^{k-1}g \quad \text{for } k = n+1, n+2, \ldots, \tag{14}$$

the function

$$v(z) = \frac{1}{2} + c_1 z^{-1} + c_2 z^{-2} + \ldots$$

is a rational function

$$v(z) = \frac{1}{2}\frac{b(z)}{a(z)}$$

where $a(z)$ and $b(z)$ are monic polynomials of degree $n$. However, there is now an additional requirement, namely that $v(z)$ is *strictly positive real*, i.e., $v(z)$ is analytic on and outside the unit circle and

$$\Phi(z) = v(z) + v(z^{-1}) > 0 \quad \text{on the unit circle.} \tag{15}$$

Then $\Phi(z)$ is a *bona fide* coercive spectral density, and therefore the spectral factorization problem

$$w(z)w(z^{-1}) = \Phi(z)$$

can be solved for the required filter, i.e., the minimum-phase, stable spectral factor

$$w(z) = C(zI - A)^{-1}B + D,$$

where we may choose coordinates so that $A = F$ and $C = h$.

Consequently we have a partial realization problem to determine a triplet $(F, g, h)$ satisfying the interpolation condition (13), but with the additional constraint that $v(z)$ is strictly positive real. In particular this requires that $F$ is Schur stable, but this is not enough. In fact, we must also have

$$a(z)b(z^{-1}) + b(z)a(z^{-1}) > 0 \quad \text{on the unit circle.}$$

These constraints make the stochastic partial realization problem considerably more complicated than the deterministic one. In particular, the infinite extension (14) is such that the Toeplitz matrix $T_k > 0$ for all $k \geq 0$.

Among all strictly positive real rational functions $v(z)$ interpolating the sequence $1, c_1, c_2, \ldots, c_n$ in the sense that

$$v(z) = \tfrac{1}{2} + \hat{c}_1 z^{-1} + \hat{c}_2 z^{-2} + \ldots \quad \text{with } \hat{c}_k = c_k \text{ for } k = 1, 2, \ldots, n,$$

find one with minimum degree. This a minimum stochastic partial realization, and its degree $p$ is called the *positive degree* of the partial covariance sequence $1, c_1, c_2, \ldots, c_n$ and is greater or equal to the algebraic degree $r$ of the same partial sequence. However, while $r$ has a generic value, $p$ does not. More precisely, whereas the generic value of the algebraic degree of (12) is $[\frac{n+1}{2}]$, it was proved in [7] that, for each $\nu = [\frac{n+1}{2}], [\frac{n+1}{2}] + 1, \ldots, n$ there is a nonempty open set of covariance data in $\mathbb{R}^\kappa$ for which the positive degree $p$ is precisely $\nu$. Thus the possibility that $p > r$ is nonrare. The positive degree can be characterized [7], but there is no easy way to compute it.

# 3 What are the solutions of the stochastic partial realization problem?

If we remove the requirement that $v(z)$ be rational of at most degree $n$ and allow it to be only meromorphic, the stochastic partial realization problem is reduced to a classical interpolation problem going back to Carathéodory [14, 15], Toeplitz [35] and Schur [34]. Schur developed a complete parameterization of the class of such meromorphic and strictly positive real interpolants in terms of what we now know as the Schur parameters $\gamma_0, \gamma_1, \gamma_2, \ldots$. In fact, to say that $v(z)$ is positive real is to say that

$$|\gamma_k| < 1 \quad \text{for } k = 0, 1, 2, \ldots. \tag{16}$$

Moreover, $\gamma_0, \gamma_1, \ldots, \gamma_{n-1}$ are uniquely determined (via, for example, the Levinson algorithm) by $c_1, c_2, \ldots, c_n$. Therefore, all meromorphic and strictly positive real interpolants are completely parameterized by the free Schur parameters $\gamma_n, \gamma_{n-1}, \ldots$ satisfying (16).

Making the very natural choice that $\gamma_k = 0$ for $k = n, n+1, \ldots$, we do obtain a rational $v(z)$ of degree $n$, which is a solution of precisely the type we are looking for. This is the *maximum entropy solution,* and the corresponding filter $w(z)$ is

$$w(z) = \sqrt{r_n}\frac{z^n}{\varphi_n(z)}, \tag{17}$$

where $\varphi_n(z)$ is the $n$:th Szegö polynomial and $\sqrt{r_n}$ is the corresponding normalization factor, which both can be determined by solving the normal equations, a linear system of equations.

Obviously the maximum entropy solution has no nontrivial zeros, since the ones that are cancel when the spectral density $\Phi(z)$ is formed. A natural question, therefore, is as follows: Given any Schur stable monic polynomial $\sigma(z)$ of degree $n$, is it possible to find a Schur stable polynomial $a(z)$ so that

$$w(z) = \frac{\sigma(z)}{a(z)} \tag{18}$$

is a solution to the stochastic partial realization problem? In [20, 19] Georgiou proved that this is possible and conjectured that this choice would be unique.

In [8] we proved an amplified version of this longstanding conjecture, showing not only that this parameterization is complete but also that the problem is well-posed in a the strong sense that the bijection is a diffeomorphism. This is proved as corollary of a more general theorem on the geometric duality between filtering and interpolation. In fact, we prove that filtering and interpolation induce complementary decompositions (foliations) of the space of positive real functions of degree less or equal to $n$. The leaves of the interpolation foliation are indexed by the partial covariance sequences and the leaves of filtering foliation by the zero polynomials $\sigma(z)$. The unique pole polynomial $a(z)$ is determined via the intersection of the appropriate leaves. (For details, also see [10].) In passing we mention that the leaves of the filtering foliation are precisely the stable manifolds of the fast filtering algorithm [28, 29], properly reformulated as in [13].

In [10] and [11] we provide alternative, but simpler, proofs of the somewhat weaker statement that the stochastic partial realization problem is well-posed in the sense of Hadamard.

However, all these proofs, including the proof of existence due to Georgiou [20], are non-constructive and thus do not offer any computational procedure. It should be noted that, although the computation of the maximum entropy solution is a linear problem, obtaining the solution corresponding to an arbitrary zero polynomial $\sigma(z)$ is a *nonlinear problem*, which explains why it is much more difficult.

The first step toward finding a computational method for solving the stochastic partial realization problem was presented in [7], where the problem was reduced to solving a Riccati-type covariance extension equation. Although there is yet no general method of solving this equation, it provides some additional insight into the issue of minimality, relating the degree of the modeling filter to the rank of the unique positive semidefinite solution of the covariance extension equation.

In a recent paper [12], we present a convex optimization problem for solving the rational covariance extension problem. Given a partial covariance sequence and the desired zeros of the modeling filter (18), the poles are uniquely determined from the unique minimum of the corresponding optimization problem. In this way we obtain an algorithm for solving the covariance extension problem, as well as a constructive proof of Georgiou's existence result and his conjecture.

# 4 Why are we interested in this problem?

There are several interesting consequences and applications to what has been discussed above. We mention just two here and refer the audience to the reference list for further reading.

The maximum entropy solution is regularly used in speech coding in designing the so called LPC filter. The speech is typically broken into segments of 20 ms. An unvoiced segment can be regarded as a realization of a stationary stochastic process obtained by passing white noise through a modeling filter $w(z)$. (The voiced segments are modeled in a similar way exchanging the white noise for a periodic "pulse train".) Now, the disadvantage with the LPC filter is that the maximum entropy filter produces a rather "flat" speech, which does not represent nasals and fricative very well. For this we need to model the notches in the spectrum by placing zeros close to the unit circle. Hence a nontrivial zero polynomial $\sigma(z)$ would be desired. The more general solutions of the stochastic partial realization problem gives us the freedom to choose the zeros arbitrarily.

Recently there has been quite some interest in a new type of stochastic identification procedure known as subspace identification [1, 36]. In [32] it was pointed out that there is no guarantee that these subspace identification algorithms will actually work for generic data. This is for the reasons mention above. In fact, at least indirectly these algorithms are based on the assumption that the algebraic and positive degrees of a partial covariance sequence coincide. Also the geometry of [36] is equivalent to the splitting geometry of stationary stochastic systems [30, 31], developed for infinite strings of data, and which strictly speaking requires stronger conditions to hold here. In the context of this talk, the procedure of [36] is to perform minimal factorization of a square Hankel matrix $H_{ii}$, where $i = [n/2]$; see (6). But as pointed out above, the rank of $H_{ii}$ only provides a lower bound of the algebraic degree, which in turn may be smaller than the positive degree, in which case the identification method will fail. It is easy to generate data for which such failure will occur even for large $n$; see [16].


## References

[1] M. Aoki, *State Space Modeling of Time Series*, Springer-Verlag, 1987.

[2] C. I. Byrnes and A. Lindquist, *The stability and instability of partial realizations*, Systems and Control Letters **2** (1982), 99–105.

[3] C. I. Byrnes and A. Lindquist, *An algebraic description of the rational solutions of the covariance extension problem*, in Linear Circuits, Systems and Signal Processing, C.I. Byrnes, C.F. Martin and R.E. Saeks (editors), Elsevier 1988, 9-17.

[4] C. I. Byrnes and A. Lindquist, *On the geometry of the Kimura-Georgiou parameterization of modelling filter*, Inter. J. of Control **50** (1989), 2301-2312.

[5] C. I. Byrnes and A. Lindquist, *Some recent advances on the rational covariance extension problem*, Proc. IEEE European Workshop on Computer-Intensive Methods in Control and Signal Processing, September 1994.

[6] C. I. Byrnes and A. Lindquist, *Toward a solution of the minimal partial stochastic realization problem*, Comptes Rendus Acad. Sci. Paris, t. 319, Série I (1994), 1231-1236.

[7] C. I. Byrnes and A. Lindquist, *On the partial stochastic realization problem*, IEEE Trans. Automatic Control AC-42 (1997), to be published.

[8] C. I. Byrnes, A. Lindquist, S. V. Gusev, and A. S. Matveev, *A complete parametrization of all positive rational extensions of a covariance sequence*, IEEE Trans. Automatic Control AC-40 (1995), 1841-1857.

[9] C. I. Byrnes, A. Lindquist, S. V. Gusev, and A. S. Matveev, *The geometry of positive real functions with applications to the rational covariance extension problem*, Proc. 33rd Conf. on Decision and Control, 3883-3888.

[10] C. I. Byrnes and A. Lindquist, *On duality between filtering and interpolation*, in *Systems and Control in the Twenty-First Century*, C.I.Byrnes, B.N.Datta, D.S.Gilliam, and C.F.Martin (editors), pp. 101-136.

[11] C. I. Byrnes, H. J. Landau, and A. Lindquist, *On the well-posedness of the rational covariance extension problem*, in *Current and Future Directions in Applied Mathematics*, M. Alber, B. Hu and J. Rosenthal (editors), pp. 81-108..

[12] C. I. Byrnes, S. V. Gusev and A. Lindquist, *A convex optimization approach to the rational covariance extension problem*, submitted for publication.

[13] C. I. Byrnes, A. Lindquist, and Y. Zhou, *On the nonlinear dynamics of fast filtering algorithms*, SIAM J. Control and Optimization, **32**(1994), 744–789.

[14] C. Carathéodory, *Über den Variabilitätsbereich der Koeffizienten von Potenzreihen, die gegebene Werte nicht annehmen*, Math. Ann. **64** (1907), 95–115.

[15] C. Carathéodory, *Über den Variabilitätsbereich der Fourierschen Konstanten von positiven harmonischen Functionen*, Rend. di Palermo **32** (1911), 193–217.

[16] A. Dahlén, A. Lindquist, and J. Mari *Experimental evidence showing that stochastic subspace identification methods may fail*, to appear.

[17] Ph. Delsarte, Y. Genin, Y. Kamp and P. van Dooren, *Speech modelling and the trigonometric moment problem*, Philips J. Res. **37** (1982), 277–292.

[18] Paul A. Fuhrmann, *On the partial realizaion problem and the recursive inversion of Hankel and Toeplitz matrices*, Contemporary Mathematics **47** (1985), 249–161.

[19] T. T. Georgiou, *Partial realization of covariance sequences*, CMST, Univ. Florida, Gainesville, 1983.

[20] T. T. Georgiou, *Realization of power spectra from partial covariance sequences*, IEEE Transactions Acoustics, Speech and Signal Processing **ASSP-35** (1987), 438–449.

[21] W. B. Gragg and A. Lindquist, *On the partial realization problem*, Linear Algebra and its Applications **50** (1983), 277–319.

[22] S. Haykin, *Toeplitz forms and their applications*, Springer-Verlag, 1979.

[23] R. E. Kalman, *Realization of covariance sequences*, Proc. Toeplitz Memorial Conference (1981), Tel Aviv, Israel, 1981.

[24] R. E. Kalman, *On minimal partial realizations of a linear input/output map*,in Aspects of Network and System Theory (R. E. Kalman and N. de Claris, eds.), Holt, Reinhart and Winston, 1971, 385–408.

[25] R. E. Kalman, *On partial realizations, transfer functions and canonical forms*, Acta Polytech. Scand. **MA31** (1979), 9–39.

[26] H. Kimura, *Positive partial realization of covariance sequences*, Modelling, Identification and Robust Control (C. I. Byrnes and A. Lindquist, eds.), North-Holland, 1987, pp. 499–513.

[27] C. Lanczos, *An iteration method for the solution of the eigenvalue problem of linear differential and integral operators*, J. Res. Nat. Bur. Standards **45** (1950), 255–282.

[28] A. Lindquist, *A new algorithm for optimal filtering of discrete-time stationary processes*, SIAM J. Control **12** (1974), 736–746.

[29] A. Lindquist, *Some reduced-order non-Riccati equations for linear least-squares estimation: the stationary, single-output case*, Int. J. Control **24** (1976), 821–842.

[30] A. Lindquist and G. Picci, *Realization theory for multivariate stationary Gaussian processes*, SIAM J. Control and Optimization **23** (1985), 809–857.

[31] A. Lindquist and G. Picci, *A geometric approach to modelling and estimation of linear stochastic systems*, Journal of Mathematical Systems, Estimation and Control **1** (1991), 241–333.

[32] A. Lindquist and G. Picci, *Canonical correlation analysis, aprroximate covariance extension, and identification of stationary time series*, Automatica **32** (1996), 709–733.

[33] S. L.Marple, Jr., *Digital Spectral Analysis and Applications*, Prentice-Hall, 1987.

[34] I. Schur, *On power series which are bounded in the interior of the unit circle I and II*, Journal für die reine und angewandte Mathematik **148** (1918), 122–145.

[35] O. Toeplitz, *Über die Fouriersche Entwicklung positiver Funktionen*, Rendiconti del Circolo Matematico di Palermo **32** (1911), 191–192.

[36] P. van Overschee and B. De Moor, *Subspace algorithms for stochastic identification problem*, IEEE Trans. Automatic Control **AC-27** (1982), 382–387.

Anders Lindquist
Division of Optimization and Systems Theory
Department of Mathematics
Royal Institute of Technology
100 44 Stockholm, Sweden
E-mail: alq@math.kth.se

# Generalized Partial Realizations

W. Manthey[1]
Universität Bremen, Bremen, Germany

U. Helmke[1]
Universität Würzburg, Würzburg, Germany

D. Hinrichsen
Universität Bremen, Bremen, Germany

*This paper is dedicated to Paul Fuhrmann on the occasion of his 60th birthday.*

**Abstract**: In this paper we extend Kalman's concept of partial realization and define generalized partial realizations of finite matrix sequences by descriptor type systems. The aim is to prove a counterpart of Kalman's main theorem of realization theory for generalized partial realizations. The paper ends with some results concerning topological aspects of generalized partial realizations.

## 1  Introduction

One of the main achievements of the structure theory of linear state space systems has been the development of a satisfactory realization theory for infinite sequences of Markov parameters. The problem is to find, for any given recursive sequence $(H_1, H_2, \ldots)$ of $p \times m$ matrices over a field $\mathbb{K}$, a linear system of minimal dimension

$$
\begin{aligned}
x(t+1) &= Ax(t) + Bu(t) \\
y(t) &= Cx(t)
\end{aligned}
\tag{1.1}
$$

such that $(H_1, H_2, \ldots)$ is the sequence of Markov parameters of the system, i.e. $H_k = CA^{k-1}B$ for all $k \in \mathbb{N}$. Kalman's main theorem of realization theory [14] states that any two minimal realizations are similar, and a realization is minimal if and only if it is controllable and observable. An elegant coordinate free method of constructing such realizations is via Fuhrmann's functional models (shift realizations), see [5], [7].

The problem becomes more complicated if minimal realizations (1.1) are sought for *finite* sequences $(H_1, \ldots, H_\tau)$. This is the *partial realization problem* which was

---

[1]Research partially supported by the German-Israeli Foundation under grant GIF I 078-304.01/91

introduced and solved by Kalman in the special case of *scalar* Markov parameters, see [15]. In fact the problem had already been considered in an implicit form by Kronecker [16] and Frobenius [4]. In his paper Kalman pointed out connections of the partial realization problem to continued fractions, the Euclidean algorithm, Padé approximations and canonical forms, and these subjects were pursued subsequently by several authors [9], [6], [1]. Topological aspects of the partial realization problem were investigated in an early paper by Brockett [2]. He proved that the finite scalar sequences which have a minimal partial realization of order $n$ form an analytical manifold. Motivated by Kalman's paper and Brockett's earlier work on $\mathrm{Rat}(n)$ Fuhrmann and Krishnaprasad [8] analyzed the parametrization of rational functions via continued fractions and conjectured that this method yields a cell decomposition of the space $\mathrm{Rat}(n)$. Their conjecture was proved in [12] and in [18] a closely related cell decomposition was applied to Brockett's sequence spaces in the context of partial realization.

The theory of finite block Hankel matrices is considerably more difficult than the theory of infinite Hankel matrices of finite rank. In particular, there is, even in the scalar case, no simple expression for the McMillan degree, i.e. the minimal dimension of a partial realization (1.1). For infinite sequences the McMillan degree is given by the rank of the associated infinite Hankel matrix. An analogous characterization is not available in the finite data case and the McMillan degree is no longer easily related to the rank of an associated Hankel matrix. For example, consider the case of the finite scalar sequence $(0, \ldots, 0, 1) \in \mathbb{K}^{2N-1}$. The associated $N \times N$ Hankel matrix has rank 1 whereas the minimal realizations of the form (1.1) are $2N - 1$-dimensional.

The reason why such complications occur is due to the fact that only realizations by state space systems of the form (1.1) are considered. In this paper we extend the concept of partial realization by considering a wider class of systems. Let $\mathrm{S}_{n,m,p}$ be the set of all quadruples $(E, A, B, C) \in \mathbb{K}^{n \times n} \times \mathbb{K}^{n \times n} \times \mathbb{K}^{n \times m} \times \mathbb{K}^{p \times n}$ which are *in standard form*, i.e. satisfy

$$I_n \in \mathbb{K} \cdot E + \mathbb{K} \cdot A, \tag{1.2}$$

where $I_n$ denotes the $n \times n$ identity matrix. Condition (1.2) implies that the pencil $sE - A$ is nondegenerate, i.e.

$$\det(sE - A) \not\equiv 0, \tag{1.3}$$

and, moreover,

$$EA = AE. \tag{1.4}$$

Conversely, any quadruple $(E, A, B, C)$ satisfying (1.3) can be transformed to standard form by an equivalence transformation $(E, A) \mapsto (LER, LAR)$ with nonsingular matrices $L$ and $R$. Specifically, choose $L = I_n$, $R = (\alpha E + \beta A)^{-1}$, where $\alpha, \beta \in \mathbb{K}$ are such that $\det(\alpha E + \beta A) \neq 0$.

With any $(E, A, B, C) \in \mathrm{S}_{n,m,p}$ we associate the (non-degenerate) descriptor system

$$\begin{aligned} Ex(t+1) &= Ax(t) + Bu(t) \\ y(t) &= Cx(t) \end{aligned}, \qquad t \in \mathbb{N}.$$

Alternatively, we may follow Nikoukhah, Willsky and Levy [21] and associate with $(E, A, B, C)$ the two–point boundary–value descriptor system

$$\begin{aligned} Ex(t+1) &= Ax(t) + Bu(t), & 0 \le t \le N-1 \\ y(t) &= Cx(t), & 0 \le t \le N \end{aligned} \tag{1.5}$$

with mixed homogeneous boundary constraint

$$V_i x(0) + V_f x(N) = 0. \tag{1.6}$$

Here $V_i$, $V_f$ are considered as given matrices. Our concept of generalized partial realization is in fact quite close to the realization theory developed in [21] for boundary value descriptor systems.

We say that a system $(E, A, B, C) \in S_{n,m,p}$ is an $n$-dimensional *generalized partial realization* of a given finite sequence $H^\tau = (H_1, \ldots, H_\tau)$ of $p \times m$-matrices if

$$H_i = CE^{\tau-i}A^{i-1}B, \qquad i = 1, \ldots, \tau.$$

The smallest $n \in \mathbb{N}$ such that there exists an $n$-dimensional generalized partial realization of $H^\tau$ is called the *generalized McMillan degree* of $H^\tau$ and denoted by $\delta(H^\tau)$. A generalized partial realization of dimension $\delta(H^\tau)$ is called *minimal*.

Clearly $S_{n,m,p}$ is invariant under similarity transformations of the form

$$(E, A, B, C) \mapsto (TET^{-1}, TAT^{-1}, TB, CT^{-1}), \quad T \in \mathrm{Gl}(n, \mathbb{K}).$$

Moreover, if two systems $(E, A, B, C), (\tilde{E}, \tilde{A}, \tilde{B}, \tilde{C}) \in S_{n,m,p}$ are similar then

$$\tilde{C}\tilde{E}^{\tau-i}\tilde{A}^{i-1}\tilde{B} = CE^{\tau-i}A^{i-1}B, \qquad i = 1, \ldots, \tau.$$

Hence if $(E, A, B, C)$ is any generalized partial realization of $H^\tau = (H_1, \ldots, H_\tau)$ then every similar system is also a generalized partial realization of $H^\tau$. Note that we only consider generalized partial realizations by systems *in standard form* and that this is an essential feature of our realization concept.

The main objective of this paper is to characterize the minimal generalized partial realizations of a given finite sequence $H^\tau = (H_1, \ldots, H_\tau)$. With any such sequence we associate a finite family of Hankel matrices $H = (H_{i+j-1})_{i=1,j=1}^{M,N}$ where $M = 1, \ldots, \tau$, $N = \tau+1-M$. We show that provided one of the associated Hankel matrices, say $H$, satisfies

$$\operatorname{rank} H < \min\{M, N\} \tag{1.7}$$

there exists a minimal generalized partial realization of $H^\tau$ of dimension $\delta(H^\tau) = \operatorname{rank} H$ and these realizations are uniquely determined modulo *generalized similarity* (see Section 3). Moreover, each minimal generalized partial realization is controllable and observable.

The above problem of generalized partial realization has a close connection with a classical problem from algebra, viz. Waring's problem for binary forms, see [11]. In fact, a homogeneous matrix polynomial in two variables of the form $f(X, Y) =$

$\sum_{i=0}^{\tau-1}\binom{\tau-1}{i}H_{i+1}X^{\tau-1-i}Y^i$ can be written as $f(X,Y) = C(XE + YA)^{\tau-1}B$ with $(E, A, B, C) \in S_{n,m,p}$ if and only if $(E, A, B, C)$ is a generalized partial realization of the finite sequence $H^\tau = (H_1, \ldots, H_\tau)$. This shows that the tasks of parametrizing solutions to Waring's problem and parametrizing minimal generalized partial realizations are equivalent. We also mention that the partial realization problem is closely related to Vandermonde factorizations of block Hankel matrices and refer the reader to [3] and the references therein for further information on this subject.

The paper is organized as follows. In Section 2 some basic facts about the Fischer–Frobenius transformation of rectangular block Hankel matrices are briefly reviewed. This material is crucial for our approach to the generalized partial realization problem. Section 3 contains our main results on minimal generalized realizations. Finally, Section 4 discusses some topological aspects. The space of controllable and observable systems $(E, A, B, C) \in S_{n,m,p}$ and the associated quotient spaces are shown to be analytic manifolds.

## 2 The Fischer–Frobenius transformation

In this section we present, for later use, some basic facts about the *Fischer–Frobenius transformation* of rectangular block Hankel matrices. For details and proofs the reader is referred to [20].

The Fischer–Frobenius transformation has its origin in the theory of *scalar* Hankel and Toeplitz matrices as well as in the theory of quadratic and Hermitian forms, see [13, §19]. Brockett [2] was the first to use the Fischer–Frobenius transformation in a system theoretic context. The Fischer–Frobenius transformation of *block* Hankel matrices was studied in [10], [20].

Throughout the paper, let $\mathbb{K} = \mathbb{R}$ or $\mathbb{C}$. For integers $m, p, M, N \in \mathbb{N}$ we denote by $\text{Hank}_{p,m}(M \times N)$ the vector space of all $M \times N$ block Hankel matrices of the form $H = (H_{i+j-1})_{i=1,j=1}^{M,N}$, $H_k \in \mathbb{K}^{p \times m}$, $k = 1, \ldots, M + N - 1$. If $n \leq \min\{Mp, Nm\}$ we set

$$\text{Hank}_{p,m}(n, M \times N) := \{H \in \text{Hank}_{p,m}(M \times N); \quad \text{rank } H = n\}.$$

The Fischer–Frobenius transformation of block Hankel matrices is defined as follows. For any $d \geq 2$, the general linear group $\text{Gl}(2, \mathbb{K})$ of all invertible $2 \times 2$ matrices $g = (g_{ij})$ over $\mathbb{K}$ acts on the $d$-dimensional $\mathbb{K}$-vector space $B(d)$ of binary forms $\gamma(X, Y) = \sum_{j=0}^{d-1} \gamma_{j+1}X^{d-1-j}Y^j$, $\gamma_{j+1} \in \mathbb{K}$ of degree $d - 1$ via the linear transformation

$$\tau_d(g) : \ \gamma(X, Y) \mapsto \gamma(g_{11}X + g_{21}Y, g_{12}X + g_{22}Y).$$

The matrix representation of the linear operator $\tau_d(g)$ with respect to the canonical basis $(X^{d-1}, X^{d-2}Y, \ldots, Y^{d-1})$ in $B(d)$ is called *Möbius matrix of order d* and is again denoted by $\tau_d(g)$. It is easily established that the mapping

$$\tau_d : \ g \mapsto \tau_d(g), \quad d \geq 2, \tag{2.1}$$

is an isomorphism between the group $\mathrm{Gl}(2, \mathbb{K})$ and the group of Möbius matrices of order $d$. Every $g \in \mathrm{Gl}(2, \mathbb{K})$ can be written as a product of elementary transformations of the following three types

$$g(a) = \begin{bmatrix} a & 0 \\ 0 & 1 \end{bmatrix}, \; a \in \mathbb{K} \setminus \{0\}, \quad w = \begin{bmatrix} 0 & 1 \\ 1 & 0 \end{bmatrix}, \quad g^t = \begin{bmatrix} 1 & t \\ 0 & 1 \end{bmatrix}, t \in \mathbb{K}. \quad (2.2)$$

An elementary calculation yields

$$\tau_d\left(g(a)\right) = \mathrm{diag}(a^{d-1}, a^{d-2}, \ldots, a, 1), \quad a \neq 0, \tag{2.3}$$

$$\tau_d(w) = \begin{bmatrix} 0 & \cdots & 0 & 1 \\ 0 & \cdots & 1 & 0 \\ \vdots & \ddots & \vdots & \vdots \\ 1 & \cdots & 0 & 0 \end{bmatrix} \in \mathrm{Gl}(d, \mathbb{K}), \tag{2.4}$$

$$\tau_d(g^t) = \begin{bmatrix} \binom{0}{0} & \binom{1}{0}t & \binom{2}{0}t^2 & \cdots & \binom{d-2}{0}t^{d-2} & \binom{d-1}{0}t^{d-1} \\ 0 & \binom{1}{1} & \binom{2}{1}t & \cdots & \binom{d-2}{1}t^{d-3} & \binom{d-1}{1}t^{d-2} \\ 0 & 0 & \binom{2}{2} & \ddots & \vdots & \vdots \\ \vdots & \vdots & \vdots & \ddots & \binom{d-2}{d-2} & \binom{d-1}{d-2}t \\ 0 & 0 & 0 & \cdots & 0 & \binom{d-1}{d-1} \end{bmatrix}, \quad t \in \mathbb{K}. \tag{2.5}$$

Setting

$$g_t = \begin{bmatrix} 1 & 0 \\ t & 1 \end{bmatrix} = w g^t w \tag{2.6}$$

we see from (2.4)–(2.6) that $\tau_d(g^t)$ and $\tau_d(g_t)$ are upper and lower triangular matrices, respectively, with all diagonal entries equal to 1.

Using the vector space isomorphism $\mathrm{Hank}_{p,m}(M \times N) \cong (\mathbb{K}^{p \times m})^{M+N-1}$, the group isomorphism $\tau_{M+N-1}$ (2.1) induces a *right* action of $\mathrm{Gl}(2, \mathbb{K})$ on the space $\mathrm{Hank}_{p,m}(M \times N)$:

$$\begin{aligned} \mathrm{Hank}_{p,m}(M \times N) \times \mathrm{Gl}(2, \mathbb{K}) &\to \mathrm{Hank}_{p,m}(M \times N) \\ (H, g) &\mapsto H \cdot g := [H_1, \ldots, H_{M+N-1}] \left(\tau_{M+N-1}(g) \otimes I_m\right) \end{aligned} \tag{2.7}$$

where $\otimes$ denotes the *Kronecker product*.

**Definition 2.1** For every $H \in \mathrm{Hank}_{p,m}(M \times N)$, $g \in \mathrm{Gl}(2, \mathbb{K})$, the block Hankel matrix $H \cdot g$ is said to be a *Fischer–Frobenius transform* of $H$.

By straight forward calculations it can be shown that an equivalent description of the $\mathrm{Gl}(2,\mathbb{K})$-action (2.7) is

$$H \cdot g = \left[ \tau_M(g)^\top \otimes I_p \right] H \left[ \tau_N(g) \otimes I_m \right], \quad H \in \mathrm{Hank}_{p,m}(M \times N), \ g \in \mathrm{Gl}(2,\mathbb{K}).$$
$$(2.8)$$

It follows that the Fischer–Frobenius transformation (2.7) leaves the rank of Hankel matrices invariant, i.e. for arbitrary $n \leq \min\{Mp, Nm\}$,

$$\mathrm{Hank}_{p,m}(n, M \times N) \cdot g = \mathrm{Hank}_{p,m}(n, M \times N), \quad g \in \mathrm{Gl}(2,\mathbb{K}).$$

For any $n \leq \min\{M, N\}$ we set

$$\mathrm{Hank}^*_{p,m}(n, M \times N) := \{H \in \mathrm{Hank}_{p,m}(n, M \times N); \quad \mathrm{rank}\, H(n,n) = n\}, \quad (2.9)$$

where $H(n,n) := (H_{i+j-1})^n_{i,j=1}$. The following theorem is a key tool for our analysis. A proof of the result can be found in [20] (see also [2], [18] for the scalar case).

**Theorem 2.2** *If $H \in \mathrm{Hank}_{p,m}(n, M \times N)$ and $n < \min\{M, N\}$, then $H \cdot g_t \in \mathrm{Hank}^*_{p,m}(n, M \times N)$ except for at most finitely many $t \in \mathbb{K}$.*

There are examples of Hankel matrices $H \in \mathrm{Hank}_{p,m}(n, M \times N)$ with $n > \min\{M, N\}$ for which the assertion of Theorem 2.2 does not hold. Hence the rank condition $n < \min\{M, N\}$ can in general not be dispensed with. However, it can be shown [17] that Theorem 2.2 also holds for $n = \min\{M, N\}$. For the scalar case ($m = p = 1$) the following result has been proved in [18] (Corollary 2.9).

**Proposition 2.3** *If $H \in \mathrm{Hank}_{1,1}(n, M \times N)$ is a scalar Hankel matrix, then $H \cdot g_t \in \mathrm{Hank}^*_{1,1}(n, M \times N)$ except for at most finitely many $t \in \mathbb{K}$.*

## 3 Minimal generalized partial realizations

In this section we determine the McMillan degree of finite matrix sequences and characterize their minimal generalized partial realizations via associated Hankel matrices under the assumption that the rank condition (1.7) is satisfied.

Given a finite sequence $H^\tau = (H_1, \ldots, H_\tau)$ of $p \times m$-matrices there are $\tau$ associated block Hankel matrices of the form

$$H = (H_{i+j-1})^{M,N}_{i=1,j=1} \in \mathrm{Hank}_{p,m}(M \times N), \qquad (3.1)$$

where $M + N - 1 = \tau$, $M = 1, \ldots, \tau$. For the sake of concise formulation we apply the following terminology directly to Hankel matrices (instead of the underlying matrix sequences).

**Definition 3.1** A linear system $(E, A, B, C) \in S_{n,m,p}$ is called a *generalized partial realization* of the Hankel matrix (3.1) if

$$H_i = CE^{M+N-1-i}A^{i-1}B, \qquad i = 1, \ldots, M + N - 1. \tag{3.2}$$

$(E, A, B, C)$ is said to be a *minimal generalized partial realization* of (3.1) if it is of minimal dimension. In this case its dimension is called the *generalized McMillan degree* of the finite Hankel matrix(3.1), which we denote by $\delta(H)$.

By definition, $(E, A, B, C)$ is a generalized partial realization of the Hankel $H$ (3.1) if and only if it is a generalized partial realization of the underlying sequence $H^\tau = (H_1, ..., H_{M+N-1})$. A generalized partial realization $(E, A, B, C)$ of $H$ (resp. $H^\tau$), is a partial realization if and only if $E = I_n$. If $(I_n, A, B, C)$ is a minimal generalized partial realization then it is also a minimal partial realization. In this case the generalized McMillan degree of $H$ coincides with the McMillan degree of $H$. Note that the converse is not true: A minimal partial realization is not necessarily a minimal generalized partial realization.

A system $(E, A, B, C) \in S_{n,m,p}$ is *controllable* if and only if

$$\text{rank}[\alpha E - \beta A, B] = n \quad \text{for all } \alpha, \beta \in \mathbb{C}, \ (\alpha, \beta) \neq (0, 0) \tag{3.3}$$

and *observable* if and only if

$$\text{rank} \begin{bmatrix} \alpha E - \beta A \\ C \end{bmatrix} = n \quad \text{for all } \alpha, \beta \in \mathbb{C}, \ (\alpha, \beta) \neq (0, 0). \tag{3.4}$$

**Remark 3.2** A two–point boundary–value descriptor system (1.5), (1.6) in standard form (1.2) is called strongly reachable (resp. strongly observable) if condition (3.3) (resp. (3.4)) is satisfied. $\quad\square$

The following generalization of the classical Cayley–Hamilton theorem can be found in [22].

**Lemma 3.3** *Suppose* $E, A \in \mathbb{K}^{n \times n}$ *and* $I_n \in \mathbb{K} \cdot E + \mathbb{K} \cdot A$. *Then for any pair of integers* $i, j \geq 0$ *there exist coefficients* $r_0, \ldots, r_{n-1} \in \mathbb{K}$ *such that*

$$A^j E^i = \sum_{l=0}^{n-1} r_l A^{n-l-1} E^l.$$

With any pair of integers $M, N \in \mathbb{N}$ and any system $(E, A, B, C) \in S_{n,m,p}$ we associate the matrix

$$H_{MN}(E, A, B, C) := O_M(E, A, C)R_N(E, A, B), \tag{3.5}$$

where

$$R_N(E, A, B) := \left( E^{N-1}B, E^{N-2}AB, \ldots, EA^{N-2}B, A^{N-1}B \right) \in \mathbb{K}^{n \times Nm}, \quad (3.6)$$

$$O_M(E, A, C) := \begin{pmatrix} CE^{M-1} \\ CE^{M-2}A \\ \vdots \\ CEA^{M-2} \\ CA^{M-1} \end{pmatrix} \in \mathbb{K}^{Mp \times n}. \quad (3.7)$$

Note that, in view of (1.4), $H_{MN}(E, A, B, C)$ is a $M \times N$ block Hankel matrix. Lemma 3.3 implies the following result.

**Corollary 3.4** *Let* $(E, A, B, C) \in S_{n,m,p}$. *Then*

$$\operatorname{rank} R_N(E, A, B) = \operatorname{rank} R_n(E, A, B) \qquad \textit{for all} \quad N \geq n, \quad (3.8)$$

$$\operatorname{rank} O_M(E, A, C) = \operatorname{rank} O_n(E, A, C) \qquad \textit{for all} \quad M \geq n. \quad (3.9)$$

**Proof:** Here we prove (3.8), equation (3.9) can be established in the same fashion. By Lemma 3.3 there exist coefficients $r_l^{(j)} \in \mathbb{K}$, $l = 0, \ldots, n-1$, $j = 1, \ldots, N$ such that

$$R_N(E, A, B) = \left( \sum_{l=0}^{n-1} r_l^{(1)} A^{n-l-1}E^l B, \ldots, \sum_{l=0}^{n-1} r_l^{(N)} A^{n-l-1}E^l B \right).$$

Thus $R_N(E, A, B) = R_n(E, A, B)[R \otimes I_m]$ with $R := \left( r_{n-i}^{(j)} \right)_{i,j=1}^{n,N}$, and we obtain $\operatorname{rank} R_N(E, A, B) \leq \operatorname{rank} R_n(E, A, B)$ for all $N \geq n$. On the other hand by condition (1.2)

$$\operatorname{Im} R_k(E, A, B) \subset \operatorname{Im} R_{k+1}(E, A, B), \qquad k \in \mathbb{N} \quad (3.10)$$

and so $\operatorname{Im} R_n(E, A, B) \subset \operatorname{Im} R_N(E, A, B)$ for all $N \geq n$. Hence (3.8) follows. $\quad \square$

In [22] it is proven that (3.3) is equivalent to the equation $\operatorname{rank} R_n(E, A, B) = n$. Combining this result with Corollary 3.4 and (3.10) we get the following rank tests for controllability. Analogous tests of observability are obtained similarly.

**Proposition 3.5** *A singular system* $(E, A, B, C) \in S_{n,m,p}$ *is controllable (resp. observable) if and only if one of the following two equivalent conditions is satisfied:*

*(i)* $\operatorname{rank} R_N(E, A, B) = n$ *(resp.* $\operatorname{rank} O_M(E, A, C) = n$) *for all* $N \geq n$ *(resp.* $M \geq n$).

*(ii)* $\operatorname{rank} R_k(E, A, B) = n$ *(resp.* $\operatorname{rank} O_l(E, A, C) = n$) *for some* $k \in \mathbb{N}$ *(resp.* $l \in \mathbb{N}$).

We now turn to properties of minimal generalized partial realizations.

**Lemma 3.6** *Suppose* $H \in \operatorname{Hank}_{p,m}(n, M \times N)$.

*(i) Then there is no generalized partial realization of $H$ of smaller dimension than $n$.*

*(ii) If $(E, A, B, C)$ is an $n$–dimensional generalized partial realization of $H$, then $(E, A, B, C)$ is controllable, observable and minimal.*

**Proof:** (i) Let $(E, A, B, C) \in S_{l,m,p}$ be an $l$–dimensional generalized partial realization of $H$. Condition (3.2) can be written in the form $H = H_{MN}(E, A, B, C)$, and hence $n = \operatorname{rank} H \le \operatorname{rank} O_M(E, A, C) \le l$, which implies $\delta(H) \ge n$.

(ii) If $(E, A, B, C)$ is an n–dimensional generalized partial realization of $H$, then it is minimal by (i). Equations (3.5)–(3.7) imply $n = \operatorname{rank} H \le \operatorname{rank} O_M(E, A, C) \le n$. Analogously we obtain $n = \operatorname{rank} R_N(E, A, B)$, and hence (ii) follows from Proposition 3.5 (ii). $\qquad\square$

For any $(E, A, B, C) \in S_{n,m,p}$ and $g = (g_{ij}) \in \operatorname{Gl}(2, \mathbb{K})$ we define

$$(E, A, B, C) \cdot g := (g_{11}E + g_{21}A, g_{12}E + g_{22}A, B, C).$$

$(E, A, B) \cdot g$ and $(E, A, C) \cdot g$ are defined analogously.

**Lemma 3.7** *The mapping*

$$S_{n,m,p} \times \operatorname{Gl}(2, \mathbb{K}) \longrightarrow S_{n,m,p}, \qquad [(E, A, B, C), g] \longmapsto (E, A, B, C) \cdot g \qquad (3.11)$$

*is a (right) group action of $\operatorname{Gl}(2, \mathbb{K})$ on $S_{n,m,p}$.*

**Proof:** By (1.2) there exists $(\alpha, \beta) \in \mathbb{K}^2$ such that $\alpha E + \beta A = I_n$. Hence

$$\frac{g_{22}\alpha - g_{12}\beta}{\det g}(g_{11}E + g_{21}A) + \frac{g_{11}\beta - g_{21}\alpha}{\det g}(g_{12}E + g_{22}A) = I_n,$$

i.e. $(E, A, B, C) \cdot g \in S_{n,m,p}$. Of course, if $e \in \operatorname{Gl}(2, \mathbb{K})$ is the identity matrix, then $(E, A, B, C) \cdot e = (E, A, B, C)$ for all $(E, A, B, C) \in S_{n,m,p}$. It is easily verified that $[(E, A, B, C) \cdot g] \cdot \tilde{g} = (E, A, B, C) \cdot (g\tilde{g})$ for all $(E, A, B, C) \in S_{n,m,p}$ and $g, \tilde{g} \in \operatorname{Gl}(2, \mathbb{K})$. $\qquad\square$

**Definition 3.8** For every $(E, A, B, C) \in S_{n,m,p}$, $g \in \operatorname{Gl}(2, \mathbb{K})$, the singular system $(E, A, B, C) \cdot g$ is said to be a *Moebius transform* of $(E, A, B, C)$.

**Definition 3.9** Given integers $n, m, p \in \mathbb{N}$ and $M, N \in \mathbb{N}$, the corresponding *Hankel mapping* is defined by

$$H_{MN} : S_{n,m,p} \to \operatorname{Hank}_{p,m}(M \times N), \quad (E, A, B, C) \mapsto H_{MN}(E, A, B, C).$$

The next Proposition states that the Moebius transformation is closely related to the Fischer–Frobenius transformation.

**Proposition 3.10** $H_{MN}$ *is an interwining mapping for the Fischer–Frobenius action (2.7) and the Moebius transformation (3.11), i.e. for* $g \in$ Gl$(2, \mathbb{K})$, $(E, A, B, C) \in S_{n,m,p}$

$$H_{MN}((E, A, B, C) \cdot g) = H_{MN}(E, A, B, C) \cdot g.$$

*Moreover*

$$\begin{array}{rcl} R_N((E, A, B) \cdot g) & = & R_N(E, A, B)\left[\tau_N(g) \otimes I_m\right], \\ O_M((E, A, C) \cdot g) & = & \left[\tau_M(g)^\top \otimes I_p\right] O_M(E, A, C). \end{array} \tag{3.12}$$

**Proof:** By (2.8) and (3.5) we have

$$H_{MN}(E, A, B, C) \cdot g = \left[\tau_M(g)^\top \otimes I_p\right] O_M(E, A, , C) R_N(E, A, B) \left[\tau_N(g) \otimes I_m\right]. \tag{3.13}$$

One immediately verifies that

$$\begin{array}{rcl} R_N(E, A, B)\left[\tau_N(g(a)) \otimes I_m\right] & = & R_N(aE, A, B), \\ \left[\tau_M(g(a))^\top \otimes I_p\right] O_M(E, A, C) & = & O_M(aE, A, C), \\ R_N(E, A, B)\left[\tau_N(w) \otimes I_m\right] & = & R_N(A, E, B), \\ \left[\tau_M(w)^\top \otimes I_p\right] O_M(E, A, C) & = & O_M(A, E, C), \end{array}$$

cf. (2.2)–(2.4). Hence

$$\left[\tau_M(g(a))^\top \otimes I_p\right] H_{MN}(E, A, B, C)\left[\tau_N(g(a)) \otimes I_m\right] = H_{MN}(aE, A, B, C), \tag{3.14}$$

$$\left[\tau_M(w)^\top \otimes I_p\right] H_{MN}(E, A, B, C)\left[\tau_N(w) \otimes I_m\right] = H_{MN}(A, E, B, C). \tag{3.15}$$

From (2.5) we get

$$\begin{aligned} R_N(E, A, B)\left[\tau_N(g^t) \otimes I_m\right] & = \left(E^{N-1}B, E^{N-2}(tE + A)B, \ldots, (tE + A)^{N-1}B\right) \\ & = R_N(E, tE + A, B) \end{aligned}$$

and similarly $\left[\tau_M(g^t)^\top \otimes I_p\right] O_M(E, A, C) = O_M(E, tE + A, C)$. Thus

$$\left[\tau_M(g^t)^\top \otimes I_p\right] H_{MN}(E, A, B, C)\left[\tau_N(g^t) \otimes I_m\right] = H_{MN}(E, tE + A, B, C). \tag{3.16}$$

The result follows from (3.13)–(3.16) since the matrices $g(a)$, $w$, $g^t$ generate Gl$(2, \mathbb{K})$. $\quad\square$

**Corollary 3.11** *Controllability and observability are preserved under the Moebius transformation (3.11).*

**Proof:** Suppose $(E, A, B, C) \in S_{n,m,p}$ is controllable and $g \in$ Gl$(2, \mathbb{K})$. By (3.11) we see that $(E, A, B, C) \cdot g \in S_{n,m,p}$. It follows from (3.12) that rank $R_n((E, A, B) \cdot g) =$ rank $R_n(E, A, B) = n$; hence $(E, A, B, C) \cdot g$ is controllable by Proposition 3.5. The observability result can be established in the same fashion. $\quad\square$

For any $n, m, p \in \mathbb{N}$, we denote

$$S_{n,m,p}^{c,o} := \{(E, A, B, C) \in S_{n,m,p}; \quad (E, A, B, C) \text{ is controllable and observable}\}.$$

Applying Proposition 3.5, we obtain

$$H_{MN}\left(S_{n,m,p}^{c,o}\right) \subset \text{Hank}_{p,m}(n, M \times N), \qquad n \leq \min\{M, N\}.$$

Hence, for $n \leq \min\{M, N\}$, the restriction of the Hankel mapping $H_{MN}$ to $S_{n,m,p}^{c,o}$ has its image in $\text{Hank}_{p,m}(n, M \times N)$.

**Lemma 3.12** *Let* $(E, A, B, C) \in S_{n,m,p}^{c,o}$ *and* $n \leq \min\{M, N\}$, $2n < M + N$. *Then* $E$ *is nonsingular if and only if* $H_{MN}(E, A, B, C) \in \text{Hank}_{p,m}^{*}(n, M \times N)$.

**Proof:** Suppose $E$ is nonsingular and $\chi_{E^{-1}A}(s) = s^n - r_n s^{n-1} - \ldots - r_1$ is the characteristic polynomial of $E^{-1}A$. Then by the Cayley–Hamilton theorem

$$E^{-n} A^n = \sum_{i=1}^{n} r_i E^{-(i-1)} A^{i-1}.$$

Multiplying this equation by $CE^{M+N-2-j}$ on the left and by $A^j B$ on the right for $j = 0, 1, \ldots, M + N - 2 - n$, we obtain

$$CE^{M+N-2-n-j} A^{n+j} B = \sum_{i=1}^{n} r_i CE^{M+N-2-(i-1)-j} A^{i-1+j} B, \quad 0 \leq j \leq M + N - 2 - n.$$

$$(3.17)$$

Thus all columns of the Hankel matrix $H_{MN}(E, A, B, C) \in \text{Hank}_{p,m}(n, M \times N)$ must be linear combinations of the first $mn$ columns, that is, the columns appearing in the first $n$ blocks. Moreover, we conclude from (3.17) that all rows of $H_{MN}(E, A, B, C)$ must be linear combinations of the first $pn$ rows. Hence $H_{MN}(E, A, B, C) \in \text{Hank}_{p,m}^{*}(n, M \times N)$.

Conversely, let $H := H_{MN}(E, A, B, C) \in \text{Hank}_{p,m}^{*}(n, M \times N)$ and assume that $E$ is singular. Then by (3.5), (1.4)

$$H(n, n) = \begin{pmatrix} CE^{M-1} \\ CE^{M-2}A \\ \vdots \\ CE^{M-n}A^{n-1} \end{pmatrix} \left( E^{N-1}B, E^{N-2}AB, \ldots, E^{N-n}A^{n-1}B \right)$$

$$= O_n(E, A, C) E^{M+N-2n} R_n(E, A, B).$$

Since $M + N - 2n > 0$, the last equation implies $\text{rank}\, H(n, n) \leq \text{rank}\, E < n$, which yields a contradiction to $H \in \text{Hank}_{p,m}^{*}(n, M \times N)$. $\qquad\square$

In classical realization theory it is shown that two minimal realizations of an infinite Hankel matrix are necessarily similar. We want to derive a counterpart of this uniqueness result for generalized partial realizations. We introduce an equivalence relation on the space $S_{n,m,p}^{c,o}$, denoted by $\sim_{MN}$, for each pair of integers $M, N \geq 1$.

**Definition 3.13** Let $(E, A, B, C), (\tilde{E}, \tilde{A}, \tilde{B}, \tilde{C}) \in S^{c,o}_{n,m,p}$. Then $(E, A, B, C)$ is called *similar to* $(\tilde{E}, \tilde{A}, \tilde{B}, \tilde{C})$ *modulo scaling of order* $(M, N)$:

$$(E, A, B, C) \sim_{MN} (\tilde{E}, \tilde{A}, \tilde{B}, \tilde{C})$$

if there exist $S \in \mathrm{Gl}(n, \mathbb{K})$ and $(\alpha, \beta) \in \mathbb{K}^2$ such that $U = \alpha E + \beta A \in \mathrm{Gl}(n, \mathbb{K})$ and

$$(\tilde{E}, \tilde{A}, \tilde{B}, \tilde{C}) = (SU^{-1}ES^{-1}, SU^{-1}AS^{-1}, SU^{N-1}B, CU^{M-1}S^{-1}).$$

**Remark 3.14** (i) If $(E, A, B, C)$ and $(\tilde{E}, \tilde{A}, \tilde{B}, \tilde{C})$ are similar modulo scaling of order $(M, N)$ then the associated Hankel matrices coincide:

$$(E, A, B, C) \sim_{MN} (\tilde{E}, \tilde{A}, \tilde{B}, \tilde{C}) \;\Rightarrow\; H_{MN}(E, A, B, C) = H_{MN}(\tilde{E}, \tilde{A}, \tilde{B}, \tilde{C}).$$
$$(3.18)$$

In other words, the homogeneous matrix polynomial $C(xE + yA)^{M+N-2}B$ is an invariant for similarity modulo scaling of order $(M, N)$.

(ii) The associated transfer functions are related in the following way:

$$\tilde{C}(s\tilde{E} - \tilde{A})^{-1}\tilde{B} = C(\alpha E + \beta A)^{M+N-1}(sE - A)^{-1}B.$$

(iii) Let $(E, A, B, C), (\tilde{E}, \tilde{A}, \tilde{B}, \tilde{C}) \in S^{c,o}_{n,m,p}$, $S, U \in \mathrm{Gl}(n, \mathbb{K})$ and suppose

$$(\tilde{E}, \tilde{A}, \tilde{B}, \tilde{C}) = (SU^{-1}ES^{-1}, SU^{-1}AS^{-1}, SU^{N-1}B, CU^{M-1}S^{-1}).$$

Then $U = \alpha E + \beta A$ holds, if and only if $\alpha\tilde{E} + \beta\tilde{A} = I_n$. Therefore if

$$S^{c,o}_{n,m,p}(\alpha, \beta) = \{(E, A, B, C) \in S^{c,o}_{n,m,p}; \alpha E + \beta A = I_n\}, \quad (\alpha, \beta) \neq (0, 0)$$

then $\sim_{MN}$ reduces to similarity on $S^{c,o}_{n,m,p}(\alpha, \beta)$: Two systems in this space are similar modulo scaling of order $(M, N)$ if and only if they are similar. $\square$

The proof of the following lemma is straight forward and is omitted.

**Lemma 3.15** $\sim_{MN}$ *is an equivalence relation on* $S^{c,o}_{n,m,p}$.

The next lemma shows that the equivalence relation $\sim_{MN}$ on $S^{c,o}_{n,m,p}$ is compatible with the Moebius transformation (3.11).

**Lemma 3.16** *Assume that* $(E, A, B, C)$ *and* $(\tilde{E}, \tilde{A}, \tilde{B}, \tilde{C})$ *are two systems in* $S^{c,o}_{n,m,p}$. *If* $(E, A, B, C) \sim_{MN} (\tilde{E}, \tilde{A}, \tilde{B}, \tilde{C})$, *then*

$$(E, A, B, C) \cdot g \sim_{MN} (\tilde{E}, \tilde{A}, \tilde{B}, \tilde{C}) \cdot g \qquad \text{for all} \quad g \in \mathrm{Gl}(2, \mathbb{K}).$$

**Proof:** Suppose that $(\tilde{E}, \tilde{A}, \tilde{B}, \tilde{C}) = (SU^{-1}ES^{-1}, SU^{-1}AS^{-1}, SU^{N-1}B, CU^{M-1}S^{-1})$ with $U = \alpha E + \beta A \in \mathrm{Gl}(n, \mathbb{K})$, $(\alpha, \beta) \in \mathbb{K}^2$ and $S \in \mathrm{Gl}(n, \mathbb{K})$. Then

$$(g_{11}\tilde{E} + g_{21}\tilde{A}, g_{12}\tilde{E} + g_{22}\tilde{A}, \tilde{B}, \tilde{C}) =$$

$$(SU^{-1}(g_{11}E + g_{21}A)S^{-1}, SU^{-1}(g_{12}E + g_{22}A)S^{-1}, SU^{N-1}B, CU^{M-1}S^{-1}),$$

i.e. $(g_{11}\tilde{E} + g_{21}\tilde{A}, g_{12}\tilde{E} + g_{22}\tilde{A}, \tilde{B}, \tilde{C}) \sim_{MN} (g_{11}E + g_{21}A, g_{12}E + g_{22}A, B, C).$ $\square$

We are now in a position to prove the main result of this paper.

**Theorem 3.17 (Generalized Partial Realization Theorem)** *Suppose*
$n < \min\{M, N\}$. *Then:*

*(i) Every $H \in \mathrm{Hank}_{p,m}(n, M \times N)$ has a minimal generalized partial realization
of dimension $n$. In particular, $\delta(H) = \mathrm{rank}\, H$.*

*(ii) Every minimal generalized partial realization of $H \in \mathrm{Hank}_{p,m}(n, M \times N)$ is
controllable and observable.*

*(iii) Two minimal generalized partial realizations $(E, A, B, C), (\tilde{E}, \tilde{A}, \tilde{B}, \tilde{C}) \in
\mathrm{S}^{c,o}_{n,m,p}$ of $H \in \mathrm{Hank}_{p,m}(n, M \times N)$ are similar modulo scaling of order $(M, N)$.*

**Proof:** Suppose $H \in \mathrm{Hank}_{p,m}(n, M \times N)$, $n < \min\{M, N\}$.
(i) In view of Lemma 3.6 (i) it suffices to show that there exists $(E, A, B, C) \in
\mathrm{S}_{n,m,p}$ with $H = H_{MN}(E, A, B, C)$. If $H \in \mathrm{Hank}^*_{p,m}(n, M \times N)$ (cf. (2.9)) then
choose $(I_n, A, B, C) \in \mathrm{S}^{c,o}_{n,m,p}$ such that $H = H_{MN}(I_n, A, B, C)$ holds. This is
possible by standard realization theory, since every $H \in \mathrm{Hank}^*_{p,m}(n, M \times N)$,
$n < \min\{M, N\}$ can be (uniquely) extended to an infinite block Hankel matrix
$(H_{i+j-1})^{\infty}_{i,j=1}$, $H_{i+j-1} \in \mathbb{K}^{p \times m}$ of rank $n$, cf. [23]. If $H \notin \mathrm{Hank}^*_{p,m}(n, M \times N)$ there
exists by Theorem 2.2 a scalar $t \in \mathbb{K} \setminus \{0\}$ such that $H \cdot g_t \in \mathrm{Hank}^*_{p,m}(n, M \times N)$,
where $g_t$ is given by (2.6). Now using the first step, choose $(I_n, A, B, C) \in \mathrm{S}^{c,o}_{n,m,p}$
such that $H \cdot g_t = H_{MN}(I_n, A, B, C)$. Then $H = H_{MN}(I_n, A, B, C) \cdot g_{-t}$. Hence $H =
H_{MN}(I_n - tA, A, B, C)$ by Proposition 3.10 and so $(E, A, B, C) := (I_n - tA, A, B, C)$
is an n–dimensional generalized partial realization of $H$. It follows from Lemma
3.6 (i) that $\delta(H) = n$.
(ii) By (i) every minimal generalized partial realization of $H$ is $n$-dimensional.
Hence (ii) follows from Lemma 3.6 (ii).
(iii) Suppose that

$$H_{MN}(E, A, B, C) = H_{MN}(\tilde{E}, \tilde{A}, \tilde{B}, \tilde{C}), \qquad (E, A, B, C), (\tilde{E}, \tilde{A}, \tilde{B}, \tilde{C}) \in \mathrm{S}^{c,o}_{n,m,p}.$$

By (1.2), Proposition 3.10, Corollary 3.11 and by the compatibility of $\sim_{MN}$ with
the Fischer–Frobenius transformation (see Lemma 3.16) we can assume that $E =
I_n$. It follows from Lemma 3.12 that the invertibility of $E = I_n$ implies that of $\tilde{E}$,
and therefore

$$H_{MN}(I_n, A, B, C) = H_{MN}(I_n, \tilde{E}^{-1}\tilde{A}, \tilde{E}^{N-1}\tilde{B}, \tilde{C}\tilde{E}^{M-1}).$$

By standard realization theory there exists a unique $S \in \mathrm{Gl}(n, \mathbb{K})$ with

$$(I_n, \tilde{E}^{-1}\tilde{A}, \tilde{E}^{N-1}\tilde{B}, \tilde{C}\tilde{E}^{M-1}) = (I_n, S^{-1}AS, S^{-1}B, CS),$$

and this implies

$$(\tilde{E}, \tilde{A}, \tilde{B}, \tilde{C}) \sim_{MN} (I_n, S\tilde{E}^{-1}\tilde{A}S^{-1}, S\tilde{E}^{N-1}\tilde{B}, \tilde{C}\tilde{E}^{M-1}S^{-1}) = (I_n, A, B, C)$$

$(\alpha = 1, \beta = 0)$. $\qquad\qquad\square$

**Corollary 3.18** *Suppose* $n < \min\{M, N\}$. *Then the McMillan degree of* $H \in$ $\mathrm{Hank}_{p,m}(n, M \times N)$ *coincides with the generalized McMillan degree of* $H$ *if and only if* $H \in \mathrm{Hank}^*_{p,m}(n, M \times N)$.

**Proof:** Let $H \in \mathrm{Hank}_{p,m}(n, M \times N)$. Then there exists a generalized partial realization $(E, A, B, C)$ of dimension $n = \mathrm{rank}\, H = \delta(H)$. If $H \in \mathrm{Hank}^*_{p,m}(n, M \times N)$, then $E$ is nonsingular by Lemma 3.12, hence there exists an $n$–dimensional state space realization (1.1) of $H$. Therefore the McMillan degree of $H$ coincides with the generalized McMillan degree of $H$. Conversely if $H \notin \mathrm{Hank}^*_{p,m}(n, M \times N)$, then $E$ is singular and there does not exist an $n$–dimensional state space realization (1.1) of $H$. $\qquad\square$

The rank condition $n < \min\{M, N\}$ has been used at two points in the proof of the existence statement (i) of Theorem 3.17: Firstly, in order to conclude that every $H \in \mathrm{Hank}^*_{p,m}(n, M \times N)$ can be extended to an infinite block Hankel matrix $(H_{i+j-1})_{i,j=1}^{\infty}$ of rank $n$, and secondly in order to make use of Theorem 2.2. If we wish to apply Theorem 3.17 to a given sequence $H^{\tau} = (H_1, \ldots, H_{\tau})$ of $p \times m$ matrices we have to find an associated Hankel matrix (3.1) which satisfies the rank condition. If none of the associated Hankel matrices satisfies this condition the theorem is not applicable. Since in the nonscalar case the condition $n < \min\{M, N\}$ is *generically not* satisfied for a matrix sequence $H^{\tau} = (H_1, \ldots, H_{\tau})$, a weakening of the assumption $n < \min\{M, N\}$ would be highly desirable.
We will now see that in the scalar case the rank condition is not needed. With every scalar sequence $h^{\tau} = (h_1, \ldots, h_{\tau})$ we associate the $M \times N$ Hankel matrix with $M = [(\tau + 1)/2] := \max\{k \in \mathbb{N};\ k \le (\tau + 1)/2\}$ rows and $N = \tau + 1 - M$ columns:

$$H(h^{\tau}) = (h_{i+j-1})_{i,j=1}^{M,N}. \tag{3.19}$$

**Corollary 3.19 (Scalar case)** *Let* $h^{\tau} = (h_1, \ldots, h_{\tau})$ *be a finite scalar sequence and let $n$ be the rank of the associated Hankel matrix $H(h^{\tau})$ defined by (3.19) where* $M = [(\tau + 1)/2]$ *and* $N = \tau + 1 - M$ .

   (i) *$H(h^{\tau})$ has a minimal generalized partial realization of dimension $n$. In particular, $\delta(H(h^{\tau})) = \mathrm{rank}\, H(h^{\tau})$.*

   (ii) *Every minimal generalized partial realization of $H(h^{\tau})$ is controllable and observable.*

  (iii) *Assume $2n < M + N$. Then any two minimal generalized partial realizations $(E, A, B, C), (\tilde{E}, \tilde{A}, \tilde{B}, \tilde{C}) \in S^{c,o}_{n,1,1}$ of $H(h^{\tau})$ are similar modulo scaling of order $(M, N)$.*

**Proof:** (i) Since $|N - M| \le 1$ every scalar Hankel matrix $H \in \mathrm{Hank}^*_{1,1}(n, M \times N)$ can be extended to an infinite Hankel matrix of rank $n$, see [13]. Therefore, making use of Proposition 2.3 instead of Theorem 2.2, step (i) of the proof of Theorem 3.17 can be applied to $H(h^{\tau})$ and yields (i).

(ii) Step (ii) of the proof of Theorem 3.17 can verbally be applied to $H(h^\tau)$ and yields (ii).

(iii) We have $n = \operatorname{rank} H(h^\tau) \leq \min\{M, N\}$. Since by assumption $2n < M + N$, Lemma 3.12 can be applied and (iii) follows by the same arguments as statement (iii) of Theorem 3.17. $\qquad\square$

The rank condition $2n < M + N$ is necessary for the validity of the uniqueness statement (iii) of Corollary 3.19. In fact, otherwise $n = N = M$ and every $H \in \operatorname{Hank}^*_{1,1}(n, n \times n)$ has an infinite number of nonsimilar partial realizations.

The following scalar example illustrates Theorem 3.17.

**Example 3.20** Let $M, N > 1$ and consider the following family of finite scalar sequences

$$(t^{M+N-2}, t^{M+N-3}, \ldots, 1) \in \mathbb{K}^{M+N-1}, \quad t \in \mathbb{K}. \tag{3.20}$$

By Theorem 3.17 (i), each sequence in the family (3.20) has the property that its generalized McMillan degree is equal to the rank of the associated Hankel matrix

$$H(t) = \begin{bmatrix} t^{M+N-2} & \cdots & t^{M-1} \\ \vdots & & \vdots \\ t^{N-1} & \cdots & 1 \end{bmatrix} \in \operatorname{Hank}_{1,1}(M \times N), \quad t \in \mathbb{K},$$

i.e.

$$\delta(H(t)) = \operatorname{rank} H(t) = 1, \quad t \in \mathbb{K}. \tag{3.21}$$

It is easy to see that $(t, 1, 1, 1) \in S^{c,o}_{1,1,1}$ is a minimal generalized partial realization of $H(t)$ for all $t \in \mathbb{K}$. By Theorem 3.17 (iii), $(t, 1, 1, 1)$ is unique up to similarity modulo scaling of order $(M, N)$. For $t \neq 0$, there exists a 1–dimensional minimal partial realization $(t^{-1}, t^{N-1}, t^{M-1})$ of $H(t)$. But for $t = 0$ the McMillan degree of $H(0)$ is $M + N - 1$ whereas the generalized McMillan degree of $H(0)$ is one. $\qquad\square$

We conclude this section with an example of a block Hankel matrix $H \in \operatorname{Hank}^*_{p,m}(n, M \times N)$ with $n = \min\{M, N\}$ which cannot be extended to an infinite block Hankel of the same rank. Hence step (i) of the proof of Theorem 3.17 cannot be applied. Nevertheless $H$ admits an n-dimensional generalized partial realization. This indicates that there is scope for improving Theorem 3.17, but this will require new proofs.

**Example 3.21** Consider the block Hankel matrix

$$H = \begin{bmatrix} \begin{bmatrix} 1 & 0 \\ 0 & 0 \end{bmatrix} & \begin{bmatrix} 0 & 0 \\ 0 & 0 \end{bmatrix} \\ \begin{bmatrix} 0 & 0 \\ 0 & 0 \end{bmatrix} & \begin{bmatrix} 0 & 0 \\ 0 & 1 \end{bmatrix} \end{bmatrix} \in \operatorname{Hank}^*_{2,2}(2, 2 \times 2).$$

For this matrix ($n = 2 = M = N$) there does not exist a rank preserving infinite extension. Nevertheless $H$ admits a (minimal) generalized partial realization of dimension rank $H = 2$, e.g.

$$(E, A, B, C) = \left( \begin{bmatrix} 1 & 0 \\ 0 & 0 \end{bmatrix}, \begin{bmatrix} 0 & 0 \\ 0 & 1 \end{bmatrix}, I_2, I_2 \right).$$

$\square$

# 4  Manifolds of partial realizations

In this section we investigate some topological aspects of the partial realization problem. In particular it is shown that $S_{n,m,p}^{c,o}$ and the quotient space of the equivalence relation $\sim_{MN}$ are analytic manifolds, provided that $\min\{m,p\} + 1 \leq n$ and $\min\{m,p\} + 1 \leq n < \min\{M,N\}$, respectively. Because of space limitations we omit the proofs. Except for Proposition 4.1 the proofs can be found in [19]. Throughout the section $S_{n,m,p}$ is endowed with the subspace topology inherited from $\mathbb{K}^{n\times n} \times \mathbb{K}^{n\times n} \times \mathbb{K}^{n\times m} \times \mathbb{K}^{p\times n} \cong \mathbb{K}^{2n^2 + n(m+p)}$. For $n = 1$ we have

$$S_{1,m,p} \cong (\mathbb{K}^2 \setminus \{(0,0)\}) \times \mathbb{K}^m \times \mathbb{K}^p, \quad S_{1,m,p}^{c,o} \cong (\mathbb{K}^2 \setminus \{(0,0)\}) \times (\mathbb{K}^m \setminus \{0\}) \times (\mathbb{K}^p \setminus \{0\}),$$

and hence $S_{1,m,p}$, $S_{1,m,p}^{c,o}$ are analytic manifolds. For $n \geq 2$, the situation is more complicated. Let

$$U_{n,m,p} := \{(E, A, B, C) \in S_{n,m,p}; \quad E, A \in \mathbb{K}I_n\},$$

$$U_{n,m,p}^{c,o} := \{(E, A, B, C) \in S_{n,m,p}^{c,o}; \quad E, A \in \mathbb{K}I_n\},$$

where $\mathbb{K}I_n = \{rI_n; \; r \in \mathbb{K}\}$. Note that by Proposition 3.5 a system of the form $(rI_n, sI_n, B, C) \in S_{n,m,p}$ can only be controllable and observable if

$$n = \operatorname{rank} R_n(rI_n, sI_n, B) = \operatorname{rank} B \leq m, \quad n = \operatorname{rank} O_n(rI_n, sI_n, C) = \operatorname{rank} C \leq p.$$

Hence

$$U_{n,m,p}^{c,o} \text{ is empty if } n > \min\{m,p\}.$$

**Proposition 4.1**   *(i) For $n \geq 2$, $S_{n,m,p} \setminus U_{n,m,p}$ is an $n^2 + 2 + n(m + p)-$ dimensional analytic manifold, and $S_{n,m,p}^{c,o} \setminus U_{n,m,p}^{c,o}$ is an open submanifold of $S_{n,m,p} \setminus U_{n,m,p}$.*

*(ii) If $n \geq \min\{m,p\} + 1$, then $S_{n,m,p}^{c,o}$ is an analytic submanifold of $S_{n,m,p} \setminus U_{n,m,p}$.*

It has been shown in [20] that when $n < \min\{M,N\}$, $\operatorname{Hank}_{p,m}(n, M \times N)$ is an $n(m + p)-$dimensional analytic manifold for which the underlying topology is the subspace topology inherited from $(\mathbb{K}^{p\times m})^{M+N-1}$. In the sequel we endow $\operatorname{Hank}_{p,m}(n, M \times N)$ with the analytic atlas described in [20].

**Proposition 4.2** *The Hankel mapping $H_{MN} : S_{n,m,p}^{c,o} \to \operatorname{Hank}_{p,m}(n, M \times N)$ is an analytic submersion, if $\min\{m,p\} + 1 \leq n < \min\{M,N\}$.*

The similarity relation modulo scaling of order $(M, N)$ partitions $S^{c,o}_{n,m,p}$ into equivalence classes

$$[(E, A, B, C)]_{\sim_{MN}} := \{(\tilde{E}, \tilde{A}, \tilde{B}, \tilde{C}) \in S^{c,o}_{n,m,p}; \quad (\tilde{E}, \tilde{A}, \tilde{B}, \tilde{C}) \sim_{MN} (E, A, B, C)\}.$$

The quotient space of $\sim_{MN}$ is the set of all these equivalence classes

$$S^{c,o}_{n,m,p}/_{\sim_{MN}} := \left\{[(E, A, B, C)]_{\sim_{MN}}; \quad (E, A, B, C) \in S^{c,o}_{n,m,p}\right\}$$

provided with the quotient topology, i.e. the finest topology for which the projection

$$\pi_{\sim_{MN}} : S^{c,o}_{n,m,p} \longrightarrow S^{c,o}_{n,m,p}/_{\sim_{MN}}, \qquad (E, A, B, C) \mapsto [(E, A, B, C)]_{\sim_{MN}}$$

is continuous. By (3.18) there exists a function

$$\bar{H}_{MN} : S^{c,o}_{n,m,p}/_{\sim_{MN}} \longrightarrow \mathrm{Hank}_{p,m}(n, M \times N), \qquad n < \min\{M, N\} \qquad (4.1)$$

for which $H_{MN} = \bar{H}_{MN} \circ \pi_{\sim_{MN}}$ and this function is uniquely determined. We now derive a topological version of the Realization Theorem 3.17.

**Corollary 4.3** *The quotient mapping* $\bar{H}_{MN} : S^{c,o}_{n,m,p}/_{\sim_{MN}} \to \mathrm{Hank}_{p,m}(n, M \times N)$ *is a homeomorphism if* $\min\{m, p\} + 1 \leq n < \min\{M, N\}$.

Using the homeomorphism $\bar{H}_{MN}$ (4.1) to transport the analytic manifold structure from $\mathrm{Hank}_{p,m}(n, M \times N)$ to $S^{c,o}_{n,m,p}/_{\sim_{MN}}$ we obtain the following

**Proposition 4.4** *Let* $\min\{m, p\} + 1 \leq n < \min\{M, N\}$. *There exists a unique structure of analytic manifold on the quotient space* $S^{c,o}_{n,m,p}/_{\sim_{MN}}$ *(compatible with the quotient topology of* $S^{c,o}_{n,m,p}/_{\sim_{MN}}$*) such that the projection* $\pi_{\sim_{MN}}$ *is an analytic submersion. The dimension of* $S^{c,o}_{n,m,p}/_{\sim_{MN}}$ *is* $n(m + p)$.

We conclude the paper with a brief description of the connectedness properties of the analytic manifolds $S^{c,o}_{n,m,p}/_{\sim_{MN}}$. These properties depend on the groundfield $\mathbb{K}$ and follow from analogous properties of the homeomorphic manifolds of block Hankel matrices.

**Corollary 4.5** *Suppose* $\min\{m, p\} + 1 \leq n < \min\{M, N\}$. *The manifold* $S^{c,o}_{n,m,p}/_{\sim_{MN}}$ *is always connected if* $\mathbb{K} = \mathbb{C}$. *In the real case it is connected if either* $\max\{m, p\} > 1$ *or* $p = m = 1$ *and* $M + N$ *is odd. If* $\mathbb{K} = \mathbb{R}$, $p = m = 1$ *and* $M + N$ *is even, then* $S^{c,o}_{n,m,p}/_{\sim_{MN}}$ *has* $n + 1$ *connected components.*

# References

[1] A. C. Antoulas. On recursiveness and related topics in system theory. *IEEE Transactions on Automatic Control*, 31:1121–1135, 1986.

[2] R. W. Brockett. The geometry of the partial realization problem. In *Proceedings of the 1978 IEEE Conference on Decision and Control, San Diego, Calif.*, pages 1048–1052, 1978.

[3] S. Feldmann and G. Heinig. Vandermonde factorization and canonical representations of block Hankel matrices. *Linear Algebra and its Applications*, 241/242/243:247–278, 1996.

[4] G. Frobenius. Über Relationen zwischen den Näherungsbrüchen von Potenzreihen. *Journal für die reine und angewandte Mathematik*, 90:1–17, 1881.

[5] P. A. Fuhrmann. On strict system equivalence and similarity. *International Journal Control*, 25:5–10, 1977.

[6] P. A. Fuhrmann. On the partial realization problem and the recursive inversion of Hankel and Toeplitz matrices. In B. N. Datta, editor, *Linear Algebra and its Role in Systems Theory*, volume 47 of *Contemporary Mathematics*, pages 149–161. Amer. Math. Soc., 1985.

[7] P. A. Fuhrmann. Functional models in linear algebra and its applications. *Linear Algebra and its Applications*, 162–164:107–151, 1992.

[8] P. A. Fuhrmann and P. S. Krishnaprasad. Towards a cell decomposition for rational functions. *IMA Journal of Mathematical Control and Information*, 3:137–150, 1986.

[9] W. Gragg and A. Lindquist. On the partial realization problem. *Linear Algebra and its Applications*, 50, 1983.

[10] G. Heinig and K. Rost. Matrices with displacement structure, generalized Bezoutians, and Möbius transformations. In *Operator Theory: Advances and Applications*, volume 40, pages 203–230. Birkhäuser, 1989.

[11] U. Helmke. Waring's problem for binary forms. *Journal of Pure and Applied Algebra*, 80:29–45, 1992.

[12] D. Hinrichsen and W. Manthey. On a cell decomposition for Hankel matrices and rational functions. *Journal für die reine und angewandte Mathematik*, 451:15–50, 1994.

[13] I. S. Iohvidov. *Hankel and Toeplitz matrices and forms*. Birkhäuser, 1982.

[14] R. E. Kalman. Mathematical description of linear dynamical systems. *SIAM Journal on Control and Optimization*, 1:152–192, 1963.

[15] R. E. Kalman. On partial realizations, transfer functions and canonical forms. *Acta Polytechn. Scand.*, 31:9–32, 1979.

[16] L. Kronecker. Zur Theorie der Elimination einer Variablen aus zwei algebraischen Gleichungen. *Monatsber. der Königl. Preuss. Akad. der Wiss., Berlin*, pages 535–600, 1881. see also Collected Works, Vol. 2.

[17] W. Manthey. A leading minor formula for finite block Hankel matrices. Technical Report 396, Institut für Dynamische Systeme, Universität Bremen, 1997.

[18] W. Manthey, U. Helmke, and D. Hinrichsen. Topological aspects of the partial realization problem. *Mathematics of Control, Signals and Systems*, 5:117–149, 1992.

[19] W. Manthey, D. Hinrichsen, and U. Helmke. Partial realizations via singular systems. Technical Report 338, Institute for Dynamical Systems, University of Bremen, 1995.

[20] W. Manthey, D. Hinrichsen, and U. Helmke. On Fischer–Frobenius transformations and the structure of rectangular block Hankel matrices. *Linear and Multilinear Algebra*, 41(3):255–288, 1996.

[21] R. Nikoukhah, B. C. Levy, and A. S. Willsky. Realization of acausal weighting patterns with boundary–value descriptor systems. *SIAM J. Control and Optimization*, 29(2):420–444, 1991.

[22] R. Nikoukhah, A. S. Willsky, and B. C. Levy. Boundary–value descriptor systems: well–posedness, reachability and observability. *Int. J. Control*, 46(5):1715–1737, 1987.

[23] E. D. Sontag. *Mathematical Control Theory*. Springer–Verlag, 1990.

W. Manthey
Institut für Dynamische Systeme
Universität Bremen
28334 Bremen

U. Helmke
Mathematisches Institut
Universität Würzburg
97074 Würzburg

D. Hinrichsen
Institut für Dynamische Systeme
Universität Bremen
28334 Bremen

# Oblique Splitting Subspaces and Stochastic Realization with Inputs


Giorgio Picci
Università di Padova, Padova, Italy


Dedicated to Paul Fuhrmann in occasion of his 60th birthday


**Abstract**: Until recently stochastic realization theory has been primarily addressed to modeling of random processes in the absence of exogenous signals (inputs). However stochastic realization of models with exogenous inputs is also of interest. In particular it is of interest for the new class of " subspace " type identification algorithms. These algorithms can be formulated as stochastic realization algorithms in an appropriate data Hilbert space [23, 24] (as it is well-known subspace methods have important advantages over the traditional parametric optimization approach to identification). We discuss here procedures for constructing minimal state-space models in presence of inputs, based on a generalization of the idea of Markovian splitting subspace which is central in stochastic realization theory for random processes. In particular we discuss a geometric procedure for constructing the minimal state-space (the predictor space) of a process with inputs which leads to an interesting identification algorithm.


## 1   Introduction

Stochastic Realization theory deals with modeling of random processes. Given a vector (say $m$-dimensional) process $y = \{y(t)\}$, construct models of $y$ representing it in terms of simpler and more basic random processes, such as white noise, Markov processes etc. In particular it deals with procedures for constructing models of (wide-sense) stationary processes, of the form

$$\begin{cases} x(t+1) & = & Ax(t) + Bw(t) \\ y(t) & = & Cx(t) + Dw(t) \end{cases} \tag{1.1}$$

where $\{w(t)\}$ is some vector normalized white noise, i.e.

$$E\{w(t)w(s)'\} = I\delta(t-s) \quad E\{w(t)\} = 0.$$


---

[1]*Key words and phrases:* Stochastic Realization, Identification, Systems with exogenous inputs, Subspace-methods Identification, Splitting Subspaces.

[2]Supported in part by the Italian ministery of higher education (MURST).

where $\delta$ is the Kronecker delta function. This representation is called a *state-space realization* of the process $y$. It involves auxiliary variables (i.e. random quantities which are not given as a part of the original data) such as the *state process* $x$ (a stationary Markov process) and the *generating white noise* $w$, whose peculiar properties lead to representations of $y$ by models having the desired structure. Constructing these auxiliary processes is part of the realization problem.

Assume that $y$ and $u$ are two jointly stationary purely-non-deterministic processes of dimensions $m$ and $p$. As we shall see in a moment, state-space modeling of $y$ in the presence of an exogenous input signal $\{u(t)\}$ can be reduced to the problem of characterizing representations of the type

$$
\begin{aligned}
x(t+1) &= Ax(t) + Bu(t) & \text{(1.2a)} \\
y_u(t) &= Cx(t) + Du(t) & \text{(1.2b)}
\end{aligned}
$$

where $\{y_u(t)\}$ is the orthogonal projection of the output $y(t)$ onto the past input space. We shall call a model of the type (1.2) a *deterministic realization of $y_u$* with input process $u$. Models of this kind reduce to the standard (Markovian) stochastic models when, of course, $u$ is white noise.

As usual a realization is called *minimal* if the dimension of the state vector is as small as possible. For minimal realizations it must necessarily hold that $(A, B, C)$ is a minimal triplet. If $A$ has all eigenvalues inside the unit circle $(|\lambda(A)| < 1)$, both $x(t)$ and $y(t)$ can be expressed as functionals of the infinite past of $u$ (i.e. their components belong to $\mathbf{U}_t^-$). Realizations with this property will be called *causal*.

An early approach ([20] and [2]) to stochastic realization with inputs, was based on the principle of constructing splitting subspaces for the future of the process $y_u$ and the past of the input process $u$. However this approach leads to state-space models which are driven by white noise and "include" also the dynamics of the process $u$, which is not interesting for applications to identification and we do not want to appear explicitly in the model. For example it was shown in [20] and in [2] that one may choose as state space for $y_u$ the predictor space of future outputs given past inputs. However this choice leads to an innovation model for $y_u$ where the state process is driven by the forward innovation process of $u$. This model includes as a cascade subsystem a state-space innovation representation for $u$ and therefore depends on the particular dynamics of the input process. This last feature is undesirable for the applications and to get models of the form (1.2) an alternative approach is needed.

We shall see in this paper that state-space realization of processes driven by a non-white process $u$, can be approached by generalizing the classical geometric procedure of stochastic realization theory for "undriven" signals [11, 13, 12]. The key idea is a generalization of the concept of splitting (and conditional orthogonality) which will be called *oblique splitting*.


**The Hilbert space formulation.** We shall review the Hilbert space setting of the stochastic realization problem below.

Assume that $y$ and $u$ are zero-mean second-order p.n.d processes defined in some probability space $\{\Omega, \mathcal{A}, \mu\}$ and consider the vector subspace of the $L^2$ space of square integrable real-valued random variables on $\{\Omega, \mathcal{A}, \mu\}$, constructed by taking finite linear combinations of the components of $y$ and $u$,

$$\mathcal{U} := \{\sum a'_k u(t_k) \qquad a_k \in \mathbb{R}^p, \ t_k \in \mathbb{Z}\} \tag{1.3}$$

$$\mathcal{Y} := \{\sum a'_k y(t_k) \qquad a_k \in \mathbb{R}^m, \ t_k \in \mathbb{Z}\} \tag{1.4}$$

Closing these vector spaces in the metric of the inner product

$$\langle \xi, \eta \rangle := E\{\xi \eta\}. \tag{1.5}$$

where $E$ denotes expectation, we obtain two Hilbert subspaces $\mathbf{U}$ and $\mathbf{Y}$ of linear functionals of the "history" of the processes $u$ and $y$.

In exactly the same manner we introduce $\mathbf{U}_t^-$, $\mathbf{Y}_t^-$, $\mathbf{U}_t^+$, $\mathbf{Y}_t^+$, the *past* and *future* subspaces of the processes $u$ and $y$ at time $t$. These are defined as the closure of the linear vector spaces spanned by the relative past or future random variables $u(t)$ and $y(t)$, in the metric of the Hilbert space $L^2\{\Omega, \mathcal{A}, \mu\}$. We shall use the notations

$$\begin{aligned}
\mathbf{U}_t^- &:= \overline{\operatorname{span}}\{\, u(s) \mid s < t \,\} \\
\mathbf{Y}_t^- &:= \overline{\operatorname{span}}\{\, y(s) \mid s < t \,\} \\
\mathbf{U}_t^+ &:= \overline{\operatorname{span}}\{\, u(s) \mid s \geq t \,\} \\
\mathbf{Y}_t^+ &:= \overline{\operatorname{span}}\{\, y(s) \mid s \geq t \,\}
\end{aligned}$$

Note that, according to a widely accepted convention, the present is included in the future only and not in the past. All these subspaces are contained in a fixed ambient space $\mathbf{H}$. For all purposes of this paper we may take $\mathbf{H} := \mathbf{U} \vee \mathbf{Y}$, where the wedge denots (closed) vector sum.

We shall define a *shift operator* $\sigma$ on $\mathbf{H}$ by setting

$$\sigma a' u(t) = a' u(t+1) \quad t \in \mathbb{Z}, \quad a \in \mathbb{R}^p \qquad \sigma a' y(t) = a' y(t+1) \quad t \in \mathbb{Z}, \quad a \in \mathbb{R}^m,$$

defining a linear map which is isometric with respect to the inner product (1.5) and extendable by linearity to all of $\mathbf{H}$.

Modeling and estimation of stationary processes on infinite or semi-infinite time intervals, naturally involve various linear operations on the variables of the process which are *time-invariant*, i.e. independent of the particular instant of time chosen as a "present". In this setting it is possible (and convenient) to fix the present instant of time to an arbitrary value say $t = 0$ and work as if time was "frozen" at $t = 0$. At the occurrence one then "shifts" the operations in time by the action of the unitary operator $\sigma^t$ on the data. In particular, the future and past subspaces of the processes $y$ and $u$ will often be considered referred at time $t = 0$ and denoted $\mathbf{Y}^+$ and $\mathbf{Y}^-$. For an arbitrary present instant $t$ we have

$$\mathbf{Y}_t^+ = \sigma^t \mathbf{Y}^+, \quad \mathbf{Y}_t^- = \sigma^t \mathbf{Y}^-.$$

At this point we need to set notations. In what follows the symbols $\vee$, $+$ and $\oplus$ will denote vector sum, *direct* vector sum and *orthogonal* vector sum of subspaces, the symbol $\mathbf{X}^\perp$ will denote the orthogonal complement of a (closed) subspace $\mathbf{X}$

of a Hilbert space with respect to some predefined ambient space. The orthogonal projection onto the subspace $\mathbf{X}$ will be denoted by the symbol $E(\cdot|\mathbf{X})$ or by the shorthand $E^{\mathbf{X}}$. The notation $E(z|\mathbf{X})$ will be used also when $z$ is vector-valued. The symbol will then just denote the vector with components $E(z_k|\mathbf{X})$, $k = 1, \ldots$. For vector quantities, $|v|$ will denote Euclidean length (or absolute value in the scalar case). Normally we shall allow $E\{\cdot\}$ to operate on matrices, taking inner products row by row.

**Definitions and Facts from geometric stochastic realization theory.** The notation $\mathbf{A} \perp \mathbf{B}|\mathbf{X}$ means that the two subspaces $\mathbf{A}, \mathbf{B} \subset \mathbf{H}$ are *conditionally orthogonal* given a third subspace $\mathbf{X}$, i.e.

$$\langle \alpha - E^{\mathbf{X}}\alpha, \beta - E^{\mathbf{X}}\beta \rangle = 0 \quad \text{for} \quad \alpha \in \mathbf{A},\ \beta \in \mathbf{B}. \tag{1.6}$$

When $\mathbf{X} = 0$, this reduces to the usual orthogonality $\mathbf{A} \perp \mathbf{B}$. Conditional orthogonality is orthogonality after subtracting the orthogonal projections onto $\mathbf{X}$. This concept is discussed in depth in [11, 13].

Let $\mathbf{X}$ be a subspace of some large stationary Hilbert space $\mathbf{H}$ of random variables containing $\mathbf{Y}$. Define

$$\mathbf{X}_t := \sigma^t \mathbf{X}, \quad \mathbf{X}_t^- := \vee_{s \le t} \mathbf{X}_s, \quad \mathbf{X}_t^+ := \vee_{s \ge t} \mathbf{X}_s.$$

**Definition 1.1.** A *Markovian Splitting Subspace* $\mathbf{X}$ for the process $y$ is a subspace of $\mathbf{H}$ making the vector sums $\mathbf{Y}^- \vee \mathbf{X}^-$ and $\mathbf{Y}^+ \vee \mathbf{X}^+$ conditionally orthogonal (i.e. uncorrelated) given $\mathbf{X}$, denoted,

$$\mathbf{Y}^- \vee \mathbf{X}^- \perp \mathbf{Y}^+ \vee \mathbf{X}^+ \,|\, \mathbf{X}. \tag{1.7}$$

The conditional orthogonality condition (1.7) can be equivalently written as

$$E[\mathbf{Y}^+ \vee \mathbf{X}^+ \,|\, \mathbf{Y}^- \vee \mathbf{X}^-] = E[\mathbf{Y}^+ \vee \mathbf{X}^+ \,|\, \mathbf{X}] \tag{1.8}$$

which gives the intuitive meaning of the splitting subspace $\mathbf{X}$ as a dynamic memory of the past for the purpose of predicting the joint future.

The subspace $\mathbf{X}$ is called *proper*, or *purely-non-deterministic* if

$$\cap_t \mathbf{Y}_t^- \vee \mathbf{X}_t^- = \{0\}, \quad \text{and} \quad \cap_t \mathbf{Y}_t^+ \vee \mathbf{X}_t^+ = \{0\}.$$

Obviously for the existence of proper splitting subspaces $y$ must also be purely non deterministic [25]. Properness is, by the Wold decomposition theorem, equivalent to the existence of two vector white noise processes $w$ and $\bar{w}$ such that,

$$\mathbf{Y}^- \vee \mathbf{X}^- = \mathbf{H}^-(w), \quad \mathbf{Y}^+ \vee \mathbf{X}^+ = \mathbf{H}^+(\bar{w})$$

Here the symbols $\mathbf{H}^-(w)$, $\mathbf{H}^+(w)$ etc. denote the Hilbert subspaces linearly generated by the of past and future of the process $w$. The spaces

$$\mathbf{S} := \mathbf{Y}^- \vee \mathbf{X}^- \quad \text{and} \quad \bar{\mathbf{S}} := \mathbf{Y}^+ \vee \mathbf{X}^+ \tag{1.9}$$

associated to a Markovian Splitting subspace $\mathbf{X}$, play an important role in the geometric theory of stochastic systems. They are called the *scattering pair* of $\mathbf{X}$ as they can be seen to form an incoming-outgoing pair in the sense of Lax-Phillips Scattering Theory [8].

**Definition 1.2.** Given a stationary Hilbert space $(\mathbf{H}, \sigma)$ containing $\mathbf{Y}$, a *scattering pair* for the process $y$ is a pair of subspaces $(\mathbf{S}, \overline{\mathbf{S}})$ satisfying the following conditions,

1. $\sigma^*\mathbf{S} \subset \mathbf{S}$ and $\sigma\overline{\mathbf{S}} \subset \overline{\mathbf{S}}$, i.e. $\mathbf{S}$ and $\overline{\mathbf{S}}$ are invariant for the left and right shift semigroups. (this means that $\mathbf{S}_t$ is increasing and $\overline{\mathbf{S}}_t$ is decreasing with time).
2. $\mathbf{S} \vee \overline{\mathbf{S}} = \mathbf{H}$
3. $\mathbf{S} \supset \mathbf{Y}^-$ and $\overline{\mathbf{S}} \supset \mathbf{Y}^+$
4. $\mathbf{S}^\perp \subset \overline{\mathbf{S}}$ or, equivalently, $\overline{\mathbf{S}}^\perp \subset \mathbf{S}$

The following representation Theorem provides the link between Markovian splitting subspaces and scattering pairs.

**Theorem 1.3.** *The intersection*

$$\mathbf{X} = \mathbf{S} \cap \overline{\mathbf{S}} \tag{1.10}$$

*of any scattering pair of subspaces of* $\mathbf{H}$ *is a Markovian splitting subspace. Conversely every Markovian splitting subspace can be represented as the intersection of a scattering pair. The correspondence* $\mathbf{X} \leftrightarrow (\mathbf{S}, \overline{\mathbf{S}})$ *is one-to-one, the scattering pair corresponding to* $\mathbf{X}$ *being given by*

$$\mathbf{S} = \mathbf{Y}^- \vee \mathbf{X}^- \qquad \overline{\mathbf{S}} = \mathbf{Y}^+ \vee \mathbf{X}^+ \tag{1.11}$$

The process of forming scattering pair associated to $\mathbf{X}$ should be thought of as an "extension" of the past and future spaces of $y$. The rationale for this extension is that scattering pairs have an extremely simple splitting geometry due to the fact that

$$\mathbf{S} \perp \overline{\mathbf{S}} \,|\, \mathbf{S} \cap \overline{\mathbf{S}} \tag{1.12}$$

which is called *perpendicular intersection*. It is easy to show that Property 4. in the definition of a scattering pair is actually equivalent to perpendicular intersection. This property of conditional orthogonality given the intersection can also be seen as a natural generalization of the Markov property[1]. Note that $\mathbf{A} \perp \mathbf{B} \,|\, \mathbf{X} \Rightarrow \mathbf{A} \cap \mathbf{B} \subset \mathbf{X}$ but the inclusion of the intersection in the splitting subspace $\mathbf{X}$ is only *proper* in general. For perpendicularly intersecting subspaces, the intersection is actually the *unique minimal subspace* making them conditionally orthogonal.

Any basis vector $x(0) := [x_1(0), x_2(0), \ldots, x_n(0)]'$ in a (finite-dimensional) Markovian splitting subspace $\mathbf{X}$ generates a stationary Markov process $x(t) := \sigma^t x(0)$, $t \in \mathbb{Z}$ which serves as a *state* of the the process $y$. If $\mathbf{X}$ is proper, the Markov process is purely non deterministic.

Denote by $\mathbf{W}_t$, $\overline{\mathbf{W}}_t$ the spaces spanned by the components, at time $t$, of the generating noises $w(t)$ and $\bar{w}(t)$, of the scattering pair of $\mathbf{X}$. Since

$$\mathbf{S}_{t+1} = \mathbf{S}_t \oplus \mathbf{W}_t, \tag{1.13}$$

we can write,

$$\mathbf{X}_{t+1} \subset \mathbf{S}_{t+1} \cap \overline{\mathbf{S}}_t = (\mathbf{S}_t \cap \overline{\mathbf{S}}_t) \oplus (\mathbf{W}_t \cap \overline{\mathbf{S}}_t) \tag{1.14}$$

---

[1]In which case $\mathbf{S} = \mathbf{X}^-$, $\overline{\mathbf{S}} = \mathbf{X}^+$ and $\mathbf{X} = \mathbf{X}^- \cap \mathbf{X}^+$ .

( since $\overline{\mathbf{S}}_t$ is decreasing in time, we have $\mathbf{S}_{t+1} \cap \overline{\mathbf{S}}_t \supset \mathbf{X}_{t+1}$) and by projecting the shifted basis $\sigma x(t) := x(t+1)$, onto the last orthogonal direct sum above, the time evolution of any basis vector $x(t) := [x_1(t), x_2(t), \ldots, x_n(t)]'$ in $\mathbf{X}_t$ can be represented by a linear equation of the type $x(t+1) = Ax(t) + Bw(t)$. It is also easy to see that by the p.n.d. property, $A$ must have all its eigenvalues strictly inside of the unit circle. Naturally, by decomposing instead $\overline{\mathbf{S}}_{t-1} = \overline{\mathbf{S}}_t \oplus \overline{\mathbf{W}}_t$ we could have obtained a *backward difference equation* model for the Markov process $x$, driven by the backward generator $\bar{w}$.

Note also that by definition of the past space, we have $y(t) \in (\mathbf{S}_{t+1} \cap \overline{\mathbf{S}}_t)$. Inserting the decomposition (1.13) and projecting $y(t)$ leads to a state-output equation of the form $y(t) = Cx(t) + Dw(t)$. Here again we could have equivalently obtained an equation driven by the backward noise $\bar{w}$ instead.

As we have seen, any basis in a Markovian splitting subspaces produces a stochastic realization of $y$. It is easy to reverse the implication. We have in fact the following characterization.

**Theorem 1.4.** [9, 13] *The state space $\mathbf{X} = span\{ x_1(0), x_2(0), \ldots, x_n(0)\}$ of any stochastic realization (1.1) is a* Markovian Splitting Subspace *for the process $y$.*

*Conversely, given a finite-dimensional Markovian splitting subspace $\mathbf{X}$, to any choice of basis $x(0) = [x_1(0), x_2(0), \ldots, x_n(0)]'$ in $\mathbf{X}$ there corresponds a stochastic realization of $y$ of the type (1.1).*

Once a basis in $\mathbf{X}$ is available, there are obvious formulas expressing the coefficient matrices $A, C$ and $\bar{C}$ in terms of the processes $x$ and $y$.

$$A = Ex(t+1)x(t)'P^{-1} \tag{1.15}$$

$$C = Ey(t)x(t)'P^{-1} \tag{1.16}$$

$$\bar{C} = Ey(t-1)x(t)' \tag{1.17}$$

where $P$ is the Gramian matrix of the basis (equal to the state covariance matrix). The matrices $B$ and $D$ however are related to the (unobservable) generating white noise $w$ and their computation requires the solution of a Linear Matrix Inequality.

Stochastic realizations are called *internal* when $\mathbf{H} = \mathbf{Y}$, i.e. the state space is built from the Hibert space made just of the linear statistics of the process $y$. For identification the only realizations of interest are the internal ones.

A central problem of geometric realization theory is to construct and to classify the *minimal* state spaces, i.e. the minimal Markovian splitting subspaces for the process $y$.

The obvious ordering by inclusion of the subspaces of $\mathbf{H}$ induces an ordering on the family of Markovian splitting subspaces. The notion of minimality is most naturally defined with respect to this ordering.

**Definition 1.5.** A Markovian splitting subspace is *minimal* if it doesn't contain (properly) other Markovian splitting subspaces.

This definition applies also to infinite-dimensional Markovian splitting subspaces, i.e. to situations where comparing dimension would not make much sense.

Contrary to the deterministic situation minimal Markovian splitting subspaces are *non unique*. Two very important examples are the *forward and backward predictor spaces* (at time zero):

$$\mathbf{X}_- := E^{\mathbf{H}^-}\mathbf{H}^+ \quad \mathbf{X}_+ := E^{\mathbf{H}^+}\mathbf{H}^- \tag{1.18}$$

for which we have the following characterization [11].

**Proposition 1.1.** *The subspaces* $\mathbf{X}_-$ *and* $\mathbf{X}_+$ *are the unique minimal splitting subspaces contained in the past* $\mathbf{H}^-$, *and, respectively, in the future* $\mathbf{H}^+$, *of the process* $y$.

The study of minimality forms an elegant chapter of stochastic system theory. There are several known geometric and algebraic characterizations of minimality of splitting subspaces and of the corresponding stochastic state-space realizations. Since however the discussion of this topic would take us too far from the main theme of the paper we shall refer the reader to the literature [11, 13].

## 2 Input-Output models

One says that there is *absence of feedback* from $y$ to $u$ in the sense of Granger, if the future of $u$ is conditionally uncorrelated[2] from the past of $y$, given the past of $u$ itself. In our Hilbert space setup this is written as,

$$\mathbf{U}_t^+ \perp \mathbf{Y}_t^- \mid \mathbf{U}_t^- \tag{2.1}$$

where $\mathbf{U}_t^-$, $\mathbf{Y}_t^-$, $\mathbf{U}_t^+$, $\mathbf{Y}_t^+$ are the past and future subspaces of the processes $u$ and $y$ at time $t$. This conditional orthogonality condition will be another basic assumption throughout this paper. It is immediate to see that in conditions of absence of feedback, the "causal estimation error" process

$$y_s(t) := y(t) - E[y(t)|\mathbf{U}_{t+1}^-] \tag{2.2}$$

coincides with $y(t) - E[y(t)|\mathbf{U}]$ and hence is uncorrelated with the whole history of the input process $u$,

$$y_s(t) \perp \mathbf{U} \text{ for all } t$$

see [21]. Hence the process $y_s$ may be called the *noisy* or *stochastic component of* $y$.

It also follows that the stochastic process $y_u$ defined by the complementary projection

$$y_u(t) := E[y(t) \mid \mathbf{U}] \qquad t \in \mathbb{Z}$$

is uncorrelated with $y_s$. It will be named the *deterministic component of* $y$.

Hence, under the feedback-free assumption we have a natural linear model of $y(t)$

$$y(t) = y_u(t) + y_s(t) = E[y(t)|\mathbf{U}_{t+1}^-] + y_s(t) \tag{2.3}$$

---

[2]Recall that uncorrelated is the same as independent in the Gaussian case.

where $E[y(t)|\mathbf{U}_{t+1}^-] = E[\,y(t)\,|\,u(s); s \leq t\,]$ is a causal "input-output" relation: the best (in the sense of minimum variance of the error) estimate of the output $y(t)$ based on the past of $u$ up to time $t$. Under some regularity conditions on the input process to be made precise later on, this estimate is described by a stable linear convolution operator.

It is obvious that state-space descriptions for the process $y$ can be obtained by combining two separate state-space models for $y_s$ and $y_u$. For example, assuming that the two processes are finite-dimensional, a (forward) innovation representation of $y$ is obtained by combining together the (forward) innovation representation of $y_s$

$$\begin{align}
x_s(t+1) &= A_s x_s(t) + B_s e_s(t) \tag{2.4a}\\
y_s(t) &= C_s x_s(t) + e_s(t) \tag{2.4b}
\end{align}$$

where $e_s(t)$ is the one-step prediction error of the process $y_s$ based on its own past, i.e. the (forward) innovation process of $y_s$, and the "deterministic" state-space model for $y_u$

$$\begin{align}
x_u(t+1) &= A_u x_u(t) + B_u u(t) \tag{2.5a}\\
y_u(t) &= C_u x_u(t) + D_u u(t). \tag{2.5b}
\end{align}$$

The process $e_s$ has then the meaning of conditional innovation of $y$ [21].

By combining together (2.4) and (2.5), the state-space innovation model of the process $y$ "with inputs" has the following structure,

$$\begin{bmatrix} x_s(t+1) \\ x_u(t+1) \end{bmatrix} = \begin{bmatrix} A_s & 0 \\ 0 & A_u \end{bmatrix} \begin{bmatrix} x_s(t) \\ x_u(t) \end{bmatrix}$$
$$+ \begin{bmatrix} 0 \\ B_u \end{bmatrix} u(t) + \begin{bmatrix} B_s \\ 0 \end{bmatrix} e_s(t)$$

$$y(t) = \begin{bmatrix} C_s & C_u \end{bmatrix} \begin{bmatrix} x_s(t) \\ x_u(t) \end{bmatrix} + D_u u(t) + e_s(t) \tag{2.6}$$

Models of this kind are naturally interpreted as state-space realizations of the familiar ARMAX-type "input-output" relations $\hat{y} = W(z)\hat{u} + G(z)\hat{e}$ (here we have $W(z) = D_u + C_u(zI - A_u)^{-1}B_u$ and $G(z) = D_s + C_s(zI - A_s)^{-1}B_s$ ) often used in the identification literature.

Hence, state-space modeling of $y$ in the presence of inputs splits into two completely separate subproblems: the realization of the stochastic component (which is well understood and thoroughly discussed in the literature) and the realization of the deterministic component $y_u$.

As we shall see, in order to construct state-space descriptions of $y_u$ driven by a non-white process $u$, it is necessary to generalize the geometric theory of stochastic realization of the previous section.

In order to streamline notations, from now on we shall *drop the subscript* $u$ throughout and whenever possible fix $t = 0$. Recall that $y(t) \in \mathbf{U}_{t+1}^-$ for all $t$ ( this is actually the feedback-free property) and for this reason we may as well agree to fix the ambient space $\mathbf{H}$ equal to $\mathbf{U}$ hereafter.

We shall need to introduce a technical assumption of "sufficient richness" of the input process.

**Assumption 2.1.** For each $t$ the input space $\mathbf{U}$ admits the direct sum decomposition

$$\mathbf{U} = \mathbf{U}_t^- + \mathbf{U}_t^+ \tag{2.7}$$

An analogous condition ( namely $\mathbf{U}_t^- \cap \mathbf{U}_t^+ = 0$) is discussed in [15] where it is shown that it is equivalent to strict positivity (coercivity) of the spectral density matrix of $u$ on the unit circle, i.e. $\Phi_u(e^{j\omega}) > cI$, $c > 0$, or to all canonical angles between the past and future subspaces of $u$ being strictly positive (or, in turn, to all canonical correlation coefficients between past and future of the input process being strictly less than one). A slightly stronger version of this condition will be needed to discuss the frequency-domain approach of the next section. Coercivity of the spectrum (and of its inverse) is discussed in [25] Chapter II, Sect. 7.

In virtue of assumption 2.1, $y(t)$ has a unique representation as a causal functional

$$y(t) = \sum_{-\infty}^{t} W_{t-k}\, u(k). \tag{2.8}$$

where $\hat{W}(z) = \sum_0^{+\infty} W_k z^{-k}$ is analytic in $\{|z| > 1\}$. Indeed, $\hat{W}(z)$ is just the transfer function of the Wiener filter $y(t) = E[y(t)|\, \mathbf{U}_{t+1}^-]$ and can be expressed as

$$\hat{W}(z) = [\Phi_{yu}(z)G(1/z)^{-T}]_+ G(z)^{-1}$$

where $G(z)$ is the outer (or minimum-phase) spectral factor of $\Phi_u$ and the symbol $[\cdot]_+$ means "analytic part", see e.g. [25] Chapter II. It is evident that $\hat{W}(z)$ is analytic and, because of nonsingularity of $\Phi_u$ on the unit circle, unique almost everywhere.

Since the input-output map relating $u$ and $y$ must be causal, it follows that in our case the only realizations of interest are the causal ones.

The *oblique* projection of a random variable $\eta \in \mathbf{U}$ onto $\mathbf{U}_t^-$ along $\mathbf{U}_t^+$ will be denoted by $E_{\|\mathbf{U}_t^+}[\eta\,|\,\mathbf{U}_t^-]$. Clearly, if $u$ is a white noise process, this is the ordinary orthogonal projection onto $\mathbf{U}_t^-$.

**Definition 2.2.** We shall call a subspace $\mathbf{X} \subset \mathbf{U}^-$ a *(causal) oblique splitting subspace* for the pair $(\mathbf{Y}^+, \mathbf{U}^-)$ if

$$E_{\|\mathbf{U}^+}[\mathbf{Y}^+ \vee \mathbf{X}^+\,|\,\mathbf{U}^-] = E_{\|\mathbf{U}^+}[\mathbf{Y}^+ \vee \mathbf{X}^+\,|\,\mathbf{X}]. \tag{2.9}$$

Note that this condition is a generalization of the conditional orthogonality condition (1.8) of the Markovian case. For the reasons explained a moment ago, we shall only consider causal splitting subspaces in this paper so that the incoming subspace $\mathbf{S}$ is now fixed equal to $\mathbf{U}^-$.

The *oblique predictor space*

$$\mathbf{X}^{+/-} := E_{\|\mathbf{U}^+}[\mathbf{Y}^+\,|\,\mathbf{U}^-]$$

is obviously contained in $\mathbf{U}^-$. Later on we shall show that $\mathbf{X}^{+/-}$ is oblique splitting. Since

$$\mathbf{X} \supset E_{\|\mathbf{U}^+}[\mathbf{Y}^+ \vee \mathbf{X}^+ \,|\, \mathbf{X}] \supset E_{\|\mathbf{U}^+}[\mathbf{Y}^+ \,|\, \mathbf{U}^-] = \mathbf{X}^{+/-}$$

every oblique splitting subspace $\mathbf{X}$ contains $\mathbf{X}^{+/-}$ and therefore the oblique predictor space is *minimal splitting.*

Write

$$y(t) = (H_W u)(t) + (W^+ u)(t) \tag{2.10}$$

where,

$$(H_W u)(t) := \sum_{-\infty}^{-1} W_{t-k} u(k), \quad (W^+ u)(t) := \sum_{0}^{t} W_{t-k} u(k). \tag{2.11}$$

Evidently $(H_W u)(t) \in \mathbf{U}^-$ and $(W^+ u)(t) \in \mathbf{U}^+$ for $t \geq 0$. This obvious fact is formally recorded in the following Lemma.

**Lemma 2.1.** *For $t \geq 0$ the random variable $(H_W u)(t)$ is the oblique projection of $y(t)$ onto $\mathbf{U}^-$ and $(W^+ u)(t)$ the oblique projection of $y(t)$ onto $\mathbf{U}^+$.*

**Lemma 2.2.** *Let $\mathbf{X}$ be an oblique splitting subspace and define*

$$\bar{\mathbf{S}} := \mathbf{Y}^+ \vee \mathbf{X}^+$$

*Then $\mathbf{X} = E_{\|\mathbf{U}^+}[\bar{\mathbf{S}} \,|\, \mathbf{U}^-]$ ( hence every $\mathbf{X}$ is the oblique predictor space for $\bar{\mathbf{S}}$, given $\mathbf{U}^-$ ).*

*Proof.* Every element $\bar{s}$ of $\bar{\mathbf{S}}$ has the form $\bar{s} = y + x$, $y \in \mathbf{Y}^+$, $x \in \mathbf{X}^+$ so that $E_{\|\mathbf{U}^+}[y \,|\, \mathbf{U}^-] \in \mathbf{X}^{+/-} \subset \mathbf{X} \subset \mathbf{U}^-$. On the other hand, by definition of oblique splitting we have

$$E_{\|\mathbf{U}^+}[\mathbf{X}^+ \,|\, \mathbf{U}^-] = E_{\|\mathbf{U}^+}[\mathbf{X}^+ \,|\, \mathbf{X}] \qquad x \in \mathbf{X}^+,$$

therefore

$$\overline{\mathrm{span}}\{E_{\|\mathbf{U}^+}[\bar{s} \,|\, \mathbf{U}^-] \,|\, \bar{s} \in \bar{\mathbf{S}}\} = \mathbf{X}.$$

This implies that $\mathbf{X}$ is the oblique predictor space of $\bar{\mathbf{S}}$ with respect to $\mathbf{U}^-$. $\qquad \square$

**Proposition 2.1.** *The predictor space $\mathbf{X}^{+/-}$ is oblique splitting. In fact, it is the minimal causal oblique splitting susbspace for $\mathbf{Y}^+$ and $\mathbf{U}^-$.*

*Proof.* The proof is given in the discussion following Theorem 3.2 in the next section. $\qquad \square$

The following is the "deterministic" analog of perpendicular intersection.

**Lemma 2.3.** *Let the symbols have the same meaning as in Lemma 2.2. Then*

$$\bar{\mathbf{S}} \cap \mathbf{U}^- = \mathbf{X}.$$

*Proof.* First note that $\mathbf{X}$ contains the intersection $\bar{\mathbf{S}} \cap \mathbf{U}^-$. For if $\eta \in \bar{\mathbf{S}} \cap \mathbf{U}^-$ then clearly it belongs to $E_{\|\mathbf{U}^+}[\bar{\mathbf{S}} \,|\, \mathbf{U}^-]$ which is equal to $\mathbf{X}$ in force of the previous Lemma.

Then just observe that, conversely, the intersection contains $\mathbf{X}$, since $\bar{\mathbf{S}} \supset \mathbf{X}$ and $\mathbf{U}^- \supset \mathbf{X}$. This proves the Lemma. $\qquad \square$

The result in particular applies to the extended future space $\bar{\mathbf{Y}}^+ = \mathbf{Y}^+ \vee (\mathbf{X}^{+/-})^+$ (this is in a sense the "minimal" $\bar{\mathbf{S}}$).

The following result is useful in subspace identification [21].

**Theorem 2.3.** *The oblique predictor space can be computed as the intersection*

$$\mathbf{X}^{+/-} = (\mathbf{Y}^+ \vee \mathbf{U}^+) \cap \mathbf{U}^-. \tag{2.12}$$

The proof is straightforward using spectral representation theory and will be given in the next section. In the same occasion it will be made clear that

$$E_{\|\mathbf{U}^+}[\mathbf{Y}^+ \,|\, \mathbf{U}^-] \supset \mathbf{Y}^+ \cap \mathbf{U}^-$$

properly and it is in general *not true* that $\mathbf{X}^{+/-} = \mathbf{Y}^+ \cap \mathbf{U}^-$, unless some special conditions are satisfied. A claim which looks suspiciously similar to this last equality is made in [18, Thm. 3].

**Lemma 2.4.** *Let the symbols have the same meaning as in Lemma 2.2. Then*

$$\bar{\mathbf{S}} = (\bar{\mathbf{S}} \cap \mathbf{U}^-) + (\bar{\mathbf{S}} \cap \mathbf{U}^+). \tag{2.13}$$

*Proof.* That

$$\bar{\mathbf{S}} \supset (\bar{\mathbf{S}} \cap \mathbf{U}^-) + (\bar{\mathbf{S}} \cap \mathbf{U}^+)$$

is obvious since both terms in the right hand side are subspaces of $\bar{\mathbf{S}}$. We shall show that the opposite inclusion also holds.

We shall first show that $\mathbf{Y}^+ \subset (\bar{\mathbf{S}} \cap \mathbf{U}^-) + (\bar{\mathbf{S}} \cap \mathbf{U}^+)$. In effect, decomposing $y(t)$ for $t \geq 0$ as in (2.10) i.e. $y(t) = (H_W u)(t) + (W^+ u)(t)$, from Lemma 2.1 above we have $(H_W u)(t) \in \mathbf{X} \subset \bar{\mathbf{S}}$, so that for $t \geq 0$ necessarily $(W^+ u)(t) = y(t) - (H_W u)(t) \in \bar{\mathbf{S}}$ as well. In fact $(H_W u)(t) \in \bar{\mathbf{S}} \cap \mathbf{U}^-$ and $(W^+ u)(t) \in \bar{\mathbf{S}} \cap \mathbf{U}^+$, given the explicit dependence on the past and future of $u$. Taking finite linear combinations of the form $\sum a'_k y(t_k), a_k \in \mathbb{R}^m, t_k \geq 0$ and then closing in the Hilbert space norm of second order random variables gives immediately the inclusion we want. Second, by projecting obliquely $x^+ \in \mathbf{X}^+$ onto the direct sum (2.7), we obtain

$$x^+ = E_{\|\mathbf{U}^+}[x^+ \,|\, \mathbf{U}^-] + E_{\|\mathbf{U}^-}[x^+ \,|\, \mathbf{U}^+].$$

The first term belongs to $\mathbf{X} = (\bar{\mathbf{S}} \cap \mathbf{U}^-)$ in view of the splitting property (2.9), so since $x^+ \in \bar{\mathbf{S}}$ by definition, the second term in the sum must belong to the same subspace. Evidently, then $E_{\|\mathbf{U}^-}[x^+ \,|\, \mathbf{U}^+] \in (\bar{\mathbf{S}} \cap \mathbf{U}^+)$. Hence $\mathbf{X}^+$ satisfies the same subspace inclusion as $\mathbf{Y}^+$. This concludes the proof. $\square$

This intersection representation extends the formula $\bar{\mathbf{S}} = (\bar{\mathbf{S}} \cap \mathbf{S}) \oplus (\bar{\mathbf{S}} \cap \mathbf{S}^\perp)$, known for "orthogonal" splitting subspaces [11, 13].

The following argument shows how state space realizations can be constructed by a procedure based on the geometry of oblique splitting subspaces.

Denote by $\mathbf{U}_t$ the $p$-dimensional subspace of $\mathbf{U}_t^+$ spanned by the components of $u(t)$. By Assumption 2.1

$$\mathbf{U}_{t+1}^- = \mathbf{U}_t^- + \mathbf{U}_t$$

and since $\bar{\mathbf{S}}_{t+1} \subset \bar{\mathbf{S}}_t$, we can then write

$$\bar{\mathbf{S}}_{t+1} \cap \mathbf{U}_{t+1}^- \subset (\bar{\mathbf{S}}_t \cap \mathbf{U}_t^-) + (\bar{\mathbf{S}}_t \cap \mathbf{U}_t). \tag{2.14}$$

Now pick a basis vector $x(t)$, say of dimension[3] $n$ in $\mathbf{X}_t$ and let $x(t+1)$ be the corresponding vector shifted by one unit of time. The $n$ scalar components of $x(t+1)$ span $\bar{\mathbf{S}}_{t+1} \cap \mathbf{U}_{t+1}^-$ so, by projecting $x(t+1)$ onto the two components of the direct sum decomposition (2.14) we obtain a unique representation of the type

$$x(t+1) = Ax(t) + Bu(t).$$

Similarly, since $y(t) \in \cap \mathbf{U}_{t+1}^-$, we have

$$y(t) \in \bar{\mathbf{S}}_t \cap \mathbf{U}_{t+1}^- = (\bar{\mathbf{S}}_t \cap \mathbf{U}_t^-) + (\bar{\mathbf{S}}_t \cap \mathbf{U}_t)$$

and by projecting $y(t)$ onto the two components of the direct sum above we immediately obtain the state-output equation

$$y(t) = Cx(t) + Du(t).$$

This leads to the following Theorem.

**Theorem 2.4.** *Assume the joint spectral density of $y$ and $u$ is rational and that the input process satisfies Assumption 2.1. Then the oblique predictor subspace $\mathbf{X}^{+/-}$ is finite dimensional. To any choice of a basis vector $x(t)$ in a finite-dimensional oblique splitting subspace $\mathbf{X}_t$, there correspond unique matrices $(A, B, C, D)$ such that*

$$\begin{aligned} x(t+1) &= Ax(t) + Bu(t) && \text{(2.15a)} \\ y(t) &= Cx(t) + Du(t) && \text{(2.15b)} \end{aligned}$$

*and the realization (2.15) is causal, i.e. $|\lambda(A)| < 1$.*

*Conversely, the state space of any other causal realization of $y$ is an oblique splitting subspace.*

*Proof.* We shall give for granted that rationality implies finite dimensionality of $\mathbf{X}^{+/-}$.

Now, it is evident that the state process $x(t)$ of the representation obtained by the geometric argument above, is stationary by construction. Hence, since $x(t)$ is a functional of the past history $\mathbf{U}_t$, there must be a $n \times p$ matrix function $F(z)$ analytic in $\{|z| > 1\}$, with rows in the space $L_p^2(\Phi_u)$ of functions square integrable on the unit circle with respect to the matrix measure $\Phi_u(e^{j\omega})d\omega/2\pi$, such that

$$x(t) = \int_{-\pi}^{+\pi} e^{j\omega t} F(e^{j\omega})\, d\hat{u},$$

where $\hat{u}$ denotes the Fourier transform (random orthogonal measure) of the process $u$ [25]. By substituting this into the state equation for $x$ derived above, we see, by uniqueness of the spectral representation, that $F(z) = (zI - A)^{-1}B$. Note that $F(z)$ is rational and actually analytic also on the unit circle, since poles of modulus 1 would prevent integrability of the spectrum of $x$, $F(z)\Phi_u(z)F(1/z)'$, on the unit circle. [Recall that $\Phi_u(z)$ has no zeros on the unit circle, so there cannot be cancellations with the zeros of $\Phi_u(z)$ ]. One easily deduces from the analiticity of $F(z)$ that the eigenvalues of the reachable subsystem of $(A, B)$ must lie inside the unit disk. On the other hand there cannot be eigenvalues of $A$ with $|\lambda(A)| \geq 1$,

---

[3]Here for the sake of illustration we assume that $\mathbf{X}_t$ is finite-dimensional.

since eigenvalues with absolute value larger than one would contradict stationarity and eigenvalues on the unit circle (necessarily unreachable) would imply that $x$ has a purely deterministic component. This is impossible since $\mathbf{X}_t^- \subset \mathbf{U}_t^-$ and $u$ is purely-non-deterministic by Assumption 2.1.

The proof of the last statement is a simple verification and will be omitted. $\square$

## 3   Spectral Theory

In this section we shall make the slightly stronger assumption that the inverse of the spectral density of $u$ is also coercive, i.e.

$$c_1 I < \Phi_u(e^{j\omega}) < c_2 I, \quad c_1, c_2 > 0 \tag{3.1}$$

whose significance is discussed in [25, p. 78]. This will allow us to develop a "frequency-domain" treatment of the input/output modelling problem without too much of technicalities.

Let $L_p^2(\Phi_u)$ be the Hilbert space of $p$-dimensional row functions $f$ defined on the unit circle, such that

$$\|f\|_{\Phi_u}^2 := \int f \Phi_u f^* \frac{d\omega}{2\pi} < \infty$$

and notice that, by (3.1) $f \in L_p^2(\Phi_u) \leftrightarrow f \in L_p^2$ so, as sets of functions, the two spaces coincide. Let $G$ and $\bar{G}$ be the outer and conjugate outer spectral factors of $\Phi_u$, i.e. let

$$\Phi_u(e^{j\omega}) = G(e^{j\omega})G(e^{j\omega})^* = \bar{G}(e^{j\omega})\bar{G}(e^{j\omega})^* .$$

It follows from (3.1) and spectral factorization theory that $G$ is a square $p \times p$ matrix function belonging to the Hardy space $H_{p \times p}^\infty$ and, similarly, $\bar{G}$ is a square $p \times p$ matrix function belonging to the conjugate Hardy space[4] $\bar{H}_{p \times p}^\infty$. The outer factor has the property that

$$\overline{\text{span}} \left\{ z^{-t} G \mid t \geq 0 \right\} = H_p^2 \tag{3.2}$$

and, dually, the conjugate outer factor is such that

$$\overline{\text{span}} \left\{ z^t \bar{G} \mid t \geq 0 \right\} = \bar{H}_p^2 . \tag{3.3}$$

Recall that $H_p^2$ and $\bar{H}_p^2$ overlap as both contain the p-dimensional subspace of constant functions. We introduce the the subspace of $H_p^2$ made of functions whose Laurent expansion has zero constant term, i.e.

$$K_p^2 := H_p^2 \ominus \mathbb{R}^p. \tag{3.4}$$

so that, for example $L_p^2 = K_p^2 \oplus \bar{H}_p^2$, $L_p^2$ being the Hilbert space of p-dimensional square integrable functions on the circle.

Let $d\hat{\nu}$ and $d\hat{\bar{\nu}}$ be the Fourier transforms of the forward and backward innovation processess of $u$

$$d\hat{u} = G\,d\hat{\nu} = \bar{G}\,d\hat{\bar{\nu}}.$$

---

[4]By definition the conjugate Hardy spaces, e.g. $\bar{H}_p^2$, $\bar{H}_p^\infty$, etc. are made of functions analytic *inside* the unit disc.

It is well known that the space of causal functionals of the past *plus present* history of the process $u$ at time zero,

$$\mathbf{U}_0^- := \overline{\operatorname{span}} \{u(t) \mid t \leq 0\},$$

has the spectral representation

$$\mathbf{U}_0^- = \{\int \hat{f} d\hat{\nu} \mid \hat{f} \in H_p^2\} := \int H_p^2 d\hat{\nu},$$

and a similar representation holds for the anticausal functionals. In fact for functionals belonging to the future

$$\mathbf{U}^+ = \{\int \bar{f} d\hat{\bar{\nu}} \mid \bar{f} \in \bar{H}_p^2\} := \int \bar{H}_p^2 d\hat{\bar{\nu}},$$

Now the past, $\mathbf{U}^-$, by convention does not contain the present $u(0)$ among its generators and it is easy to see that the spectral representative of $\mathbf{U}^-$ is actually $K_p^2$. In other words we have

$$\mathbf{U}^- = \{\int \hat{f} d\hat{\nu} \mid \hat{f} \in K_p^2\} := \int K_p^2 d\hat{\nu} . \tag{3.5}$$

The following Theorem provides more useful representations of the same spaces which are in a sense "model-free".

**Theorem 3.1.** *Assume that the spectral density, $\Phi_u$, of the input process $u$ satisfies the condition (3.1) and let $G$ and $\bar{G}$ be its outer and conjugate outer spectral factors. Then $G^{-1}$ and $\bar{G}^{-1}$ are also outer and conjugate outer, respectively. The subspaces $K_p^2$ and $\bar{H}_p^2$ are closed complementary subspaces of $L_p^2(\Phi_u)$, i.e.*

$$L^2(\Phi_u) = K_p^2 + \bar{H}_p^2, \tag{3.6}$$

*and the following spectral representations hold*

$$\mathbf{U}_0^- = \{\int \hat{f} d\hat{u} \mid \hat{f} \in H_p^2\} := \int H_p^2 d\hat{u} \tag{3.7a}$$

$$\mathbf{U}^- = \{\int \bar{f} d\hat{u} \mid \bar{f} \in K_p^2\} := \int K_p^2 d\hat{u} \tag{3.7b}$$

$$\mathbf{U}^+ = \{\int \bar{f} d\hat{u} \mid \bar{f} \in \bar{H}_p^2\} := \int \bar{H}_p^2 d\hat{u} \tag{3.7c}$$

*In other words, the spectral representation map $I_u$, mapping $\hat{f} \in L_p^2$ into the random variable $\eta = \int \hat{f} d\hat{u} \in \mathbf{U}$, maps $K_p^2$ onto $\mathbf{U}^-$ and $\bar{H}_p^2$ onto the complementary subspace $\mathbf{U}^+$.*

*Proof.* We first show that

**Lemma 3.1.** *Under the asumption (3.1) the operator $M_G : H_p^2 \to H_p^2$ of multiplication (from the right) by the matrix function $G$ has closed range.*

*Proof.* Let $\{h_n\}$ be a Cauchy sequence in $M_G H_p^2$ and let $\{f_n\}$ be the corresponding sequence of pre-images. Note that $\{f_n\}$ is also Cauchy in $H_p^2$, since by (3.1)

$$\|f_n - f_m\|^2 \leq \frac{1}{c_1} \|(f_n - f_m)G\|^2 = \frac{1}{c_1} \|h_n - h_m\|^2$$

Then $f_n \to f \in H_p^2$ and by continuity $h_n \to fG \in \text{Range} M_G$. So $\text{Range} M_G$ is closed. $\qquad\square$

Hence multiplication by $G \in H_{p \times p}^\infty$ maps $H_p^2$ onto itself. By taking the preimages of the functions of $H_p^2$ identically equal to the canonical basis vectors $e_1, \dots, e_p$ in $\mathbb{R}^p$, and arranging them as row vectors of a $p \times p$ matrix $F$, one finds immediately that $F = G^{-1}$ since $FG = I_p = GF$ and hence

$$H_p^2 = H_p^2 FG = H_p^2 GF = \left(H_p^2 G\right) F = H_p^2 F = \text{closure}\{H_p^2 F\}$$

so that $F = G^{-1}$ is outer and $M_F$ also has closed range.

Now

$$\mathbf{U}_0^- = \left\{ \int \hat{f} \, d\hat{\nu} \mid \hat{f} \in H_p^2 \right\} := \int H_p^2 G^{-1} d\hat{u}$$

and the first representation in (3.7) follows since in the last formula we can exchange $H_p^2$ for $H_p^2 G^{-1}$. A symmetric argument proves (3.7). The third representation (3.7) also follows from the spectral representation of $\mathbf{U}^-$ in terms of the innovation $d\hat{\nu}$ displayed above which involves the "reduced" Hardy space $K_p^2$. $\quad\square$

Hence we have spectral representatives for the past and future spaces of the process $u$ in terms of well-known Hardy spaces. A usful representation for the oblique projection operators follows directly from Theorem 3.1.

**Corollary 3.1.** *The spectral representative of the oblique projection operator onto* $\mathbf{U}^-$ *along* $\mathbf{U}^+$, $E_{\|\mathbf{U}^+}[\cdot \mid \mathbf{U}^-]$, *is the* $L_p^2$-*orthogonal projection* $P^{K_p^2}$.

*Namely, if $\eta = \int \hat{f} \, d\hat{u}$, then*

$$E_{\|\mathbf{U}^+}[\eta \mid \mathbf{U}^-] = I_u\left(P^{K_p^2} \hat{f}\right) = \int P^{K_p^2} \hat{f} \, d\hat{u}. \tag{3.8}$$

*Similarly,*

$$E_{\|\mathbf{U}^-}[\eta \mid \mathbf{U}^+] = I_u\left(P^{\bar{H}_p^2} \hat{f}\right) = \int P^{\bar{H}_p^2} \hat{f} \, d\hat{u}. \tag{3.9}$$

*Proof.* We just need to check that $P^{K_p^2}$ is idempotent and its kernel is $\bar{H}_p^2$. But this is trivial. $\qquad\square$

We shall now find the spectral representative of the oblique predictor space $\mathbf{X}^{+/-}$. Let $\hat{W}(z)$ be the transfer function of the convolution operator (2.8). Note that $\Phi_y(z) = \hat{W}(z)\,\Phi_u(z)\hat{W}(1/z)' \geq c_1 \hat{W}(z)\,\hat{W}(1/z)'$ bythe first inequalty in (3.1) and since $\hat{W}$ is analytic it follows that actually $\hat{W} \in H_{m \times p}^2$. Consider the spectral density matrix $\Phi$ defined by

$$\Phi(z) := \hat{W}(z)\,\hat{W}(1/z)'. \tag{3.10}$$

In the following we shall assume that $p \leq m$ (there are more output channels than inputs) and, without loss of generality, that $\hat{W}(z)$ is *full column rank*. For if the columns of $\hat{W}$ were linearly dependent as functions in $H_m^2$, we could eliminate some input components and substitute them by suitable linear combinations of the others.

Below we shall refer to the spectral factorization of $\Phi(z)$ and to the related family of spectral factors. Following [11], we shall denote by $\hat{W}_-$ and $\bar{W}_+$ the outer and conjugate outer factor of $\Phi(z)$ and by $\bar{W}$ the unique spectral factor such that

$$K(z) := \bar{W}(z)^{-L}\,\hat{W}(z) \tag{3.11}$$

is an *inner function*, called the *structural function* of $\hat{W}$. Obviously the structural inner function carries all the information about the pole structure of $\hat{W}$. In the inner-outer factorization

$$\hat{W}(z) = \hat{W}_-(z)Q(z) \tag{3.12}$$

the inner function $Q$ carries information about the *stable* zero structure of $\hat{W}$ while in

$$\bar{W}(z) = \bar{W}_+(z)\bar{Q}(z) \tag{3.13}$$

the conjugate inner function $\bar{Q}$ carries information about the *unstable* zero structure of $\hat{W}$. ( Indeed $\bar{W}$ and $\hat{W}$ have the same zeros). In the following we shall assume that there are no pole-zero cancellations in forming the product $\bar{Q}K$.

Since $y(t) = I_u\!\left(z^t\hat{W}\right)$ and

$$\hat{W} = \bar{W}_+\,\bar{W}_+^{-1}\bar{W}\,\bar{W}^{-1}\hat{W} = \bar{W}_+\,\bar{Q}K$$

it is clear that

$$\mathbf{Y}^+ = \overline{\mathrm{span}}\{\,y(t)\,|\,t \geq 0\,\} = I_u\!\left(\bar{H}_p^2\,\bar{Q}K\right) \tag{3.14}$$

and from this representation it follows that

**Theorem 3.2.** *Assume that the co-inner functions $(\bar{Q},\,K^*)$ are right-coprime. Then the oblique predictor space has the spectral representation*

$$\mathbf{X}^{+/-} = K_p^2 \ominus K_p^2\,K := H(K) \tag{3.15}$$

*where $K$ is the structural function of $\hat{W}$.*

*Proof.* In fact,

$$\mathbf{X}^{+/-} = \int P^{K_p^2}\bar{H}_p^2\,\bar{Q}K\,d\hat{u}$$

and in force of coprimeness of $(\bar{Q},\,K^*)$, the range of the Hankel operator $\bar{f} \to P^{K_p^2}\bar{f}\bar{Q}K$, $\bar{f} \in \bar{H}_p^2$ is precisely the right invariant subspace $H(K)$ [4]. $\qquad\square$

This result shows in particular that

$$(\mathbf{X}^{+/-})^+ = \int \bar{H}_p^2 K\,d\hat{u} \supset \int \bar{H}_p^2\bar{Q}K\,d\hat{u} = \mathbf{Y}^+$$

and hence $\mathbf{Y}^+ \vee (\mathbf{X}^{+/-})^+ = (\mathbf{X}^{+/-})^+$. Therefore $\mathbf{X}^{+/-}$ is oblique splitting by virtue of

$$P^{K_p^2}\bar{H}_p^2 K = H(K)$$

which is a well-known equivalent definition of $H(K)$.

At this point it is possible to give an economical justification of the claim made in Theorem 2.3.

*Proof of Theorem 2.3*

*Proof.* It is easy to check that the generalized future space $\bar{\mathbf{Y}}^+ := \mathbf{Y}^+ \vee \mathbf{U}^+$ has the spectral representation

$$\mathbf{Y}^+ \vee \mathbf{U}^+ = I_u\left(\bar{H}_p^2 K\right)$$

This is so because $\bar{H}_p^2 \bar{Q} K \vee \bar{H}_p^2 = \left(\bar{H}_p^2 \bar{Q} \vee \bar{H}_p^2 K^*\right) K$ and in the vector sum between parentheses the conjugate inner functions $\bar{Q}$ and $K^*$ are right coprime. It is on the other hand very well-known (see e.g. [6] or [4, Lemma 13.5]) that

$$\bar{H}_p^2 K \cap K_p^2 = H(K)$$

which is just (2.12) written in the spectral domain. $\qquad\square$

ACKNOWLEDGMENT

It is a pleasure to thank György Michaletzky for very helpful discussions.

REFERENCES


[1] H. Akaike, Markovian representation of stochastic processes by canonical variables, *SIAM J. Control*, **13** , pp. 162–173, (1975).

[2] H. Akaike, Stochastic Theory of Minimal Realization. *IEEE Trans. Automat. Contr.*, vol. AC-19, no. 6, pp. 667–674, 1974.

[3] H. Akaike, Canonical Correlation Analysis of Time Series and the Use of an Information Criterion. *System Identification: Advances and Case Studies* (R. Mehra and D. Lainiotis, Eds.), Academic, 1976, pp. 27–96.

[4] P.A. Fuhrmann, *Linear Operators and Systems in Hilbert Space*, McGraw Hill 1981.

[5] C.W.J. Granger, Economic processes involving feedback, *Information and Control* **6**, (1963), pp. 28-48.

[6] H. Helson, *Lectures on Invariant Subspaces*, Academic Press, N.Y. 1961.

[7] W. E. Larimore, System identification, reduced-order filtering and modeling via canonical variate analysis, *Proc. American Control Conference*, 1983, pp. 445–451.

[8] P.D.Lax and R.S.Phillips, *Scattering Theory*, Academic Press, New York, 1967.

[9] A. Lindquist, G. Picci and G. Ruckebusch On minimal splitting subspaces and Markovian representation, *Math. System Theory*, **12**: 271-279, 1979.

[10] A. Lindquist and G. Picci, On the stochastic realization problem *SIAM J. Control and Optimization*, **17**: 365–389, 1979.

[11] A. Lindquist and G. Picci, Realization theory for multivariate stationary Gaussian processes, *SIAM J. Contr. & Optimiz.* **23** (1985), pp. 809–857.

[12] A. Lindquist and M. Pavon, On the structure of state space models of discrete-time vector processes, *IEEE Tr. on Automatic Control*, **AC-29**, p.418-432, 1984.

[13] A. Lindquist and G. Picci, A geometric approach to modelling and estimation of linear stochastic systems, *Journal of Mathematical Systems, Estimation and Control*, **1**:241–333, 1991.

[14] A. Lindquist and G. Picci, On "subspace-methods" identification and stochastic model reduction, *Proc. 10th IFAC Symposium on System Identification*, Copenhagen, DK, **2** (1994), pp. 397–403.

174


[15] A. Lindquist and G. Picci, Canonical correlation analysis approximate covariance extension and identification of stationary time series, *Automatica*, vol. 32, pp. 709-733, 1996.

[16] A. Lindquist, G. Michaletzky and G. Picci, Zeros of Spectral Factors, the Geometry of Splitting Subspaces, and the Algebraic Riccati Inequality, *SIAM J. Control & Optimization*, **33**,p. 365-401, ( 1995).

[17] A. Lindquist and G. Picci, Geometric Methods for State Space Identification, in *Identification, Adaptation, Learning*, (Lectures given at the NATO-ASI School, *From Identifiation to Learning* held in Como, Italy, Aug.1994), Springer Verlag, 1996.

[18] M. Moonen, B. De Moor, L. Vanderberghe and J. Vandewalle, On- and Off-Line Identification of Linear State-Space Models. *Int. J. Control* **49** (1989), pp. 219–232.

[19] M. Moonen and J. Vandewalle, QSVD Approach to On- and Off-Line State-Space Identification. *Int. J. Control* **51** (1990), pp. 1133–1146.

[20] G. Picci, Stochastic realization of Gaussian processes *Proc. of the IEEE*, **64** (1976), pp. 112-122.

[21] G. Picci and T. Katayama, Stochastic realization with exogenous inputs and "Subspace Methods" Identification, *Signal Processing*, 1996, (in press).

[22] G. Picci and T. Katayama: "A simple "subspace" identification algorithm with exogenous inputs", *Proceedings of the 1996 triennial IFAC Congress*, San Francisco, Ca., paper n. 0916, session 3a-06-5.

[23] G. Picci, Geometric Methods in Stochastic Realization and System Identification *CWI Quarterly* special Issue on System Theory, **9**, pp. 205-240, 1996.

[24] G. Picci, Stochastic Realization and System Identification, in *Statistical Methods in Control and Signal Processing*, T. Katayama and I. Sugimoto eds, M. Dekker, N.Y. 1997.

[25] N. I. Rozanov, *Stationary Random Processes*. Holden-Day (1963).

[26] G. Ruckebusch, Répresentations markoviennes de processus gaussiens stationnaires, *C.R.Acad.Sc.Paris* , Series A, **282**, p. 649-651, 1976.

[27] G. Ruckebusch, A state space approach to the stochastic realization problem, *Proc. 1978 IEEE Intern. Symp. Circuits and Systems*, p. 972–977, 1978.

[28] P. Van Overschee and B. De Moor, Subspace algorithms for the stochastic identification problem, *Automatica* **29** (1993), pp. 649–660.

[29] P. Van Overschee and B. De Moor, N4SID: Subspace algorithms for the identification of combined deterministic– stochastic systems, *Automatica* **30** (1994), pp. 75–93.

[30] P. Van Overschee and B. De Moor, A unifying theorem for subspace system identification algorithms and its interpretation. *Proc. 10th IFAC Symposium on System Identification* **2** (1994), pp. 145–156.

[31] M. Verhaegen, Identification of the deterministic part of MIMO State Space Models given in Innovations form from Input-Output data, *Automatica* **30** (1994), pp. 61–74.



Giorgio Picci
Dipartimento di Elettronica e Informatica
Università di Padova e LADSEB-CNR
via Gradenigo 6/a, 35131 Padova, Italy
Email: picci@dei.unipd.it

# Lyapunov Revisited:
# Variations on a Matrix Theme

Hans Schneider[1]
University of Wisconsin, Madison, USA

**Dedicated to Paul A. Fuhrmann
on the occasion of his 60th birthday**

**Abstract**: In this expository note it is shown that a cone version of the Perron–Frobenius theorem implies various generalizations of a matrix form of Lyapunov's famous theorem:

Si les équations différentielles du mouvement troublé sont telles qu'il est possible de trouver une fonction définie $V$, dont la dérivée $\dot{V}$ soit une fonction de signe fixe et contraire à celui de $V$, ou se réduise identiquement à zéro, le mouvement non troublé est stable.

Lyapunov's basic result on the stability of solutions of differential equations [Lyap, Ch.I, §16, Th.I] is here quoted from the French translation of his 1892 memoir. Lyapunov also investigated a more restrictive concept, that of asymptotic stability, in the case of linear differential equations with constant coefficients where $V$ is a homogeneous form of degree $m$, see [Lyap, Ch. II]. Gantmacher [Gant, Ch.XV, §5] considered the case of constant coefficients $\dot{x} = Ax$, $x \in \mathbb{C}^n$, $A \in \mathbb{C}^{nn}$, and restricted $V$ to a homogeneous quadratic form. Thus

$$V(x) = x^* H x, \quad H^* = H$$

and

$$\dot{V}(x) = \dot{x}^* H x + x^* H \dot{x} = x^* (A^* H + H A) x.$$

Putting

$$W(x) = x^* K x, \quad K \succ 0,$$

where

$$K \succ 0 := K \text{ positive definite},$$

and noting that

$$x(t) \to 0 \text{ as } t \to \infty \Leftrightarrow \Re(\lambda) < 0, \text{ all } \lambda \in \operatorname{spec}(A),$$

---

[1]Work supported by NSF Grant DMS-9424346.

he obtained a result [Gant, Ch.XV, Th.3'] that is usually called "Lyapunov's theorem" by matrix theorists which I state in a slightly more general form:

*Theorem 0* : Let $A \in \mathbb{C}^{nn}$ and let $K \succ 0$. Then there exists $H \succ 0$ such that $AH + HA^* = K$ if and only if $A$ is positive stable (i.e. has all eigenvalues in the *right* half plane).

Gantmacher's reformulation, see also [Hahn, Kap. II, §8], had a deep influence on the inertia theory of matrices as developed in the 1960's and subsequently, but I shall pursue this topic no further. Note that in Lyapunov's original formulation the theorem concerned the *existence* of a function $V$ with certain properties, while Gantmacher's version concerns *solving* a matrix equation. The two formulations are equivalent for we have:

$$\forall K \succ 0, \ \exists H \succ 0, \ AH + HA^* = K \iff \exists H \succ 0, \ AH + HA^* \succ 0.$$

I had met this situation before in Perron–Frobenius theory. Thus we define, for $P \in \mathbb{R}^{mn}$,

$$P > 0 := p_{ij} > 0, \ \text{all } (i,j),$$

$$P \geq 0 := p_{ij} \geq 0, \ \text{all } (i,j)$$

and employ the spectral radius $\rho(P)$ defined as usual by

$$\rho(P) = \max\{|\lambda| : \lambda \in \text{spec}(P)\}.$$

If $P \geq 0$ it follows by Perron–Frobenius that $\rho(P)$ is an eigenvalue of $P$. We further have, see e.g. [BePl, Theorem 6.2.],

*Theorem 1* : Let $A = \sigma I - P$ where $P \geq 0$. Then the following are equivalent:

1. $\sigma > \rho(A)$.

2. For all $y > 0$, there exists $x > 0$ such that $Ax = y$. (viz. $A^{-1} > 0$).

3. There exists $x > 0$ such that $Ax > 0$.

Again we have

$$\forall \iff \exists .$$

In [Schn] I found a unified treatment and generalized Lyapunov's theorem. The key is a generalization of Perron–Frobenius to cones which is due to Krein–Rutman [KrRu] in a Banach space. We consider only the finite dimensional case here.

*Definition*: A subset $C$ of a (finite dimensional) space $V$ over $\mathbb{R}$ is a (pointed, full, closed)) *cone* if

1. $C + C \subseteq C$, viz. $x + y \in C$, $\forall x, y \in C$.

2. $\mathbb{R}_+ C \subseteq C$, viz. $\alpha x \in C$, $\forall \alpha \geq 0, x \in C$.

3. $C \cap -C = \{0\}$, viz. $x, -x \in C \Rightarrow x = 0$.

4. $C - C = V$ viz. $\forall z \in V$, $\exists x, y \in C$, $z = x - y$,
   equivalently, the interior $C^0 \neq \phi$.

5. $C$ is closed.

We now redefine for $x \in V$:
$$x \geq 0 \; : \; x \in C,$$
$$x > 0 \; : \; x \in C^0,$$
and for $T \in \mathrm{Hom}(V)$:
$$T \geq 0 : \; TC \subseteq C.$$

Again, Perron–Frobenius (Krein–Rutman) applies: If $T \geq 0$ then $\rho(T) \in \mathrm{spec}(T)$, and as a consequence we obtain

*Theorem* 2 : Let $C$ be a cone. Let $T = R - S \in \mathrm{Hom}(V)$:
$$T = R - S, \; R^{-1} \geq 0, \; S \geq 0.$$

Then the following are equivalent:

1. $\rho(R^{-1}S) < 1$.

2. $T^{-1}C^0 \subseteq C^0 \; (T^{-1} \geq 0)$.

3. $TC^0 \cap C^0 \neq \phi$.

The obvious model is $C = \mathbb{R}_+^n$, the set of all vectors with nonnegative components in $\mathbb{R}^n$. In this case Theorem 2 is a slight generalization of Theorem 1 to splittings of type $T = R - S$ which Varga exploited and called *regular splittings*, see [Varg, p. 88]. We, however, are interested in the following set up:
$$V = \mathcal{H}_n, \; C = \mathcal{P}_n,$$

where
$$\mathcal{H}_n = \text{real space of Hermitians in} \mathbb{C}^{nn},$$
$$\mathcal{P}_n = \text{cone of positive semidefinite Hermitians in } \mathcal{H}_n.$$

If $R \in \mathrm{Hom}(\mathcal{H}_n)$ is defined by $R(H) = AHA^*$, where $A \in C^{nn}$ is nonsingular, then $R \geq 0$ and $R^{-1} \geq 0$. If $S \in \mathrm{Hom}(\mathcal{H}_n)$ is defined by $S(H) = \Sigma_k C_k^* H C_k$ then $S \geq 0$. The operator $R^{-1}S$ in (1.) of Theorem 2 now becomes
$$R^{-1}S \; = \; \Sigma_{k=1}^s \, (A^{-1}C_k \times \bar{A}^{-1}\bar{C}_k)$$

where $\times$ is the Kronecker (tensor) product. Thus Theorem 2 specializes to:

*Theorem* 3 : Let $A$, $C_k$, $k = 1, \ldots, s$, be complex $n \times n$ matrices. Let $H$ be Hermitian. Then the following are equivalent:

1. $A$ is nonsingular and

$$\rho(\Sigma_{k=1}^{s}A^{-1}C_k \times \bar{A}^{-1}\bar{C}_k) < 1.$$

2. For all $K \succ 0$, there exists a unique $H \succ 0$ such that

$$AHA^* - \Sigma_{k=1}^{s}C_kHC_k^* = K.$$

3. There exists an $H \succ 0$ such that

$$AHA^* - \Sigma_{k=1}^{s}C_kHC_k^* \succ 0.$$

4. $A$ is nonsingular and there exists an $H \succ 0$ such that

$$\rho((\Sigma_{k=1}^{s}A^{-1}C_kHC_k^*A^{*-1})H^{-1}) < 1.$$

Condition (4.) of Theorem 3 is a consequence of (3.) and was pointed out to me by S. Friedland.

The spectral radius of the operator in (1.) of Theorem 3 can be evaluated in terms the eigenvalues of its constituent matrices (only) under special assumptions. One such assumption is that the matrices $A$, $C_k$, $k = 1, \ldots, s$, are *simultaneously triangulable*, viz. there exists $Q \in \mathbb{C}^{nn}$ such that $Q^{-1}AQ$, $Q^{-1}C_kQ$, $k = 1, \ldots, s$, are (upper) triangular. In this case there exists an obvious *natural correspondence* $(\alpha_i, \gamma_i^{(1)}, \ldots, \gamma_i^{(s)})$, $i = 1, \ldots, n$, of the eigenvalues of $A$, $C_k$, $k = 1, \ldots, s$, such that every (noncommutative) polynomial $p(A, C_1, \ldots, C_s)$ has eigenvalues $p(\alpha_i, \gamma_i^{(1)}, \ldots, \gamma_i^{(s)})$, $i = 1, \ldots, n$. (A theorem of McCoy's assert that this latter property is equivalent to simultaneous triangulability). It was known to Frobenius that a set of pairwise commutative matrices is simultaneously triangulable. In particular, if $C_k = C^k$, $k = 0, \ldots, s$, then the $C_k$ can be simultaneously triangulated. See [Taus] for more information and references on this topic.

If $A$, $C_k$, $k = 1, \ldots, s$, are simultaneously triangulable, then for the operator $R^{-1}S$ in Theorem 3 we have

$$\text{spec}(R^{-1}S) = \{\Sigma_{k=1}^{s}\alpha_i^{-1}\gamma_i^{(k)}\bar{\alpha}_j^{-1}\bar{\gamma}_j^{(k)} : i, j = 1, \ldots, n\}.$$

We may apply Cauchy's inequality to obtain [Schn, Theorem 1]:

*Theorem 4* : Let $A$, $C_k$, $k = 1, \ldots, s$, be complex $n \times n$ matrices which can be simultaneously triangulated. Suppose the eigenvalues of $A$, $C_k$ under a natural correspondence are $\alpha_i$, $\gamma_i^{(k)}$, $i = 1, \ldots, n$, $k = 1, \ldots, s$. For Hermitian $H$, let

$$T(H) = AHA^* - \Sigma_{k=1}^{s}C_kHC_k^*.$$

Then the following are equivalent:

1. $\epsilon_i := |\alpha_i|^2 - \Sigma_{k=1}^{s}|\gamma_i^{(k)}|^2 > 0$, $i = 1, \ldots, n$.

2. For all $K \succcurlyeq 0$, there exists a unique $H \succcurlyeq 0$ such that T(H) = K.

3. There exists an $H \succcurlyeq 0$ such that $T(H) \succcurlyeq 0$.

We note the following special cases:

If

$$T(H) = (B + I)H(B + I)^* - (BHB^* + IHI^*) = BH + HB^*$$

then $\epsilon_i = \beta_i + \bar{\beta}_i$ and thus we obtain Lyapunov's Theorem.

If

$$T(H) = IHI^* - CHC^*$$

then $\epsilon_i = 1 - |\gamma_i|^2$ and thus we have a result due to Stein.

We now turn to a generalization due to D.H. Carlson, published in [Hill]. Let

$$\Phi = \Phi^* \in \mathbb{C}^{s+1,s+1}$$

and consider the operator $T$ defined by

$$T(H) = \Sigma_{h,k=0}^{s} \varphi_{hk} C_h H C_k^*$$

for $H \in \mathcal{H}_n$. We define the $n \times (s+1)n$ matrix

$$\underline{C} = [C_0, \ldots, C_s]$$

and we obtain

$$T(H) = \underline{C}(\Phi \times H)\underline{C}^* = \underline{C}(U \times I)(\Delta \times H)(U^* \times I)\underline{C}^* = \underline{B}(\Delta \times H)\underline{B}^*,$$

where $U$ is a unitary matrix, $\Phi = U\Delta U^*$ and $\underline{B} = \underline{C}U \in \mathbb{C}^{n,(s+1)n}$. If $C_0, \ldots, C_s$ are simultaneously triangulable and we put

$$\underline{\gamma_i} = [\gamma_i^{(0)}, \ldots, \gamma_i^{(s)}], \; i = 1, \ldots, n,$$

where $(\gamma_i^{(0)} \ldots \gamma_i^{(s)})$, $i = 1, \ldots, n$, is a natural correspondence of the eigenvalues, $k = 0, \ldots, s$. Since the eigenvalues of $B_k$, $k = 0, \ldots, s$, are $\Sigma_{h=0}^{s}\gamma_h^{(i)}u_{hk}$, $i = 1, \ldots, n$, Theorem 4 can be generalized to the following result, where by $\pi(\Phi)$ we denote the number of positive eigenvalues of the Hermitian matrix $\Phi$.

*Theorem 5* : Let $C_k$, $0 = 1, \ldots, s$, be complex $n \times n$ matrices which can be simultaneously triangulated. Suppose the eigenvalues of $C_0, \ldots, C_s$ under a natural correspondence are $\gamma_i^{(0)}, \ldots \gamma_i^{(s)}$, $i = 1, \ldots, n$. Let $\Phi = \Phi^* \in \mathbb{C}^{s+1,s+1}$, where $\pi(\Phi) = 1$. For Hermitian $H$, let

$$T(H) = \Sigma_{h,k=0}^{s} \varphi_{hk} C_h H C_k^*.$$

Then the following are equivalent:

1. $\underline{\gamma}_i \Phi \underline{\gamma}_i^* > 0$, $i = 1, \ldots, n$.

2. For all $K \succ 0$, there exists a unique $H \succ 0$ such that $T(H) = K$.

3. There exists an $H \succ 0$ such that $T(H) \succ 0$.

Clearly the assumptions of Theorem 5 are satisfied if $A \in \mathbb{C}^{nn}$ and $C_k = A^k$, $k = 1, \ldots, n$. Thus we derive a result independently due to Kharitonov [Khar], see also [Gutm, Theorem 6.1].

*Theorem 6* : Let $A \in \mathbb{C}^{nn}$ have eigenvalues $\alpha_i$, $i = 1, \ldots, n$. Let $K \in \mathbb{C}^{nn}$ be positive definite and suppose that $\Phi$ is a Hermitian matrix in $\mathbb{C}^{s+1,s+1}$ with $\pi(\Phi) = 1$. Then the following are equivalent:

1. $\Sigma_{h,k=0}^s \alpha_i^h \varphi_{hk} \bar{\alpha}_i^k > 0$, $i = 1, \ldots, n$.

2. The (unique) solution $H$ of $\Sigma_{h,k=0}^s \varphi_{hk} A^h H A^{k*} = K$ is positive definite.

Theorems 5 and 6 do not hold without the assumption that $\pi(\Phi) = 1$, as is shown by the following example with $\pi(\Phi) = 2$. Let

$$A = \begin{bmatrix} 2 & 0 \\ 0 & 2 \end{bmatrix}, \ C = \begin{bmatrix} 1 & 0 \\ 0 & -1 \end{bmatrix}, \ H = \begin{bmatrix} 1 & 1 \\ 1 & 1 \end{bmatrix}.$$

Then

$$AHA^* + CHC^* = \begin{bmatrix} 5 & 3 \\ 3 & 5 \end{bmatrix}.$$

A perturbation argument shows that we can find $H$ with $\pi(H) = 1$ or $\pi(H) = 2$ such that $AHA^* + CHC^* \succ 0$. However, a weaker result holds, see [Hill] for remarks on this topic. We need an additional assumption, which is again satisfied if the $C_k$, $k = 0, \ldots, s$, commute pairwise and hence if $C_k = A^k$, $k = 0, \ldots s$, see [CaPi] for information. Dropping the assumption that $\pi(\Phi) = 1$, we still have

*Theorem 7* : Let $C_k$, $k = 0, \ldots, s$, be complex $n \times n$ matrices which can be simultaneously triangulated . Assume that for each distinct sequence of corresponding eigenvalues $(\gamma_i^{(0)}, \ldots, \gamma_i^{(s)})$, $i = 1, \ldots, n$, of $C_0, \ldots, C_s$ there exists a common eigenvector. Let $\Phi = \Phi^* \in \mathbb{C}^{s+1,s+1}$. Then the following are equivalent:

1. $\underline{\gamma}_i \Phi \underline{\gamma}_i^* > 0$, $i = 1, \ldots, n$.

3. There exists an $H \succ 0$ such that $\Sigma_{h,k=0}^s \varphi_{hk} C_h H C_k^* \succ 0$.

As a special case of Theorem 7 we state

*Theorem 8* : Let $A \in \mathbb{C}^{nn}$ have eigenvalues $\alpha_i$, $i = 1, \ldots, n$. Suppose that $\Phi$ is a Hermitian matrix in $\mathbb{C}^{s+1,s+1}$. Then the following are equivalent:

1. $\Sigma_{h,k=0}^{s} \alpha_i^h \varphi_{hk} \bar{\alpha}_i^k > 0$, $i = 1, \ldots, n$.

3. There exists $H \succ 0$ such that $\Sigma_{h,k=0}^{s} \varphi_{hk} A^h H A^{k*} \succ 0$.

Proofs of the last two theorems are implicit in [Hill] or [Khar].

# References


[BePl]  Berman, A. & Plemmons, R.J., *Nonnegative matrices in the Mathematical Sciences*, Academic (1979) and SIAM (1994)

[CaPi]  Carlson, D.H. & Pierce, S., Common eigenvectors and quasicommutativity of simultaneously triangulable matrices, *Lin. Alg. Appl.* 71:49–55 (1985).

[Gant]  Gantmacher, F.R., (i) *Teoriya Matrits*, Gosd. Isd. Tech-teoret. (1953), (ii) *Theory of Matrices*, Chelsea (1959).

[Gutm]  Gutman, S., Root clustering in parameter space, *Lecture Notes in Control and Information Sciences*, Springer (1990).

[Hahn]  Hahn, W., Theorie und Anwendung der zweiten Methode von Ljapunov, *Ergeb. Math. Grenzg.*, N.F. 22, Springer (1959),

[Hill]  Hill, R.D., Inertia theory for simultaneously triangulable complex matrices, *Lin. Alg. Appl.* 2:131–142 (1969).

[Khar]  Kharitonov, V.L., Distribution of the roots of an autonomous system, *Avtomatika i Telmekhanika* 5:42–47 (1981).

[KrRu]  Krein, M.G. & Rutman M.A., Linear operators leaving invariant a cone in Banach space, (i) *Uspehi Mat. Nauk* 3(23):3–95 (1948), (ii) *Trans. Amer. Math. Soc.* Ser. 1, 10:199–325 (1952).

[Lyap]  Lyapunov (Liapunoff) M.A., Problème général de la stabilité du mouvement, (i) *Comm. Math. Soc. Kharkov* (1892) (ii) *Ann. Fac. Sci. Toulouse* 9(2)(1907) (iii) *Ann. Math. Studies*, 17(1947), Princeton U.P.

[Schn]  Schneider, H., Positive operators and an inertia theorem, *Num. Math.* 7:11–17 (1965).

[Taus]  Taussky, O., Commutativity in finite matrices, *Amer. Math. Month.* 64:229–235 (1957).

[Varg]  Varga, R.S., *Matrix Iterative Analysis*, Prentice–Hall (1962).



Hans Schneider
Department of Mathematics
University of Wisconsin
Madison, Wisconsin 53706, USA

# Linear Systems and Discontinuous Dynamics

J. M. Schumacher
CWI, Amsterdam, The Netherlands, and
Tilburg University, Tilburg, The Netherlands

Dedicated to Paul Fuhrmann at the occasion of his sixtieth birthday

**Abstract:** A dynamical system may be called *discontinuous* if it is subject to abrupt changes in its dynamic characteristics. Such changes may be induced either externally ("time events") or internally ("state events"). Here we discuss so-called *linear complementarity systems*. These are piecewise linear systems in which state events occur due to transitions from one branch of an ideal-diode-type characteristic to the other. We present a precise definition of the dynamics of such systems and give sufficient conditions for well-posedness.

## 1   Introduction

Recent years have seen a growing interest in systems that are subject to abrupt changes in their dynamic characteristics. There are actually several different motivations for this interest, coming from various domains of science. A discontinuous dynamical system arises for instance when a continuous system is coupled to a controller that switches between a finite number of operating modes (for instance *on* and *off*); the switching may be induced by external inputs or by measurements taken from the system itself. A second source of discontinuous dynamics lies in processes that are subject to unilateral constraints; think for instance of a robot arm that comes into contact with a rigid surface, or a tank in a chemical plant that at some point becomes completely filled. Sudden changes are sometimes inherent in the physics of a situation; examples could be a fluid that starts to boil, or a crane that overreaches. In quite another context, one may be interested in computing or communication devices whose main purpose is to switch between discrete states but which are nevertheless subject to disturbances of a continuous nature, such as clock drift. Impetus for the study of discontinuous dynamical systems therefore comes both from control theory and from computer science, as well as from the modeling and simulation of mechanical and electrical systems and of chemical processes.

Of course the different backgrounds of researchers in discontinuous dynamical systems give rise to different angles from which the field is studied. For those whose background is in differential equations, the main modeling alternative is to use a traditional smooth dynamical model. An important reason not to choose alternative can for instance lie in speed of simulation; certain phenomena that happen on a timescale much faster than the timescale of interest are then described as instantaneous transitions in an attempt to avoid time-consuming calculations. Fast (real-time) simulation of complicated physical systems is crucial for instance in the training of pilots but also for testing of software that interacts with a physical environment. If this is taken as a motivation to study discontinuous dynamical systems, then naturally aspects related to *simulation* are of prime interest. For those whose background is in computer science, the main modeling alternative is to use a traditional completely discrete model such as a transition graph, in which time does not appear at all. Such researchers will be motivated to introduce continuous time by applications in which time is for some reason critical; hence they will tend to be interested in *verification* aspects. It remains to be seen whether the field of discontinuous dynamical systems with its disparate approaches will evolve into a coherent area of study; for the moment however, the possibility of combining ideas from various directions certainly adds to the attractiveness of the subject.

There should be no suggestion that the study of discontinuous dynamical systems is new; in fact the history of the subject goes back at least several decades. One should mention the fundamental work of Filippov [5], the work that has been done on bang-bang control [3, 14] and on systems with relays or bistable elements [18, §22], and the classical treatments of mechanical systems with unilateral constraints [17, 12]; all of this refers to the fifties and sixties. The term *hybrid*, which is currently often used for systems that incorporate both continuous dynamics and discrete switching, appears to have been employed in this way for the first time by Witsenhausen in 1966 [22]. The resurgence of interest in hybrid systems, as indicated for instance in [2], can be attributed to several factors. Among those are the fact that computer scientists are extending their interest from the computer itself to its environment, and the fact that control theorists are more and more aware that in many industrial applications the continuous format that has long been standard in control theory is not quite appropriate. Actually it is somewhat embarrassing for control theorists to note the discrepancy between the high state of development of continuous control theory and the relatively small amount of theory that is available on switching control, despite the fact that programmable logic controllers dominate the market in many corners of industry.

The above doesn't imply that the systematic study that has been made of continuous control systems in the past decades is irrelevant. On the contrary, it will be argued in this paper that, at least for some types of discontinuous dynamical systems, the availability of an extensive theory of continuous multivariable systems [11, 23, 6, 16, 13] will be of crucial importance in advancing the theory beyond the stage that was reached in the fifties and sixties. The reasoning behind this may briefly be explained as follows.

In order not to complicate the discussion unnecessarily, let us consider 'closed' dynamical systems, i. e. systems with no inputs. (We will see that, despite appearances, control theory *is* of relevance to such systems.) In a closed hybrid system, switching will typically take place when the continuous variables in the system reach certain threshold values. In a number of cases, it is possible to describe the switching by a so-called *complementarity* characteristic. Such a characteristic involves two continuous variables, say $u(t)$ and $y(t)$, that are both required to be nonnegative; there are two associated operating modes, one in which $u(t)$ is constantly zero, and one in which $y(t)$ is constantly zero. If in the first mode the variable $y(t)$ reaches zero and tends to become negative, then a switch to the second mode occurs; and if in the second mode $u(t)$ tends to cross zero then the first mode becomes operative. If all the switching in the system can be described in this way, then we effectively have a continuous dynamical systems with a number of pairs of continuous variables $(u_i, y_i)$ that represent canonical switching elements. Because the switching elements are canonical, all properties of the closed system as a whole (existence and uniqueness of solutions, stability, etc.) must be expressible in terms of the continuous part with its associated variables $y_i$ and $u_i$, and this is where the input/output systems theory comes in.

A simple example will be given below in which the complementarity structure described in general terms above does indeed arise naturally. Actually the reader may already have noted that what was described is essentially the ideal diode characteristic, with current through and voltage across the diode as the complementary variables. The example below is mechanical and uses position and force as complementary variables. Another obvious application is in hydraulic systems with valves for which current and pressure may be taken as complementary. There are also less obvious applications. For instance, it has been shown that the mode switching brought about by any piecewise linear element can be described in complementarity terms [21]. The fact that such a canonical description is available may make piecewise linear modeling an attractive alternative to fully nonlinear modeling; in turn this would open large new grounds for discontinuous dynamical systems.

In this paper, the following conventions will be in force. For a positive integer $k$, $\bar{k}$ denotes the set $\{1, 2, \ldots, k\}$. For $I \subset \bar{k}$, $I^c$ denotes the complementary set $\bar{k} \setminus I$. Given $M \in \mathbb{R}^{k \times l}$ and two subsets $I \subset \bar{k}$ and $J \subset \bar{l}$, the $(I, J)$-submatrix of $M$ is defined as $M_{IJ} := (m_{ij})_{i \in I, j \in J}$. Following a convention in [4], we shall also write $M_{I\bullet}$ instead of $M_{I\bar{l}}$ and $M_{\bullet J}$ instead of $M_{\bar{k}J}$. For $a \in \mathbb{R}^k$ and $I \subset \bar{k}$, we write $a_I = (a_i)_{i \in I}$. A vector $a \in \mathbb{R}^k$ is called *nonnegative*, and we write $a \geqslant 0$, if $a_i \geqslant 0$ for all $i \in \bar{k}$. A (finite or infinite) sequence of real numbers is said to be *lexicographically nonnegative* if either it is a sequence of zeros or its first nonzero element is positive; we write $(a^1, a^2, \ldots) \succeq 0$. If $(a^1, a^2, \ldots)$ is a sequence of vectors, we write $(a^1, a^2, \ldots) \succeq 0$ when $(a_i^1, a_i^2, \ldots) \succeq 0$ for all $i$.

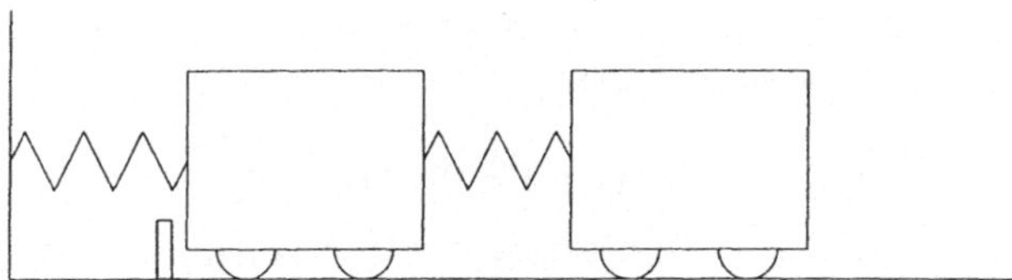

Figure 1: A discontinuous dynamical system.

## 2   An example

To motivate the development below, let us consider the following very simple example of a discontinuous dynamical system (see Fig. 1). Two carts are connected to each other and to a fixed wall by springs. The motion of the left cart is restricted by a purely non-elastic stop. For simplicity, we shall normalize all constants to 1 and let the springs be linear, and we shall assume that the stop is placed at the equilibrium position of the left cart. An 'event' takes place when the left cart hits the stop or when it is pulled away from a position at the stop. When an event takes place, the system switches from 'constrained mode' to 'unconstrained mode' or vice versa; these two modes may also be viewed as the two discrete states of the system.

Let us now discuss the dynamics of the system in the example. It is not difficult to write equations of motion for each of the two modes separately. Let $x_1(t)$ and $x_2(t)$ represent the deviations of the left and the right cart respectively from their equilibrium positions, and let $x_3(t)$ and $x_4(t)$ denote the corresponding velocities. In the unconstrained mode the equations (in first-order form) are the ones that would hold if there were no block:

$$
\begin{aligned}
\dot{x}_1(t) &= x_3(t) \\
\dot{x}_2(t) &= x_4(t) \\
\dot{x}_3(t) &= -2x_1(t) + x_2(t) \\
\dot{x}_4(t) &= x_1(t) - x_2(t).
\end{aligned}
\tag{1}
$$

The equations of motion in the constrained mode are the ones that would hold if the first cart were nailed to the block:

$$
\begin{aligned}
x_1(t) &= 0 \\
\dot{x}_2(t) &= x_4(t) \\
x_3(t) &= 0 \\
\dot{x}_4(t) &= -x_2(t).
\end{aligned}
\tag{2}
$$

To give a complete description of the hybrid system, one also needs to specify under what conditions events take place and what the effects of such events will be. Under the assumption of inelastic collision, one can argue that a transition from the unconstrained mode to the constrained mode will occur at times $t_0$ when the following Boolean expression in terms of equality and inequality conditions on the state variables evaluates to TRUE:

$$(x_1(t_0) = 0) \wedge ((x_3(t_0) < 0) \vee (((x_3(t_0) = 0) \wedge$$
$$\wedge ((x_2(t_0) < 0) \vee ((x_2(t_0) = 0) \wedge (x_4(t_0) < 0)))))).$$

Moreover, when this event takes place the variable $x_3$ is reset to zero, whereas the other continuous state variables keep the values that they had just before the event. A transition from the constrained mode to the unconstrained mode will take place at times $t_0$ when the following expression evaluates to TRUE:

$$(x_2(t_0) = 0) \wedge (x_4(t_0) > 0).$$

This event produces no jumps in the continuous state variables. Note that it is possible that the conditions for a transition from the constrained mode to the unconstrained mode are satisfied immediately after a transition from the unconstrained mode to the constrained mode has taken place; in such cases there are two events in one time instant (in a well-defined order).

The example shows that the specification of a hybrid system in the form of a "flat" differential automaton is not always very convenient. It is easier to specify the system in the form

$$
\begin{aligned}
\dot{x}_1(t) &= x_3(t) & (3)\\
\dot{x}_2(t) &= x_4(t) & (4)\\
\dot{x}_3(t) &= -2x_1(t) + x_2(t) + u(t) & (5)\\
\dot{x}_4(t) &= x_1(t) - x_2(t) & (6)\\
y(t) &= x_1(t) & (7)\\
y(t) \geq 0, \quad u(t) &\geq 0, \quad y(t)u(t) = 0 & (8)
\end{aligned}
$$

in which a *constraint force* denoted by $u(t)$ has been introduced. Note that the systems obtained by setting $u(t) = 0$ and $y(t) = 0$ in the above are indeed equivalent to (1) and (2) respectively; in the second case, note that imposing $y = 0$ forces the relation $u = -x_2$ so that $u$ is an output even though the equations are written as if it was an input. Moreover, the transition condition from constrained to unconstrained mode is exactly the one that is needed to prevent $u(t)$ from becoming negative. The above system is an example of a *linear complementarity system*. A precise definition of this class of dynamical systems will be given in the next section.

# 3 Linear complementarity systems

Consider the following system of linear differential and algebraic equations and inequalities:

$$\dot{x}(t) = Ax(t) + Bu(t) \tag{9a}$$
$$y(t) = Cx(t) + Du(t) \tag{9b}$$
$$y(t) \geqslant 0, \quad u(t) \geqslant 0, \quad y^{\mathsf{T}}(t)u(t) = 0. \tag{9c}$$

The equations (9a) and (9b) constitute a linear system in state space form; we let the number of inputs be equal to the number of outputs. The relations (9c) are called *complementarity conditions*. Because of the nonnegativity constraints, the vanishing of the inner product $y^{\mathsf{T}}(t)u(t)$ implies that actually for each index $i$ at least one of the variables $y_i(t)$ and $u_i(t)$ must be zero. The set of indices for which $y_i(t) = 0$ (we shall call this the *active index set*) need not be constant in time, so that the system may switch from one "operating mode" to another. To define the dynamics of (9) completely, we will have to specify when these mode switches occur, what their effect will be on the state variables, and how a new mode will be selected. A proposal for answering these questions (cf. [10]) will be explained below. By the specification of the complete dynamics of (9) we describe a class of dynamical systems called *linear complementarity systems*.

Let $n$ denote the length of the vector $x(t)$ in the equations (9a–9b) and let $k$ denote the number of inputs and outputs. There are then $2^k$ possible choices for the active index set. The equations of motion when the active index set is $I$ are given by

$$\begin{aligned}
\dot{x}(t) &= Ax(t) + Bu(t) \\
y(t) &= Cx(t) + Du(t) \\
y_i(t) &= 0, \quad i \in I \\
u_i(t) &= 0, \quad i \in I^c.
\end{aligned} \tag{10}$$

We shall say that these equations represent the system in *mode $I$*. An equivalent and somewhat more explicit form is given by the (generalized) state equations

$$\begin{aligned}
\dot{x}(t) &= Ax(t) + B_{\bullet I}u_I(t) \\
0 &= C_{I\bullet}x(t) + D_{II}u_I(t)
\end{aligned} \tag{11}$$

together with the output equations

$$\begin{aligned}
y_{I^c}(t) &= C_{I^c\bullet}x(t) + D_{I^cI}u_I(t) \\
u_{I^c}(t) &= 0.
\end{aligned} \tag{12}$$

At this point we need to digress in order to recall some facts concerning equations of the form (11), which can be derived from the geometric theory of linear systems (see [23, 1, 13] for the general background). Denote by $V_I$ the *consistent subspace* of mode $I$, i. e. the set of initial conditions $x_0$ for which there exist smooth functions $x(\cdot)$ and $u_I(\cdot)$, with $x(0) = x_0$, such that (11) is satisfied. The space $V_I$ can be computed as the limit of the sequence defined by

$$
\begin{aligned}
V^0 &= \mathbb{R}^n \\
V^{i+1} &= \{x \in V^i \mid \exists u \in \mathbb{R}^{|I|} \text{ s.\,t. } Ax + B_{\bullet I}u \in V^i,\ C_{I\bullet}x + D_{II}u = 0\}.
\end{aligned}
\tag{13}
$$

There exists a linear mapping $F_I$ such that (11) will be satisfied for $x_0 \in V_I$ by taking $u_I(t) = F_I x(t)$. The mapping $F_I$ is uniquely determined, and more generally the function $u_I(\cdot)$ that satisfies (11) for given $x_0 \in V_I$ is uniquely determined, if the full-column-rank condition

$$
\ker \begin{bmatrix} B_{\bullet I} \\ D_{II} \end{bmatrix} = \{0\}
\tag{14}
$$

holds and moreover we have

$$
V_I \cap T_I = \{0\},
\tag{15}
$$

where $T_I$ is the subspace that can be computed as the limit of the following sequence:

$$
\begin{aligned}
T^0 &= \{0\} \\
T^{i+1} &= \{x \in \mathbb{R}^n \mid \exists \tilde{x} \in T^i,\ \tilde{u} \in \mathbb{R}^{|I|} \text{ s.\,t. } x = A\tilde{x} + B_{\bullet I}\tilde{u},\ C_{I\bullet}\tilde{x} + D_{II}\tilde{u} = 0\}.
\end{aligned}
\tag{16}
$$

In the present context, the subspace $T_I$ is best thought of as the *jump space* associated to mode $I$, that is, as the space along which fast motions will occur that take an inconsistent initial state instantaneously to a point in the consistent space $V_I$; note that under the condition (15) this projection is uniquely determined. To make the interpretation of $T_I$ as a jump space precise, introduce the class of *impulsive-smooth distributions* that was studied by Hautus [8] (see also [9, 7]). The general form of an impulsive-smooth distribution $\phi$ is

$$
\phi = p(\tfrac{d}{dt})\delta + f
\tag{17}
$$

where $p(\cdot)$ is a polynomial, $\frac{d}{dt}$ denotes the distributional derivative, $\delta$ is the delta distribution with support at zero, and $f$ is a distribution that can be identified with the restriction to $(0, \infty)$ of some function in $C^\infty(\mathbb{R})$. The class of such distributions will be denoted by $C_{\text{imp}}$. For an element of $C_{\text{imp}}$ of the form (17), we write $\phi(0^+)$ for the limit value $\lim_{t\downarrow 0} f(t)$. Having introduced the class $C_{\text{imp}}$, we can replace the system of equations (11) by its distributional version

$$
\begin{aligned}
\tfrac{d}{dt}x &= Ax + B_{\bullet I}u_I + x_0\delta \\
0 &= C_{I\bullet}x + D_{II}u_I
\end{aligned}
\tag{18}
$$

in which the initial condition $x_0$ appears explicitly, and we can look for a solution of (18) in the class of vector-valued impulsive-smooth distributions. It was shown in [9] that if the conditions (14) and (15) are satisfied, then there exists a unique solution $(x, u_I) \in C_{\text{imp}}^{n+|I|}$ to (18) for each $x_0 \in V_I + T_I$; moreover, the solution is such that $x(0^+)$ is equal to $P_{V_I}^{T_I} x_0$, the projection of $x_0$ onto $V_I$ along $T_I$. The solution is most easily written down in terms of its Laplace transform:

$$\hat{x}(s) \;=\; (sI - A)^{-1} x_0 \;+\; (sI - A)^{-1} B_{\bullet I} \hat{u}_I(s) \tag{19}$$

$$\hat{u}_I(s) \;=\; -G_{II}^{-1}(s) C_{I \bullet} (sI - A)^{-1} x_0, \tag{20}$$

where

$$G_{II}(s) \;:=\; C_{I \bullet}(sI - A)^{-1} B_{\bullet I} + D_{II}. \tag{21}$$

Note that the notation is consistent in the sense that $G_{II}(s)$ can also be viewed as the $(I, I)$-submatrix of the transfer matrix $G(s) := C(sI - A)^{-1}B + D$. It is shown in [9] (see also [15]) that the transfer matrix $G_{II}(s)$ associated to the system parameters in (11) is left invertible when (14) and (15) are satisfied. Since the transfer matrices $G_{II}(s)$ that we consider are square, left invertibility is enough to imply invertibility, and so (by duality) we also have $V_I + T_I = \mathbb{R}^n$. Summarizing, we can list the following equivalent conditions. In the formulation of the theorem, we call a matrix $M$ over a field $\mathbb{F}$ *totally invertible* if none of its principal minors vanishes.

**Theorem 3.1** *Consider a time-invariant linear system with $k$ inputs and $k$ outputs, given by standard state space parameters $(A, B, C, D)$. The following conditions are equivalent.*

1. *For each index set $I \subset \bar{k}$, the associated system (11) admits for each $x_0 \in V_I$ a unique smooth solution $(x, u)$ such that $x(0) = x_0$.*

2. *For each index set $I \subset \bar{k}$, the associated distributional system (18) admits for each initial condition $x_0$ a unique impulsive-smooth solution $(x, u)$.*

3. *The conditions (14) and (15) are satisfied for all $I \subset \bar{k}$.*

4. *The transfer matrix $G(s) = C(sI - A)^{-1}B + D$ is totally invertible (as a matrix over the field of rational functions).*

In this paper we shall be interested in conditions for existence and uniqueness of solutions, and so we shall usually consider systems $(A, B, C, D)$ for which the above conditions hold.

After this digression, we can continue to define a notion of solution for the system (9). The following preliminary definitions will be needed.

**Definition 3.2** An impulsive-smooth distribution $\phi = p(\frac{d}{dt})\delta + f$ as in (17) will be called *initially nonnegative* if the leading coefficient of the polynomial $p(\cdot)$ is positive, or, in case $p = 0$, the smooth function $f$ is nonnegative on an interval of the form $(0, \varepsilon)$ with $\varepsilon > 0$. A vector-valued impulsive-smooth distribution will be called *initially nonnegative* if each of its components is initially nonnegative in the above sense.

**Definition 3.3** A triple of vector-valued impulsive-smooth distributions $(u, x, y)$ will be called an *initial solution* to (9) with *initial state* $x_0$ and *solution mode* $I$ if

1. the triple $(u, x, y)$ satisfies the distributional equations

$$\begin{aligned} \tfrac{d}{dt}x &= Ax + Bu + x_0\delta \\ y &= Cx + Du. \end{aligned}$$

2. both $u$ and $y$ are initially nonnegative

3. $y_i = 0$ for all $i \in I$ and $u_i = 0$ for all $i \notin I$.

Given a vector $x_0$, we can consider the collection of all index sets $I$ such that there exists an initial solution to (9) with initial state $x_0$ and solution mode $I$. This collection will be denoted by $\mathcal{S}(A, B, C, D; x_0)$, or simply by $\mathcal{S}(x_0)$ if the context is clear.

We are now ready to define the concept of a solution to (9). We shall first define what might be called a "full" solution to these equations, involving all of the ingredients that play a role. Depending on which aspects of the behavior one is interested in (for instance the continuous part or the switching part), one may then define related solution concepts which emphasize the aspects of interest.

**Definition 3.4** Consider seven-tuples $(L, \tau, x_e, I, u_c, x_c, y_c)$ of the following form:

- $L$ is either $\{0, \dots, N\}$ for some $N > 0$ or $\mathbb{Z}_+$ and is called the *event label set*

- $\tau$ is a monotonous function from $L$ to $\mathbb{R}_+ \cup \{\infty\}$ whose values are called *event times*

- $x_e$ is a function from $L$ to $\mathbb{R}^n$ whose values are called *event states*

- $I$ is a function from $L$ to $2^{\bar{k}}$ whose values are called *operating modes*

- $u_c$, $x_c$, and $y_c$ are smooth functions defined on $[0, T) \setminus \tau(L)$, with values in $\mathbb{R}^k$, $\mathbb{R}^n$, and $\mathbb{R}^k$ respectively.

Such a seven-tuple is called a *(full) solution* to (9) on an interval $[0, T)$ if the following conditions hold:

1. $\tau(0) = 0$ and $\sup_{i \in L} \tau(i) = T$

2. $I(i) \in \mathcal{S}(x_e(i))$ for all $i \in L$

3. for all $i$ such that $i \in L$, $i + 1 \in L$, and $\tau(i+1) = \tau(i)$, we have $x_e(i+1) = P_{V_{I(i)}}^{T_{I(i)}} x_e(i)$

4. for all $i$ such that $i \in L$, $i + 1 \in L$, and $\tau(i+1) > \tau(i)$, we have $x_e(i) = \lim_{t \downarrow \tau(i)} x_c(t)$ and $x_e(i+1) = \lim_{t \uparrow \tau(i+1)} x_c(t)$

5. for all $i$ such that $i \in L$ and $i+1 \in L$, the triple $(u_c, x_c, y_c)$ satisfies (10) with $I = I(i)$ for $t \in (\tau(i), \tau(i+1))$

6. for all $t \in [0, T)$ we have $u_c(t) \geqslant 0$ and $y_c(t) \geqslant 0$.

The *initial state* corresponding to a given solution is the vector $x_e(0)$. The *multiplicity* of an event time $t \in \tau(L)$ is the number of labels $i \in L$ such that $\tau(i) = t$.

The above definition is more involved than would be necessary for the systems that we shall consider in this paper. Under conditions that will be formulated below, we have *dominance of the continuous state* in the sense that the evolution of the continuous state vector is completely determined by its initial value, and in particular does not depend on the choice of an initial mode. The definition as given is in a form that can be relatively easily modified to accommodate externally induced switching by redefining the set of continuation modes $\mathcal{S}(x)$.

Suppose now that we are mainly interested in the behavior of the continuous states of the system. To avoid trivial distinctions between solutions, we shall identify two piecewise continuous functions that are defined almost everywhere if they agree on their common domain of definition. A notion of solution that places emphasis on the continuous states can then be given as follows.

**Definition 3.5** The *continuous trace* of a solution $(L, \tau, x_e, I, u_c, x_c, y_c)$ to (9) on an interval $[0, T)$ is the triple $(u_c, x_c, y_c)$ of vector-valued functions defined almost everywhere on $[0, T)$.

**Definition 3.6** A triple $(u_c, x_c, y_c)$ of almost everywhere defined piecewise smooth vector-valued functions will be called a *continuous-state solution* to (9) if it is the continuous trace of a full solution.

The industrious reader may verify that, in the concrete case of the example of Section 2, the definitions above reproduce the rules that we formulated for that case.

# 4   Well-posedness

A basic issue concerns the existence and uniqueness of solutions. We understand well-posedness in the following "local" sense.

**Definition 4.1** The complementary-slackness system (9) is *(locally) well-posed* if for each initial state there exists an $\varepsilon > 0$ such that (9) admits a unique continuous-state solution on $[0, \varepsilon)$.

The definition requires that for each initial vector in $\mathbb{R}^n$ there exists a unique solution on an interval of positive length starting with at most a finite number of jumps followed by smooth continuation. Note also that the definition expresses dominance of the continuous state, since no information about an initial discrete state ("starting mode") is required. Due to the possible occurrence of phenomena such as deadlock, ill-posedness is more common in discontinuous dynamical systems than it is in classical continuous systems. In particular, examples of linear complementarity systems that are not well-posed in the above sense are easy to find, see for instance [20].

To formulate sufficient conditions for well-posedness, we again need to introduce some linear systems terminology. The *leading column indices* $\eta_1, \dots, \eta_k$ of a linear system $(A, B, C, D)$ with Markov parameters $G^i$ are defined by

$$\eta_j := \inf\{i \in \mathbb{N} \mid G^i_{\bullet j} \neq 0\},$$

where we take $\inf \varnothing = \infty$. The *leading row indices* $\rho_1, \dots, \rho_k$ of the system $(A, B, C, D)$ are defined by

$$\rho_j := \inf\{i \in \mathbb{N} \mid G^i_{j\bullet} \neq 0\}.$$

For systems with totally invertible transfer matrices, the leading row and column indices are clearly all finite; in fact, we even have $\rho_i \leqslant n$ and $\eta_i \leqslant n$ for all $i$. The *leading row coefficient matrix* $M_r(A, B, C, D)$ and the *leading column coefficient matrix* $M_c(A, B, C, D)$ of a system $(A, B, C, D)$ with $k$ inputs and $k$ outputs are defined as follows:

$$M_r(A, B, C, D) := \begin{pmatrix} G^{\rho_1}_{1\bullet} \\ \vdots \\ G^{\rho_k}_{k\bullet} \end{pmatrix}, \quad M_c(A, B, C, D) := (G^{\eta_1}_{\bullet 1} \dots G^{\eta_k}_{\bullet k}). \tag{22}$$

We also need some terminology from mathematical programming [4]: a square matrix $M \in \mathbb{R}^{k \times k}$ is said to be a *P-matrix* if all its principal minors are positive. Now we can state the following well-posedness result; see [10] for a proof.

**Theorem 4.2** *Let a linear complementarity system of the form (9) be given. Suppose that the system satisfies the conditions of Thm. 3.1, and that both the leading column coefficient matrix and the leading row coefficient matrix of the associated linear system $(A, B, C, D)$ are P-matrices. Under these conditions, the complementarity system is well-posed. Moreover, no solution requires a multiplicity of event times higher than two.*

The notion of well-posedness that is used here does not require continuous dependence on initial conditions. Actually, examples of linear complementarity systems in which the dependence on initial conditions is *discontinuous* are not hard to find (see [10]). For similar reasons of easy failure we have not included a claim on uniqueness of *full* solutions. Finally we have not claimed *global* existence of solutions since we have not discussed the possibility of event times having a finite accumulation point. Examples of complementarity systems that show a "Zeno behavior" do not seem to be easily found, so it may be conjectured that solutions will exist for all time under conditions similar to those of Thm. 4.2. Finally let us note that complementarity systems that are not well-posed in the *forward* sense may still be useful in the analysis of *two-point* boundary value problems, such as occur in the maximum principle for optimal control problems subject to unilateral state constraints.

## 5 Conclusions

In this paper we have discussed a class of hybrid systems that allow a relatively compact specification. Indeed, we can describe a system with $2^k$ discrete states essentially by giving the parameters of a linear system with $k$ inputs and $k$ outputs. Moreover, the analysis of the hybrid system can be carried out in terms of the associated linear system, which brings the results of linear multivariable systems theory to bear on a class of hybrid systems. The framework we presented is limited and needs to be extended in several directions, in particular to allow the inclusion of continuous and discrete inputs. In addition to the basic issues of well-posedness that have been discussed here, there is a need to explore for instance conditions for stability, efficient methods for simulation, compositional specification of large systems, and verification of safety properties by systematic search for worst-case situations.

## References

[1] G. Basile and G. Marro. *Controlled and Conditioned Invariants in Linear System Theory*. Prentice Hall, Englewood Cliffs, NJ, 1992.

[2] R. W. Brockett. Hybrid models for motion control systems. In H. L. Trentelman and J. C. Willems, editors, *Essays on Control. Perspectives in the Theory and its Applications* (lectures and the mini-courses of the European control conference (ECC'93), Groningen, the Netherlands, June/July 1993), pages 29–53. Birkhäuser, 1993.

[3] D. W. Bushaw. *Differential Equations with a Discontinuous Forcing Term*. PhD thesis, Dept. of Math., Princeton Univ, 1952.

[4] R. W. Cottle, J.-S. Pang, and R. E. Stone. *The Linear Complementarity Problem*. Academic Press, Boston, 1992.

[5] A. F. Filippov. Differential equations with discontinuous right-hand sides. *Matematik. Sbornik.*, 51:99–128, 1960. In Russian. English translation: *Am. Math. Soc. Transl.* 62 (1964).

[6] P. A. Fuhrmann. *Linear Systems and Operators in Hilbert Space*. McGraw-Hill, New York, NY, 1981.

[7] A. H. W. Geerts and J. M. Schumacher. Impulsive-smooth behavior in multi-mode systems. Part I: State-space and polynomial representations. *Automatica*, 32:747–758, 1996.

[8] M. L. J. Hautus. The formal Laplace transform for smooth linear systems. In G. Marchesini and S.K. Mitter, editors, *Mathematical Systems Theory*, Lect. Notes Econ. Math. Syst. 131, pages 29–47. Springer, New York, 1976.

[9] M. L. J. Hautus and L. M. Silverman. System structure and singular control. *Lin. Alg. Appl.*, 50:369–402, 1983.

[10] W. P. M. H. Heemels, J. M. Schumacher, and S. Weiland. Linear complementarity systems. In preparation.

[11] T. Kailath. *Linear Systems*. Prentice-Hall, Englewood Cliffs, N.J., 1980.

[12] C. W. Kilmister and J. E. Reeve. *Rational Mechanics*. Longmans, London, 1966.

[13] M. Kuijper. *First-Order Representations of Linear Systems*. Birkhäuser, Boston, 1994.

[14] J. P. LaSalle. Time optimal control systems. *Proc. Natl. Acad. Sci. U.S.*, 45:573–577, 1959.

[15] A. S. Morse and W. M. Wonham. Status of noninteracting control. *IEEE Trans. Automat. Contr.*, AC-16:568–581, 1971.

[16] H. Nijmeijer and A. J. van der Schaft. *Nonlinear Dynamical Control Systems*. Springer-Verlag, Berlin, 1990.

[17] J. Pérès. *Mécanique Générale*. Masson & Cie., Paris, 1953.

[18] L. S. Pontryagin, V. G. Boltyanskii, R. V. Gamkrelidze, and E. F. Mishchenko. *The Mathematical Theory of Optimal Processes*. Interscience, New York, 1962.

[19] A. J. van der Schaft and J. M. Schumacher. Complementarity modeling of hybrid systems. Report BS-R9611, CWI, Amsterdam, 1996. URL: http://www.cwi.nl/ftp/CWIreports/BS/BS-R9611.ps.Z.

[20] A. J. van der Schaft and J. M. Schumacher. The complementary-slackness class of hybrid systems. *Math. Contr. Signals Syst.*, 9:266–301, 1996.

[21] L. Vandenberghe, B. L. De Moor, and J. Vandewalle. The generalized linear complementarity problem applied to the complete analysis of resistive piecewise-linear circuits. *IEEE Trans. Circuits Syst.*, CAS-36:1382–1391, 1989.

[22] H. S. Witsenhausen. A class of hybrid-state continuous-time dynamic systems. *IEEE Transactions on Automatic Control*, AC-11:161–167, 1966.

[23] W. M. Wonham. *Linear Multivariable Control: A Geometric Approach.* Springer Verlag, New York, 3rd edition, 1985.

J.M. Schumacher
CWI
P.O. Box 94079
1090 GB  Amsterdam
The Netherlands
Tel. +31-20-5924090
Fax +31-20-5924199
E-mail Hans.Schumacher@cwi.nl

# Differential Invariants and Curvature Flows in Active Vision

Allen Tannenbaum and Anthony Yezzi, Jr.
University of Minnesota, Minneapolis, USA

**Abstract**: In this paper, we discuss the use of differential invariants, and curvature driven flows for active vision. We concentrate on three problem areas: invariant flows, active contours, and $L^1$-based methods for optical flow and stereo. The solutions to these key problems will all be based on curvature based evolutions which are obtained in a completely natural manner from geometric and physical principles.

*This paper is dedicated with much admiration and affection to Professor Paul Fuhrmann on the occasion of his 60th birthday.*

## 1   Introduction

In this paper, we will discuss some of the work that we have been conducting on the development of novel techniques for employing visual information in control systems [7, 8, 18, 40, 19, 20, 33, 34, 41]. We concentrate on the computer vision and image processing aspects of our work, in particular the use of certain geometric invariant evolution equations. Besides the control applications, these techniques are already being applied in medical imaging (MRI, CT, and ultrasound), as well as shape and object recognition problems.

More precisely, let us consider the key research area of *visual tracking* which may be employed for a number of problems in robotics, manufacturing, as well as automatic target recognition. Even though tracking in the presence of a disturbance is a classical control issue, because of the highly uncertain nature of the disturbance, this type of problem is very difficult and challenging. Visual tracking differs from standard tracking problems in that the feedback signal is measured using imaging sensors. In particular, it has to be extracted via computer vision and image processing algorithms and interpreted by a reasoning algorithm before being used in the control loop. Furthermore, the response speed is a critical aspect. We have

[1]This work was supported in part by grants from the National Science Foundation ECS-9122106, by the Air Force Office of Scientific Research AF/F49620-94-1-00S8DEF, AF/F49620-94-1-0461, by the Army Research Office DAAL03-92-G-0115, DAAH04-94-G-0054, DAAH04-93-G-0332, and by MURI Grant. Allen Tannenbaum is also a Visiting Professor at the Technion, Israel.

been developing robust control algorithms for some years now, valid for general classes of distributed parameter and nonlinear systems based on interpolation and operator theoretic methods; see [14] and the references therein. In this paper, we will indicate how such control techniques may be combined with our new approach to image processing in order to develop "state of the art" visual tracking algorithms.

Indeed, because of our interest in active vision, we have been conducting research into advanced algorithms in image processing and computer vision for a variety of uses: image smoothing and enhancement, image segmentation, morphology, denoising algorithms, shape recognition, edge detection, optical flow, shape-from-shading, and deformable contours ("snakes"); see [7, 8, 18, 40, 19, 20, 33, 34] and the references therein. Our ideas are motivated by certain types of geometric invariant flows rooted in the mathematical theory of curve and surface evolution. There are now available powerful numerical algorithms based on Hamilton-Jacobi type equations and the associated theory of viscosity solutions for the computer implementation of this methodology [28]. Specifically, we intend to consider the following problems:

1. *Differential Invariants.* We will first discuss some of our research from [7, 8]. The foundation for this will be a new paradigm for the group invariant recognition of visual objects. For objects whose boundaries are described by plane curves, the *group-invariant signature curve,* which is parametrized by the group-invariant curvature function and its first derivative with respect to the group-invariant arc-length element, provides a complete representation of the equivalence class of objects under group transformations, with two curves being mapped into each other if and only if they have identical signature curves. As such, these differential invariant signature curves hold considerable promise for the resolution of a number of key processes in low-level visual recognition systems. Invariant recognition will be one of the foundations of our developing theory of controlled active vision. We will also discuss some our related work on invariant flows [26, 27, 33, 34, 35]. These pieces will be put together for an invariant object recognition system.

2. *Active Contours.* One of the key techniques in active vision and tracking is that of *deformable contours* or *snakes.* The work we discuss in this paper is based on [18, 40]. Snakes are autonomous processes which employ image coherence in order to track features of interest over time. In the past few years, a number of approaches have been proposed for the problem of snakes. The underlying principle in these works is based upon the utilization of deformable contours which conform to various object shapes and motions. Snakes have been used for edge and curve detection, segmentation, shape modeling, and especially for visual tracking. We have developed a novel deformable contour model which is derived from a generalization of Euclidean curve shortening evolution. Our snake model is based on the geometric intuition of multiplying the Euclidean arc-length by a function

tailored to the features of interest to which we want to flow, and then writing down the resulting *gradient evolution equations.* Mathematically, this amounts to defining a new Riemannian metric in the plane intrinsically determined by the geometry of the given image, and then computing the corresponding gradient flow. This leads to some intriguing new snake models which efficiently attract the given active contour to the desired feature (which basically lies at the bottom of a *potential well*). The method also allows us to naturally write down 3-D active surface models as well. Our model can handle multiple contours as well as topological changes such as merging and breaking which classical snakes cannot. This will be one of the key ideas which we use in visual tracking.

3. *Optimal Control Algorithms in Image Processing.* A number of the algorithms we have developed are based on ideas in optimal control and the corresponding gradient flows, especially our work about estimation of optical flows and stereo disparity [19, 20]. Indeed, let us consider in a bit more detail optical flow. The computation of optical flow has proved to be an important tool for problems arising in active vision, including visual tracking. The optical flow field is defined as the velocity vector field of apparent motion of brightness patterns in a sequence of images. It is assumed that the motion of the brightness patterns is the result of relative motion, large enough to register a change in the spatial distribution of intensities on the images. We are now exploring various constrained optimization approaches for the purpose of accurately computing optical flow. In this paper, we apply an $L^1$ type minimization technique to this problem following [19]. These ideas make strong contact with viscosity theory for Hamilton-Jacobi equations from optimal control.

All of the work surveyed in the paper is the result of a collaborative effort with our various co-authors. Explicitly, the work on the signature curves and differential invariants is joint with Eugene Calabi, Peter Olver, and Cheri Shakiban [7, 8], the work on the invariant flows is based on a number of papers written with Sigurd Angenent, Peter Olver, and Guillermo Sapiro [3, 26, 27, 33, 34, 35], the active contours is joint with Satya Kichenasamy, Arun Kumar, and Peter Olver [18, 40], the optical flow work is based on our paper with Gary Balas and Arun Kumar [19], and the $L^1$ approach to disparity is from our work with Steve Haker, Arun Kumar, Curt Vogel, and Steve Zucker [20]. We would like to thank all of our co-authors for all we have learned from them.

# 2  Differential Invariants

The use of symmetry (Lie) groups and the associated theory of invariants have played an essential role in computer vision, in problems ranging from image processing to object recognition. Surveys can be found in the volume [31]. We follow [7, 8] in this section.

Our approach to differential invariants in computer vision is based upon the following set-up. We have a transformation group $G$ acting on a space $E$, representing the image space, whose subsets are the objects of interest. In computer vision applications, the group $G$ is typically the Euclidean, affine, similarity, or projective group. We are particularly interested in how the geometry induced by the transformation group $G$ applies to submanifolds contained in the space $E$. Formally, a *differential invariant $I$* of $G$ is a real-valued function, depending on the submanifold and its derivatives at a point, which is unaffected by the action of $G$. It is a classical fact (due to S. Lie) that a transformation group admits a finite number of fundamental differential invariants, $I_1, \ldots, I_N$, and a system of invariant differential operators $\mathcal{D}_1, \ldots, \mathcal{D}_n$, equal in number to the dimension of the submanifold, and such that every other differential invariant is a function of the fundamental differential invariants and their successive derivatives with respect to the invariant differential operators. For example, in the case of Euclidean curves in the plane, the group action is provided by the Euclidean group consisting of translations and rotations, and every differential invariant is a function of the Euclidean curvature and its derivatives with respect to Euclidean arc length. Similarly, for affine planar geometry, the underlying group is the *equi-affine* (or *special affine*) group of area-preserving affine transformations, and every differential invariant of a curve is a function of the affine curvature and its various derivatives with respect to affine arc length.

An important consequence of a powerful result of É. Cartan ([9], [17]) is that for transitive group actions, an object can be fully reconstructed, modulo group transformations, from a prescribed (and finite) collection of differential invariants. For example, a curve in the Euclidean plane is uniquely determined, modulo translation and rotation, from its curvature invariant $\kappa$ and its first derivative with respect to arc length $\kappa_s$. Thus, the curve is uniquely prescribed by its *Euclidean signature curve*, parametrized by the two functions $(\kappa, \kappa_s)$. Similarly, a curve in the affine plane is uniquely determined, modulo an affine transformation, by its *affine signature curve* which is the planar curve parametrized by the affine curvature and its derivative with respect to affine arc length.

For the applications of invariant theory to practical engineering problems in vision, one must compute a differential invariant, such as the curvature of a curve, by a discrete numerical approximation. A robust and efficient numerical implementation is crucial, but is a nontrivial problem in that the more important differential invariants depend on high order derivatives and are thus particularly sensitive to noise and round-off error. Although the differential invariants reflect the invariance of the image under a transformation group, most standard numerical approximation schemes fail to incorporate this symmetry. Consequently, two objects which are equivalent under a group transformation, while having the same differential invariants, may have unequal numerical versions, thereby complicating the implementation of their invariant characterization by differential invariant signatures. In our approach, the problem of invariance of the numerical approximation is solved through the introduction of an explicitly group-invariant numerical scheme, based on suitable combinations of joint invariants of the mesh points used to approximate the object in question. Thus, our schemes are *automatically* invariant under the prescribed transformation group. In order to construct a numerical approximation to a differential invariant $I$, we employ a finite difference approach, so that the approximation will be computed using appropriate combinations of the coordinates of the mesh points. The approximation will be invariant under the underlying group $G$, and hence its numerical values will not depend on the group transformations, provided it depends only on the joint invariants of the mesh points. Thus, *any G-invariant numerical approximation to a differential invariant must be governed by a function of the joint invariants of $G$.* The crucial computational issue, then, is to practically determine the appropriate joint invariant which, in the limit as the mesh size goes to zero, recovers the desired differential invariant. In our approach, one interpolates the mesh points with a curve that has a constant value for the differential invariant, using this as the relevant approximation.

In what follows, we outline the general theory, and then relate this to our proposed approach to an invariant object recognition system suitable for visual tracking.

## 2.1 Invariant Signature Manifolds

In this section, we will sketch the general theory of signature manifolds. See [7, 8] for all the details. Accordingly, let $G$ be an $r$-dimensional Lie group acting transitively on the plane $E \simeq \mathbf{R}^2$, with coordinates $x, y$. We are interested in the differential invariants of smooth curves $C \subset E$ under the group $G$. We represent the curve $C$ (locally) as a function $y = u(x)$. Let $J^n \simeq \mathbf{R}^{n+2}$ denote the $n$–th jet space of $E$ — the coordinates of $J^n$ are provided by the independent variable $x$, the dependent variable $u$ and the derivatives of $u$ with respect to $x$ up to order $n$, denoted $(x, u^{(n)}) = (x, u, u_x, u_{xx}, \ldots, u_n)$. A function $F(x, u^{(n)})$ depending on the jet space coordinates is said to have *order* $n$ provided $F$ explicitly depends on $n$–th order derivatives. The group $G$ acts on curves by transforming them pointwise, and hence induces a prolonged action $G^{(n)}$ on the jet space, which is found by determining how the derivative coordinates are transformed under the group elements. We make the technical assumption, for simplicity, that $G$ is an *ordinary* $r$-dimensional transformation group, which means that $G^{(n)}$ acts transitively on (an open subset of) $J^n$ for each $0 \leq n \leq r - 1$.

**Theorem 1** *Let $G$ be an ordinary $r$-dimensional transformation group acting on $E \simeq \mathbf{R}^2$. Then there is, up to constant multiple, a unique $G$-invariant one-form of lowest order, $ds = P(x, u^{(n)})dx$, which we call the $G$-invariant arc length element. The order $n$ of $ds$ is at most $n \leq r - 2$.*

For example, the Euclidean group has dimension $r = 3$, and admits an arc length element of order $n = 1 = r - 2$. On the other hand, the equi-affine group has dimension $r = 5$, but its arc length element has order $n = 2 < 3 = r - 2$.

**Theorem 2** *Let $G$ be an ordinary $r$-dimensional transformation group acting on $E \simeq \mathbf{R}^2$. Then there is, up to constant multiple, a unique differential invariant of lowest order, $\kappa(x, u^{(r-1)})$, having order exactly $r - 1$, which we call the $G$-invariant curvature. Moreover, the derivatives of the $G$-invariant curvature with respect to the $G$-invariant arc length, $d^m \kappa / ds^m$, $m \geq 0$, provide a complete list of differential invariants of $G$, meaning that any other differential invariant is a function of these: $I = I(\kappa, \kappa_s, \kappa_{ss}, \ldots)$.*

Of course, once one has determined the formulas for $\kappa$ and $ds$ for curves given by graphs $y = u(x)$, one can recompute them for arbitrary parametrized curves $C = (x(t), y(t))$. The result as stated still holds true; moreover the resulting differential invariants are clearly also invariant under reparametrization.

Then a complete list of differential invariant signatures associated with a plane curve is provided by the group-invariant curvature and its successive derivatives with respect to arc length. How many of these are required to uniquely characterize the curve up to a group transformation? The answer is that we only need to know the first two, namely $\kappa$ and $\kappa_s$.

**Definition.** Let $G$ be an ordinary transformation group acting on $E \simeq \mathbf{R}^2$. Then the *signature curve* associated with a parametrized plane curve $C = \{(x(t), y(t))\} \subset E$ is the curve $S \subset Z \simeq \mathbf{R}^2$ parametrized by the $G$-invariant curvature and its first derivative with respect to arc length: $S = \{(\kappa(t), \kappa_s(t))\} \subset Z$.

The importance of the signature curve lies in the fact that it uniquely characterizes the original curve up to a group transformation [8]:

**Theorem 3** *Let $G$ be an ordinary transformation group acting on $E \simeq \mathbf{R}^2$. Two smooth $(C^r)$ curves $C$ and $\bar{C}$ are equivalent up to a group transformation, $\bar{C} = g \cdot C$, if and only if their signature curves are identical: $\bar{S} = S$.*

We should note that Theorem 3 is a special case of a general theorem of É. Cartan characterizing equivalent submanifolds of a homogeneous space. Cartan's Theorem states that the signature submanifold corresponding to a submanifold $S \subset G/H$ is parametrized by its $n$–th order differential invariants, where $n$ denotes the order of a Frenet frame on $S$. The signature curve (or manifold) plays the same role in the theory of transformation groups that the classifying curve (or manifold) does in the Cartan equivalence method [24]. Two submanifolds are equivalent under a group transformation if and only if their signature submanifolds are identical; see [9], [17], for precise statements and a variety of geometric examples.

Since the invariant curvature has order $r - 1$, if we are given $r$ points $P_1, \ldots, P_r \in E$ in "general position", there exists a unique constant curvature curve passing through them. Let $\tilde{\kappa}(P_1, \ldots, P_r)$ denote its curvature. Note that $\tilde{\kappa}(P_1, \ldots, P_r)$ is a joint invariant of the $r$ points. Let $C \subset E$ be an arbitrary curve in the plane. We are interested in constructing a $G$-invariant finite difference approximation to its $G$-invariant curvature $\kappa(P_1)$ at a given point $P_1 \in C$ in the curve. Choose $r - 1$ nearby points $P_2, \ldots, P_r \in C$. Then the curvature

$$\tilde{\kappa} = \tilde{\kappa}(P_1, \ldots, P_r) \approx \kappa(P_1) \tag{1}$$

of the constant curvature curve $C_0$ passing through the points determines our approximation to $\kappa(P_1)$. This provides a general method for constructing $G$-invariant finite difference approximations to the $G$-invariant curvature of a curve and its higher derivatives.

## 2.2 Invariant Flows

We will now outline some of our work on geometric-invariant diffusion equations and the resulting scale-spaces. This section is based on [26, 27, 33, 34, 35]. Besides the connection to invariant theory, such flows are very relevant in theory of scale-spaces or multiscale representation. This was introduced by Witkin in [39] and developed after that by several authors in different frameworks; see [31] for an extensive list of references on the subject. Initially, most of the work was devoted to linear scale-spaces derived via linear filtering. In the last years, a number of non-linear and geometric scale-spaces have been investigated as well.

The combination of invariant theory with geometric multiscale analysis was first investigated in [33, 34]. There, the authors introduced an affine invariant geometric scale-space, and extended part of the work to other groups as well in [35, 26, 27]. Related work was also carried out in [1, 2]. The shape representations which we derive allow us to compute invariant signatures at different scales and in a robust way. These flows are already being used with satisfactory results in various applications.

Let $C(p, t) : S^1 \times [0, \tau) \to \mathbf{R}^2$ $(C = (x, y)^T)$ be a family of simple planar curves, where $p$ parametrizes the curve and $t$ the family ($p$ and $t$ are independent). Assume that we want to formulate an intrinsic *geometric heat flow* for plane curves which is invariant under a certain transformation group $G$. This type of flows replace the classical heat flow, which is equivalent to Gaussian smoothing, and frequently used in image analysis. As above, let $r$ denote the group arc-length, i.e., the simplest invariant parametrization of the group [24, 26]. Then, the *invariant geometric heat flow* is given by

$$\frac{\partial C(p, t)}{\partial t} = \frac{\partial^2 C(p, t)}{\partial r^2}, \tag{2}$$

$$C(p, 0) = C_0(p).$$

If $G$ acts linearly, it is easy to see that since $dr$ is an invariant of the group, so is $C_{rr}$. $C_{rr}$ is called the *group normal*. For nonlinear actions, the flow (2) is still $G$-invariant, since $\frac{\partial}{\partial r}$ is the unique *invariant derivative* [24, 26].

The flow given by (2) is non-linear, since the group arc-length $r$ is a time-dependent parametrization. This flow gives the invariant geometric heat-type flow of the group, and provides the invariant direction of the deformation. For $G$ the special affine, full affine, similarity or projective group $\mathrm{SL}(\mathbf{R}, 3)$, (2) is the simplest non-trivial $G$-invariant flow (it has the lowest order in the spatial derivatives). This follows from a general result proven in [26, 27] which shows that for subgroups of $\mathrm{SL}(\mathbf{R}, 3)$, every $G$-invariant evolution is of the form

$$\frac{\partial C(p, t)}{\partial t} = I(\kappa, \frac{\partial \kappa}{\partial r}, \ldots, \frac{\partial^n \kappa}{\partial r^n}) \frac{\partial^2 C(p, t)}{\partial r^2}, \tag{3}$$
$$C(p, 0) = C_0(p),$$

where $I(\cdot)$ is a given function, $\kappa$ is the group curvature (i.e., the simplest non-trivial differential invariant of the group), and $ds$ is the invariant arc-length (the simplest invariant one form).

For the Euclidean group, it is interesting to note that the simplest nontrivial flow is given by (constant motion)

$$u_t = c\sqrt{1 + u_x^2}, \quad c \text{ a constant.}$$

(Here $g = \sqrt{1 + u_x^2}$.) In this case the curvature (the ordinary planar curvature $\kappa$) has order 2. This equation is obtained for the invariant function $I = 1/\kappa$.

Note that for each of these flows smoothing filters may be designed; see [35]. Finally, our methods are extendable to 3D images. Indeed, we have developed affine invariant volumetric smoothers in [27]. We intend to use our affine smoothers in movies as a preprocessing tool for motion estimation.

## 2.3 Invariant Object Recognition System

Let us outline how the preceding ideas can be combined to formulate an affine invariant object recognition system. (In principle, the outline given below is valid for any of the symmetry groups in use in computer vision.) Despite the extensive activity in recent years on invariant shape recognition algorithms, the corresponding problem of invariant detection of shapes has received considerably less attention. Some work along these lines has been reported in [33, 34], where the theory of geometric invariant smoothing of planar curves (boundaries of planar shapes) was initiated; see also [1]. In particular, using the methods of [1, 34], a shape can be smoothed in an affine invariant manner before the computation of invariant descriptors; see for example [13]. As we noted above, this work was partially extended for other groups and dimensions in [26, 27, 35]. Motivated by this work, efforts in the derivation of a projective invariant smoothing process was begun in the work of [12].

We are now deriving simple geometric object detection and image denoising schemes which incorporate affine invariance. These invariant edge detection and denoising algorithms should constitute the first step in a fully affine invariant system of object recognition. The second step, if necessary, would be the affine smoothing mentioned above (for curve-based systems), to be followed by the computation of affine invariant descriptors. Incorporating such affine invariant detection and denoising schemes in object recognition systems will reduce the "algorithmic" noise introduced by using non-invariant methods. Indeed, smoothing an image with a non-invariant algorithm before computing its affine curvature, will introduce "artificial noise" into the computation, and when comparing the signatures of different shapes, it would be very difficult to know if the differences correspond to the shapes or were introduced by the algorithm.

The next step in a shape recognition system is to recognize the objects bounded by the edges that have been detected by the segmentation procedure. For this purpose, as we have discussed above, a new approach incorporating either Euclidean or affine invariance, based on the general concept of a differential invariant signature curve, was recently considered in [7, 8]. These papers also derive a new, fully affine-invariant numerical algorithm for computing the affine invariants required to uniquely characterize the curve up to affine motion, and hence, in conjunction with the aforementioned algorithms, form the basis of a fully curve-based affine-invariant object detection and recognition procedure.

We have derived in [25] two different affine edge detectors. The first one is derived by weighted differences of images obtained as solutions of the affine invariant scale-space developed in [1, 34, 35]. The second one is given by the simplest affine invariant function which shows behavior similar to the magnitude of the Euclidean gradient. This affine gradient is derived from the classification of differential invariants described in [24]. These affine invariant edge maps are then used in [25] to define *affine invariant active contours,* extending the work in [18, 11, 37]. The active contours are therefore used to integrate the local information obtained by the affine edge detectors. The boundaries of the scene objects are determined as lying at the bottom of a potential well relative to a geometrically defined energy functional tailored to the affine geometry of the plane weighted by an affine invariant stopping function. (This was the philosophy of [11, 18, 37] in the Euclidean case.) Hence, in contrast with previous approaches, distances in the resulting metric space are affine invariants, and are based on the affine edge maps and classical affine differential geometry [6]. The boundaries of the desired features, are computed via a gradient descent flow based on a new notion of affine invariant curve metric introduced in [25] as well. This curve metric not only makes it possible to define a gradient flow for the active contour computation but also shows that the affine invariant heat flow introduced in [1, 2, 33, 34] is minimizing, in an affine invariant form, the area enclosed by the curve. The affine invariant edge maps are also used to extend the work in [1, 35] to obtain a completely affine invariant flow for image denoising and simplification. The framework outlined here for affine invariant edge detectors and active contours can be used in combination with other scale-spaces.

# 3  Visual Tracking

In this section, we discuss several algorithms based on gradient flows and therefore very much in the spirit of optimal control which will be key in our developing visual tracking program.

## 3.1  Geometric Active Contours

In this section, we will describe a new paradigm for *snakes* or *active contours* based on principles from geometric optimization theory. We follow [18, 40]. Active contours may be regarded as autonomous processes which employ image coherence in order to track various features of interest over time. Such deformable contours have the ability to conform to various object shapes and motions. Snakes have been utilized for segmentation, edge detection, shape modeling, and visual tracking. Active contours have also been widely applied for various applications in medical imaging. For example, snakes have been employed for the segmentation of myocardial heart boundaries as a prerequisite from which such vital information such as ejection-fraction ratio, heart output, and ventricular volume ratio can be computed. The recent book by Blake and Yuille [5] contains an excellent collection of papers on the theory and practice of deformable contours together with a large list of references.

In the classical theory of snakes, one considers energy minimization methods where controlled continuity splines are allowed to move under the influence of external image dependent forces, internal forces, and certain contraints set by the user. As is well-known there may be a number of problems associated with this approach such as initializations, existence of multiple minima, and the selection of the elasticity parameters. Moreover, natural criteria for the splitting and merging of contours (or for the treatment of multiple contours) are not readily available in this framework.

In [18], we have formulated a novel deformable contour model to successfully solve such problems, and which will become one of our key techniques for tracking. Our method is based on the Euclidean curve shortening evolution which defines the gradient direction in which a given curve is shrinking as fast as possible relative to Euclidean arc-length, and on the theory of conformal metrics. Namely, we multiply the Euclidean arc-length by a function tailored to the features of interest which we want to extract, and then we compute the corresponding gradient evolution equations. The features which we want to capture therefore lie at the bottom of a potential well to which the initial contour will flow. Further, our model may be easily extended to extract 3D contours based on motion by mean curvature [18, 40].

Let us briefly review some of the details from [18]. First of all, in [10] and [22], a snake model based on the level set formulation of the Euclidean curve shortening equation is proposed. More precisely, the model is

$$\frac{\partial \Psi}{\partial t} = \phi(x,y)\|\nabla\Psi\|(\mathrm{div}(\frac{\nabla\Psi}{\|\nabla\Psi\|}) + \nu).\tag{4}$$

Here the function $\phi(x,y)$ depends on the given image and is used as a "stopping term." For example, the term $\phi(x,y)$ may chosen to be small near an edge, and so acts to stop the evolution when the contour gets close to an edge. One may take [10, 22], $\phi := 1/(1 + \|\nabla G_\sigma * I\|^2)$ where $I$ is the (grey-scale) image and $G_\sigma$ is a Gaussian (smoothing filter) filter. The function $\Psi(x,y,t)$ evolves in (4) according to the associated level set flow for planar curve evolution in the normal direction with speed a function of curvature which was introduced in [28, 36].

It is important to note that the Euclidean curve shortening part of this evolution, namely $\Psi_t = \|\nabla\Psi\|\mathrm{div}(\frac{\nabla\Psi}{\|\nabla\Psi\|})$ is derived as a gradient flow for shrinking the perimeter as quickly as possible. As is explained in [10], the constant *inflation term* $\nu$ is added in (4) in order to keep the evolution moving in the proper direction. Note that we are taking $\Psi$ to be negative in the interior and positive in the exterior of the zero level set.

We would like to modify the model (4) in a manner suggested by Euclidean curve shortening [15]. Namely, we will change the ordinary Euclidean arc-length function along a curve $C = (x(p), y(p))^T$ with parameter $p$ given by $ds = (x_p^2 + y_p^2)^{1/2}dp$, to $ds_\phi = (x_p^2 + y_p^2)^{1/2}\phi dp$, where $\phi(x,y)$ is a positive differentiable function. Then we want to compute the corresponding gradient flow for shortening length relative to the new metric $ds_\phi$.

Accordingly set

$$L_\phi(t) := \int_0^1 \|\frac{\partial C}{\partial p}\|\phi dp.$$

Then taking the first variation of the modified length function $L_\phi$, and using integration by parts (see [18]), we get that

$$L'_\phi(t) \;=\; -\int_0^{L_\phi(t)} \langle \frac{\partial C}{\partial t}, \phi\kappa\vec{\mathcal{N}} - (\nabla\phi\cdot\vec{\mathcal{N}})\vec{\mathcal{N}}\rangle ds$$

which means that the direction in which the $L_\phi$ perimeter is shrinking as fast as possible is given by

$$\frac{\partial C}{\partial t} = (\phi\kappa - (\nabla\phi\cdot\vec{\mathcal{N}}))\vec{\mathcal{N}}. \tag{5}$$

This is precisely the gradient flow corresponding to the miminization of the length functional $L_\phi$. The level set version of this is

$$\frac{\partial\Psi}{\partial t} = \phi\|\nabla\Psi\|\mathrm{div}(\frac{\nabla\Psi}{\|\nabla\Psi\|}) + \nabla\phi\cdot\nabla\Psi. \tag{6}$$

One expects that this evolution should attract the contour very quickly to the feature which lies at the bottom of the potential well described by the gradient flow (6). As in [10, 22], we may also add a constant inflation term, and so derive a modified model of (4) given by

$$\frac{\partial\Psi}{\partial t} = \phi\|\nabla\Psi\|(\mathrm{div}(\frac{\nabla\Psi}{\|\nabla\Psi\|}) + \nu) + \nabla\phi\cdot\nabla\Psi. \tag{7}$$

Notice that for $\phi$ as above, $\nabla\phi$ will look like a doublet near an edge. Of course, one may choose other candidates for $\phi$ in order to pick out other features.

We now have very fast implementations of these snake algorithms based on level set methods [28]. Clearly, the ability of our snakes to change topology, and quickly capture the desired features will make them an indispensable tool for our visual tracking algorithms. Finally, we have also developed 3D active contour evolvers for image segmentation, shape modeling, and edge detection based on both snakes (inward deformations) and bubbles (outward deformations) in our work [18, 40].

## 3.2 Optical Flow and Stereo Disparity

The computation of optical flow has proved to be an important tool for problems arising in active vision. In this section, we discuss a method from [19] for doing this. The optical flow field is the velocity vector field of apparent motion of brightness patterns in a sequence of images [16]. One assumes that the motion of the brightness patterns is the result of relative motion, large enough to register a change in the spatial distribution of intensities on the images. Thus, relative motion between an object and a camera can give rise to optical flow. Similarly, relative motion among objects in a scene being imaged by a static camera can give rise to optical flow.

In [19], we consider a spatiotemporal differentiation method for optical flow. Even though in such an approach, the optical flow typically estimates only the isobrightness contours, it has been observed that if the motion gives rise to sufficiently large intensity gradients in the images, then the optical flow field can be used as an approximation to the real velocity field and the computed optical flow can be used reliably in the solutions of a large number of problems. Thus, optical flow computations have been used quite successfully in problems of three-dimensional object reconstruction, and in three-dimensional scene analysis for computing information such as depth and surface orientation. In object tracking and robot navigation, optical flow has been used to track targets of interest. Discontinuities in optical flow have proved an important tool in approaching the problem of image segmentation. The problem of computing optical flow is ill-posed in the sense of Hadamard. Well-posedness has to be imposed by assuming suitable *a priori* knowledge [30]. In [19], we employ a variational formulation for imposing such *a priori* knowledge.

One constraint which has often been used in the literature is the "optical flow constraint"(OFC). The OFC is a result of the simplifying assumption of constancy of the intensity, $E = E(x, y, t)$, at any point in the image [16]. It can be expressed as the following linear equation in the unknown variables $u$ and $v$

$$E_x u + E_y v + E_t = 0, \tag{8}$$

where $E_x$, $E_y$ and $E_t$ are the intensity gradients in the $x$, $y$, and the temporal directions respectively, and $u$ and $v$ are the $x$ and $y$ velocity components of the apparent motion of brightness patterns in the images, respectively. It has been shown that the OFC holds provided the scene has Lambertian surfaces and is illuminated by either a uniform or an isotropic light source, the 3-D motion is translational, the optical system is calibrated and the patterns in the scene are locally rigid; see [4].

It is not difficult to see from equation (8) that computation of optical flow is unique only up to computation of the flow along the intensity gradient $\nabla E = (E_x, E_y)^T$ at a point in the image [16]. (The superscript $T$ denotes "transpose.") This is the celebrated *aperture problem*. One way of treating the aperture problem is through the use of regularization in computation of optical flow, and consequently the choice of an appropriate constraint. A natural choice for such a constraint is the imposition of some measure of consistency on the flow vectors situated close to one another on the image.

In their pioneering work, Horn and Schunk [16] use a quadratic smoothness constraint. The immediate difficulty with this method is that at the object boundaries, where it is natural to expect discontinuities in the flow, such a smoothness constraint will have difficulty capturing the optical flow. For instance, in the case of a quadratic constraint in the form of the square of the norm of the gradient of the optical flow field [16], the Euler-Lagrange (partial) differential equations for the velocity components turn out to be *linear* elliptic. The corresponding parabolic equations therefore have a linear diffusive nature, and tend to blur the edges of a given image. In the past, work has been done to try to suppress such a constraint in directions orthogonal to the occluding boundaries in an effort to capture discontinuities in image intensities that arise on the edges.

We have proposed in [19] a novel method for computing optical flow based on the theory of the evolution of curves and surfaces. The approach employs an $L^1$ type minimization of the norm of the gradient of the optical flow vector rather than quadratic minimization as has been undertaken in most previous regularization approaches. This type of approach has already been applied to derive a powerful denoising algorithm [32]. The equations that arise are nonlinear degenerate parabolic equations. The equations diffuse in a direction orthogonal to the intensity gradients, i.e., in a direction along the edges. This results in the edges being preserved. The equations can be solved by following a methodology very similar to the evolution of curves based on the work in [28, 36]. Proper numerical implementation of the equations leads to solutions which incorporate the nature of the discontinuities in image intensities into the optical flow.

We can summarize our procedure as follows:

1. Let $E = E(x, y, t)$ be the intensity of the given moving image. Assume constancy of intensity at any point in the image, i.e.,

$$E_x u + E_y v + E_t = 0,$$

   where

$$u = \frac{dx}{dt}, \quad v = \frac{dy}{dt},$$

   are the components of the apparent motion of brightness patterns in the image which we want to estimate.

2. Consider the regularization of optical flow using the $L^1$ cost functional

$$\min_{(u,v)} \int \int \sqrt{u_x^2 + u_y^2} + \sqrt{v_x^2 + v_y^2} + \alpha^2 (E_x u + E_y v + E_t)^2 dx dy,$$

   where $\alpha$ is the smoothness parameter.

3. The corresponding Euler-Lagrange equations may be computed to be

$$\begin{aligned} \kappa_u - \alpha^2 E_x (E_x u + E_y v + E_t) &= 0, \\ \kappa_v - \alpha^2 E_x (E_x u + E_y v + E_t) &= 0, \end{aligned}$$

where the curvature $\kappa_u := \operatorname{div}(\frac{\nabla u}{\|\nabla u\|})$, and similarly for $\kappa_v$.

4. These equations are solved via "gradient descent" by introducing the system of nonlinear parabolic equations

$$
\begin{aligned}
\hat{u}_{t'} &= \kappa_{\hat{u}} - \alpha^2 E_x (E_x \hat{u} + E_y \hat{v} + E_t), \\
\hat{v}_{t'} &= \kappa_{\hat{v}} - \alpha^2 E_x (E_x \hat{u} + E_y \hat{v} + E_t),
\end{aligned}
$$

for $\hat{u} = \hat{u}(x, y, t')$, and similarly for $\hat{v}$.

The above equations have a significant advantage over the classical Horn-Schunck quadratic optimization method since they *do not blur edges*. Indeed, the diffusion equation

$$
\Phi_t = \Delta \Phi - \frac{1}{\|\nabla \Phi\|^2} \langle \nabla^2 \Phi(\nabla \Phi), \nabla \Phi \rangle = \kappa_\Phi \|\nabla \Phi\|,
$$

does not diffuse in the direction of the gradient $\nabla \Phi$; see [2]. Our optical flow equations are perturbations of the following type of equation: $\Phi_t = \frac{\kappa_\Phi}{\|\nabla \Phi\|} \|\nabla \Phi\|$. Since $\|\nabla \Phi\|$ is maximal at an edge, our optical flow equations do indeed preserve the edges. Thus the $L^1$–norm optimization procedure allows us to retain edges in the computation of the optical flow.

This approach to the estimation of motion will be one of the tools which we will employ in our tracking algorithms. The algorithm has already proven to be very reliable for various type of imagery [19]. We have also applied a similar technique to the problem of stereo disparity; [20]. We also want to apply some of the tecnhiques of Vogel [38] in order to significantly speed up our optical flow algorithms.

# References


[1] L. Alvarez, F. Guichard, P. L. Lions, and J. M. Morel, "Axioms and fundamental equations of image processing," *Arch. Rational Mechanics* **123** (1993), pp. 200-257.

[2] L. Alvarez, P. L. Lions, and J. M. Morel, "Image selective smoothing and edge detection by nonlinear diffusion," *SIAM J. Numer. Anal.* **29** (1992), pp. 845-866.

[3] S. Angenent, G. Sapiro, and A. Tannenbaum, "On the affine heat equation for non-convex curves," submitted for publication, 1996.

[4] A. D. Bimbo, P. Nesi, and J. L. C. Sanz, "Optical flow computation using extended constraints," Technical report, Dept. of Systems and Informatics, University of Florence, 1992.

[5] A. Blake, R. Curwen, and A. Zisserman, "A framework for spatio-temporal control in the tracking of visual contours," to appear in *Int. J. Compter Vision*.

[6] W. Blaschke, *Vorlesungen über Differentialgeometrie II,* Verlag Von Julius Springer, Berlin, 1923.

[7] E. Calabi, P. Olver, and A. Tannenbaum, "Affine geometry, curve flows, and invariant numerical approximations," *Advances in Mathematics* **124** (1996), pp. 154–196.

[8] E. Calabi, P. Olver, C. Shakiban, and A. Tannenbaum, "Differential and numerically invariant signature curves applied to object recognition," to appear in *Int. J. Computer Vision*, 1996.

[9] E. Cartan, *La Méthode du Repére Mobile, la Théorie des Groupes Continus, et les Espaces Généralisés; Exposés de Géométrie,* Hermann, Paris, 1935.

[10] V. Casselles, F. Catte, T. Coll, and F. Dibos, "A geomteric model for active contours in image processing," *Numerische Mathematik* **66** (1993), pp. 1-31.

[11] V. Caselles, R. Kimmel, and G. Sapiro, "Geodesic snakes," to appear in *Int. J. Computer Vision*.

[12] O. Faugeras, "On the evolution of simple curves of the real projective plane," *Comptes rendus de l'Acad. des Sciences de Paris* **317**, pp. 565-570, September 1993.

[13] O. Faugeras and R. Keriven, "Scale-spaces and affine curvature," *Proc. Europe-China Workshop on Geometric Moddeling and Invariants for Computer Vision,* edited by R. Mohr and C. Wu, 1995, pp. 17-24.

[14] C. Foias, H. Ozbay, and A. Tannenbaum, *Robust Control of Distributed Parameter Systems, Lecture Notes in Control and Information Sciences* **209**, Springer-Verlag, New York, 1995.

[15] M. Grayson, "The heat equation shrinks embedded plane curves to round points," *J. Differential Geometry* **26** (1987), pp. 285–314.

[16] B. K. P. Horn and B. G. Schunck, "Determining optical flow," *Artificial Intelligence,* 23:185–203, 1981.

[17] G. R. Jensen, *Higher order contact of submanifolds of homogeneous spaces, Lecture Notes in Math.* **610**, New York, Springer–Verlag, 1977.

[18] S. Kichenassamy, A. Kumar, P. Olver, A. Tannenbaum, and A. Yezzi, "Conformal curvature flows: from phase transitions to active vision," *Archive of Rational Mechanics and Analysis* **134** (1996), pp. 275–301.

[19] A. Kumar, A. Tannenbaum, and G. Balas, "Optical flow: a curve evolution approach," *IEEE Transactions on Image Processing* **5** (1996), pp. 598–611.

[20] A. Kumar, S. Haker, C. Vogel, A. Tannenbaum, and S. Zucker, "Stereo disparity and $L^1$ minimization," to appear in *Proceedings of CDC*, December 1997.

[21] S. Lie, "Theorie der Transformationsgruppen I," *Math. Ann.* **16** (1880), pp. 441–528.

[22] R. Malladi, J. Sethian, B. and Vermuri, "Shape modelling with front propagation: a level set approach," *IEEE PAMI* **17** (1995), pp. 158–175.

[23] D. Mumford and J. Shah, "Optimal approximations by piecewise smooth functions and associated variational problems," *Comm. on Pure and Applied Math.* **42** (1989).

[24] P. Olver, *Equivalence, Invariants, and Symmetry*, Cambridge University Press, 1995.

[25] P. Olver, G. Sapiro, and A. Tannenbaum, "Affine invariant edge maps and active contours," submitted for publication, 1997.

[26] P. Olver, G. Sapiro, and A. Tannenbaum, "Differential invariant signatures and flows in computer vision: a symmetry group approach," in *Geometry Driven Diffusion in Computer Vision*, edited by Bart ter Haar Romeny, Kluwer, 1994.

[27] P. Olver, G. Sapiro, and A. Tannenbaum, "Invariant geometric evolutions of surfaces and volumetric smoothing," *SIAM J. on Analysis*, February 1997.

[28] S. J. Osher and J. A. Sethian, "Fronts propagation with curvature dependent speed: Algorithms based on Hamilton-Jacobi formulations," *Journal of Computational Physics* **79** (1988), pp. 12-49.

[29] P. Perona and J. Malik, "Scale-space and edge detection using anisotropic diffusion," *IEEE Trans. Pattern Anal. Machine Intell.* **12** (1990), pp. 629-639.

[30] T. Poggio, V. Torre, and C. Koch, "Computational vision and regularization theory," *Nature* **317** (1985), pp. 314-319.

[31] B. ter Haar Romeny (editor), *Geometry–Driven Diffusion in Computer Vision*, Kluwer, Holland, 1994.

[32] L. I. Rudin, S. Osher, and E. Fatemi, "Nonlinear total variation based noise removal algorithms," *Physica D* **60** (1993), 259–268.

[33] G. Sapiro and A. Tannenbaum, "On affine plane curve evolution," *Journal of Functional Analysis* **119** (1994), pp. 79-120.

[34] G. Sapiro and A. Tannenbaum, "Affine invariant scale-space," *International Journal of Computer Vision* **11** (1993), pp. 25–44.

[35] G. Sapiro and A. Tannenbaum, "Invariant curve evolution and image analysis," *Indiana University J. of Mathematics* **42** (1993), pp. 985–1009.

[36] J. A. Sethian, "A review of recent numerical algorithms for hypersurfaces moving with curvature dependent speed," *J. Differential Geometry* **31** (1989), pp. 131–161.

[37] J. Shah, "Recovery of shapes by evolution of zero-crossings," *Technical Report, Math. Dept. Northeastern Univ*, Boston MA, 1995.

[38] C. Vogel, "Total variation regularization for ill-posed problems," Technical Report, Department of Mathematics, Montana State University, April 1993.

[39] A. P. Witkin, "Scale-space filtering," *Int. Joint. Conf. Artificial Intelligence*, pp. 1019-1021, 1983.

[40] A. Yezzi, S. Kichenesamy, A. Kumar, P. Olver, and A. Tannenbaum, "Geometric active contours for segmentation of medical imagery," *IEEE Trans. Medical Imaging* **16** (1997), pp. 199-209.

[41] A. Yezzi, "Modified mean curvature motion for image smoothing and enhancement, to appear in *IEEE Trans. Image Processing*, 1997.

Allen Tannenbaum and Anthony Yezzi, Jr.
Department of Electrical Engineering
University of Minnesota
Minneapolis, MN 55455

# On the Convergence of the Newton Iteration for Spectral Factorization

Jan C. Willems
University of Groningen
Groningen, The Netherlands

**Abstract**: The problem of factoring a polynomial matrix $B = B^*$ into $B(\xi) = H(\xi)^*H(\xi)$, with $H$ Hurwitz, is called the Hurwitz spectral factorization problem. We show that the Newton iteration, applied to the computation of $H$, converges.

**Keywords**: Spectral factorization, Newton iteration, quadratic differential forms, two-variable polynomial matrices.

## 1  Introduction

It is a great pleasure for me to contribute this paper for the Festschrift for Paul Fuhrmann at the occasion of his 60–th birthday. Paul was perhaps the first pure-mathematics trained researcher that I had occasion to collaborate with. At first, this collaboration was mainly in the form of exchanges of ideas, but later on, we co-authored a few papers as well [1], [2], [8]. In many ways, seeing how a mathematics trained researcher as Paul approached problems in systems and control was – well over 25 years ago – a new experience to me, and I remember admiring in a somewhat starry-eyed way, the creativity that the internal logic of mathematics could lead to in a field as systems and control. Paul's virtuosity, his impressive oeuvre and boundless creativity, and his warmth, energy and friendship brought me closer to it than I have ever been.

The purpose of this paper is to explain some ideas on the interaction of one- and two-variable polynomial matrices that have emerged in our present research program and that are more than reminiscent of Paul Fuhrmann's work.

We start with a few items of notation. Throughout, we denote one-variable polynomials with real coefficients by $\mathbb{R}[\xi]$, and two-variable polynomials by $\mathbb{R}[\zeta, \eta]$ (thus $\xi$ is usually the indeterminate in the one-variable case, and $\zeta, \eta$ are the indeterminates in the two-variable case). Polynomial matrices are denoted by $\mathbb{R}^{\bullet \times \bullet}[\xi], \mathbb{R}^{\bullet \times n}[\xi]$ (when the number of rows is not specified), etc., with similar notation in the two-variable case. We will use some special operators acting on polynomial matrices: $*, \star, \bullet$ and $\partial$ :

- $*$ maps $\mathbb{R}^{\bullet \times \bullet}[\xi]$ into itself: if $P \in \mathbb{R}^{n_1 \times n_2}[\xi]$, then $P^* \in \mathbb{R}^{n_2 \times n_1}[\xi]$ is defined as $P^*(\xi) := P^T(-\xi)$ ($^T$ denotes transposition);

- $\star$ maps $\mathbb{R}^{\bullet \times \bullet}[\zeta, \eta]$ into itself: if $P \in \mathbb{R}^{n_1 \times n_2}[\zeta, \eta]$, then $P \in \mathbb{R}^{n_2 \times n_1}[\zeta, \eta]$ is defined as $P^\star(\zeta, \eta) := P^T(\eta, \zeta)$;

- $\bullet$ maps $\mathbb{R}^{\bullet \times \bullet}[\zeta, \eta]$ into itself: if $P \in \mathbb{R}^{n_1 \times n_2}[\zeta, \eta]$, then $\overset{\bullet}{P} \in \mathbb{R}^{n_1 \times n_2}[\zeta, \eta]$ is defined as $\overset{\bullet}{P}(\zeta, \eta) := (\zeta + \eta)P(\zeta, \eta)$;

- $\partial$ maps $\mathbb{R}^{\bullet \times \bullet}[\zeta, \eta]$ into $\mathbb{R}^{\bullet \times \bullet}[\xi]$: if $P \in \mathbb{R}^{n_1 \times n_2}[\zeta, \eta]$, then $\partial P \in \mathbb{R}^{n_1 \times n_2}[\xi]$ is defined as $\partial P(\xi) := P(-\xi, \xi)$.

We call an element $P$ of $\mathbb{R}^{\bullet \times \bullet}[\xi]$ *symmetric* if $P = P^*$, and of $R^{\bullet \times \bullet}[\zeta, \eta]$ if $P = P^\star$, and *skew-symmetric* if $P = -P^*$, or $P = -P^\star$. Note that the operator $\partial$ thus maps (skew-)symmetric elements to (skew-)symmetric elements. Of course, it is possible to view, in an obvious way, the one-variable polynomial matrix $P \in \mathbb{R}^{\bullet \times \bullet}[\zeta]$ as an element of $\mathbb{R}^{\bullet \times \bullet}[\zeta, \eta]$ in which it so happens that no powers of $\eta$ appear. Thus when the "one-variable" polynomial matrix $P$ is viewed with the indeterminate $\zeta$, $P^\star$ is the "one-variable" polynomial matrix $P^T(\eta)$.

An important relation among the $\bullet$ and the $\partial$ operators is given in the following proposition (see [9] for a proof).

**Proposition 1.1** *The image of the $\bullet$ operator equals the kernel of the $\partial$ operator. In other words, for $\Phi \in \mathbb{R}^{\bullet \times \bullet}[\zeta, \eta]$,*

$$\frac{\Phi(\zeta, \eta)}{\zeta + \eta}$$

*is a polynomial iff $\Phi(\xi, -\xi) = 0$.*

The field of rational functions over $\mathbb{R}$ is denoted by $\mathbb{R}(\xi)$; $R^{\bullet \times \bullet}(\xi)$ denotes the set of matrices of rational functions. An element $P \in \mathbb{R}(\xi)$ is said to be *proper* if $P = p_1/p_2$, with the degree of $p_1 \in \mathbb{R}[\xi]$ less than or equal to that of $p_2 \in \mathbb{R}[\xi]$, *bi-proper* if these degrees are equal, and *strictly proper* if "less than" holds. Obviously any $P \in \mathbb{R}(\xi)$ can be written as $P = P_1 + P_2$, with $P_1 \in \mathbb{R}[\xi]$ the polynomial part and $P_2 \in \mathbb{R}(\xi)$ strictly proper. The polynomial part of $P$ is denoted by $P_\infty$. All this can be generalized in an obvious way to $\mathbb{R}^{\bullet \times \bullet}[\xi]$. In particular, $P \in \mathbb{R}^{\bullet \times \bullet}[\xi]$ is bi-proper iff it is square and $P_\infty \in \mathbb{R}^{\bullet \times \bullet}$ is invertible (equivalently, iff $P^{-1}$ exists and is also proper).

A square matrix $P \in \mathbb{R}^\bullet[\xi]$ is said to be *Hurwitz* if $det(P)$ is a Hurwitz polynomial, i.e. a non-zero polynomial with all its roots in the open left half of the complex plane.

We are interested in the question of factoring a polynomial matrix $B = B^*$ into the product $B = F^*F$. Such factorization questions, which go under the name of *spectral factorization*, have many applications in control and signal processing, and go back to the work of Wiener in the first half of this century. There have been literally countless articles on this problem since. Formally, let $B = B^* \in \mathbb{R}^{q \times q}[\xi]$. We call the factorization $B = H^*H$ a *Hurwitz spectral factorization* of $B$ if $H \in \mathbb{R}^{q \times q}[\xi]$ is Hurwitz. It is well known when such a factorization exists. We state this for easy reference.

**Theorem 1.1** *Let* $B = B^* \in \mathbb{R}^{q \times q}[\xi]$. *Then there exists* $H \in \mathbb{R}^{q \times q}[\xi]$ *with* $H$ *Hurwitz such that*

$$\boxed{B = H^*H} \tag{1}$$

*iff*

$$\boxed{B(i\omega) > 0 \qquad \forall \omega \in \mathbb{R}} \tag{2}$$

*where (2) means that the Hermitian matrix* $B(i\omega) \in \mathbb{C}^{q \times q}$ *is positive definite for all* $\omega \in \mathbb{R}$. *This* $H$ *is unique up to pre-multiplication by an orthogonal matrix.*

The aim of this paper is to study the convergence of the Newton iteration as an algorithm for computing a Hurwitz spectral factor. The Newton iteration for (1), studied before by Kučera and others ([5, 3, 6]), and the convergence results are stated in the following theorem.

**Theorem 1.2** *Assume that* $B = B^* \in \mathbb{R}^{q \times q}[\xi]$ *satisfies (2). Let* $X_0 \in \mathbb{R}^{q \times q}[\xi]$ *be Hurwitz and satisfy* $((X_0^*)^{-1}BX_0^{-1})_\infty = I$. *Then the Newton iteration*

$$\boxed{X_{k+1}^*X_k + X_k^*X_{k+1} = B + X_k^*X_k} \tag{3}$$

*with the normalization condition*

$$\boxed{(X_{k+1}X_k^{-1})_\infty = I} \tag{4}$$

*defines a unique sequence* $X_1, X_2, \ldots, X_k, \ldots \in \mathbb{R}^{q \times q}[\xi]$. *Moreover, each of the* $X_k$*'s is Hurwitz, and* $X_k \to H$ *as* $k \to \infty$, *with* $H$ *Hurwitz and satisfying (1). This convergence is quadratic.*

# 2    Quadratic differential forms

Let $\Phi \in \mathbb{R}^{n_1 \times n_2}[\zeta, \eta]$, i.e.

$$\Phi(\zeta, \eta) = \sum_{k,\ell} \Phi_{k,\ell} \zeta^k \eta^\ell,$$

with $\Phi_{k\ell} \in \mathbb{R}^{n_1 \times n_2}$. This two-variable polynomial matrix induces the mapping

$$L_\Phi : C^\infty(\mathbb{R}, \mathbb{R}^{n_1}) \times C^\infty(\mathbb{R}, \mathbb{R}^{n_2}) \longrightarrow C^\infty(\mathbb{R}, \mathbb{R}),$$

defined by

$$L_\Phi(v, w) := \sum_{k,\ell} \left(\frac{d^k}{dt^k} v\right)^T \Phi_{k\ell} \left(\frac{d^\ell}{dt^\ell} w\right).$$

We call $L_\Phi$ a *bilinear differential form (BLDF)*. For $\Phi \in \mathbb{R}^{n \times n}[\zeta, \eta]$, this leads to the *quadratic differential form (QDF)*

$$Q_\Phi : C^\infty(\mathbb{R}, \mathbb{R}^n) \to C^\infty(\mathbb{R}, \mathbb{R}),$$

defined by $Q_\Phi(w) := L_\Phi(w, w)$. Note that $Q_\Phi = Q_{\Phi^*} = Q_{\frac{1}{2}(\Phi + \Phi^*)}$. Accordingly, we only consider QDF's induced by symmetric $\Phi$'s. Note that $L_{\overset{\bullet}{\Phi}}(v, w) = \frac{d}{dt} L_\Phi(v, w)$ and $Q_{\overset{\bullet}{\Phi}}(w) = \frac{d}{dt} Q_\Phi(w)$. Following this correspondence of $\Phi = \Phi^* \in \mathbb{R}^{n \times n}[\zeta, \eta]$ with $Q_\Phi$, we will identify $\Phi$ and $Q_\Phi$ whenever there is no danger of confusion. We denote $C^\infty(\mathbb{R}, \mathbb{R}^n)$ by $C^\infty$ when the co-domain is obvious from the context (the domain is always $\mathbb{R}$).

We say that $Q_\Phi$ (or $\Phi$) is *nonnegative* (denoted $Q \geq 0$) if $Q_\Phi(w) \geq 0$ for all $w \in C^\infty$, and *positive* if in addition $Q_\Phi(w) = 0$ implies $w = 0$. Every $\Phi \geq 0$ can be factored as $\Phi(\zeta, \eta) = M^T(\zeta) M(\eta)$, with $M \in \mathbb{R}^{\bullet \times \bullet}[\xi]$.

We will deal with continuous-time real linear time-invariant differential systems, as discussed in [7], and amply elaborated in our recent work. Thus the time axis is $\mathbb{R}$, the signal space is $\mathbb{R}^q$ (the number of variables may of course depend on the case at hand), and the behavior $\mathfrak{B}$ is the solution set of a system of linear constant coefficient differential equations

$$R\left(\frac{d}{dt}\right)w = 0, \tag{5}$$

where $R \in \mathbb{R}^{\bullet \times q}[\xi]$. We denote the resulting system by $\Sigma_R = (\mathbb{R}, \mathbb{R}^q, \mathfrak{B}_R)$ and its behavior by

$$\mathfrak{B}_R = \{w \in C^\infty(\mathbb{R}, \mathbb{R}^q) \mid R\left(\frac{d}{dt}\right)w = 0\}. \tag{6}$$

This class of systems, or alternatively its behaviors, is denoted by $\mathfrak{L}^q$. For obvious reasons we refer to (5) as a *kernel representation* of $\Sigma_R$.

We say that $\mathfrak{B} \in \mathfrak{L}^q$ is *autonomous* if $w_1, w_2 \in \mathfrak{B}$, $w_1(t) = w_2(t)$ for $t < 0$ implies $w_1 = w_2$. Autonomous systems are those that allow a representation (5) with $R \in \mathbb{R}^{q \times q}[\xi]$ nonsingular: $\det(R) \neq 0$. An autonomous system $\mathfrak{B} \in \mathfrak{L}^q$ is said to be *asymptotically stable* if $w \in \mathfrak{B}$ implies $w(t) \to 0$ as $t \to \infty$. Asymptotically stable systems correspond to those for which $R$ can be taken to be Hurwitz.

Let $\Sigma \in \mathfrak{L}^q$ be represented by (5), and let $S \in \mathbb{R}^{\bullet \times \bullet}[\xi]$. Then

$$x = S(\frac{d}{dt})w \tag{7}$$

is called a *state map* for (5) if (5,7) jointly define a state system, i.e. if whenever $w_1, w_2 \in \mathfrak{B}_R$ satisfy $S(\frac{d}{dt})w_1(0) = S(\frac{d}{dt})w_2(0)$, then $w_1 \wedge w_2$ ($\wedge$ denotes concatenation at 0) is a weak solution of (5) and $x_1 \wedge x_2$ is absolutely continuous. A state map is *minimal* if for all $a \in \mathbb{R}^{rowdim(S)}$, there exists $w \in \mathfrak{B}_R$ such that $S(\frac{d}{dt})w(0) = a$, in other words if the map $w \in \mathfrak{B}_R \mapsto S(\frac{d}{dt})w(0)$ is surjective. For autonomous systems, the minimal state map construction can be carried out as follows. Consider the set $\{f \in \mathbb{R}^{1 \times q}[\xi] \mid fR^{-1} \text{is strictly proper}\}$. This set is actually a finite-dimensional vector space. Let $S = col(f_1, f_2, \ldots, f_n)$ be a basis for this vector space. This $S$ defines a minimal state map for (5).

We will encounter the question when two systems admit the same state map.

**Lemma 2.1** *Assume that $R_1$, $R_2 \in \mathbb{R}^{q \times q}[\xi]$, with $\det(R_1) \neq 0$, and $\det(R_2) \neq 0$, are such that $R_2 R_1^{-1}$ is bi-proper. Then $x = S(\frac{d}{dt})w$ is a state map for $R_1(\frac{d}{dt})w = 0$ iff it is a state map for $R_2(\frac{d}{dt})w = 0$.*

**Proof:** This lemma follows immediately from the fact that, with $f \in \mathbb{R}^{1 \times q}[\xi]$, $fR_1^{-1}$ is strictly proper iff $fR_2^{-1}$ is. ∎

Let $\mathfrak{B} \in \mathfrak{L}^q$. We will call the QDF $Q_\Phi$ $\mathfrak{B}$-*nonnegative* (denoted $\Phi \overset{\mathfrak{B}}{\geq} 0$ ) if $Q_\Phi(w) \geq 0$ for all $w \in \mathfrak{B}$ and $\mathfrak{B}$-*positive* (denoted $\Phi \overset{\mathfrak{B}}{>} 0$ ) if $\Phi \overset{\mathfrak{B}}{\geq} 0$ and if $Q_\Phi(w) = 0$ implies $w = 0$. In [9] $\mathfrak{B}$ positivity is studied in depth. The following proposition from there plays a role in the sequel.

**Proposition 2.1** *Let $\mathfrak{B} \in \mathfrak{L}^q$. Then $\mathfrak{B}$ is asymptotically stable if there exists $\Psi = \Psi^* \in \diagdown^{q \times q}[\zeta, \eta]$ such that $\Psi \overset{\mathfrak{B}}{\geq} 0$ and $\overset{\bullet}{\Psi} \overset{\mathfrak{B}}{<} 0$.*

# 3  The polynomial matrix Lyapunov equation

The analysis of (3) leads to the analysis of the (linear) polynomial equation

$$\boxed{X^* A + A^* X = B} \tag{8}$$

with $A$, $B = B^*$, $X \in \mathbb{R}^{q \times q}[\xi]$. In (8) we view $A$, $B$ as given and $X$ as the unknown. Because of its similarity to the case in which $A$, $B$, $X$ are ordinary matrices, we call (8) the *polynomial matrix Lyapunov equation*. It is of much interest to study this equation in its full generality. However, we only consider the case that corresponds to the situation that is encountered in the Newton iteration (3).

**Theorem 3.1** *Assume that $B = B^* \in \mathbb{R}^{q \times q}[\xi]$, and that $A \in \mathbb{R}^{q \times q}[\xi]$ is Hurwitz. Assume further that $(A^*)^{-1}BA^{-1}$ is bi-proper, normalized to $((A^*)^{-1}BA^{-1})_\infty = 2I$. Then (8) admits a solution such that $XA^{-1}$ is bi-proper and a unique such solution with*

$$\boxed{(XA^{-1})_\infty = I.} \tag{9}$$

*This unique solution satisfies $((X^*)^{-1}BX^{-1})_\infty = 2I$*

**Proof:** Consider the more general version of (8) $YA + A^*X = B$, with $X, Y \in \mathbb{R}^{q \times q}[\xi]$ unknown. To show the existence of a solution, observe that using the Smith form for polynomial matrices, we may assume with loss of generality that $A$ is diagonal. This equation then reduces to $q^2$ equations of the form $ax + b^*y = c$, with $a, b, c \in \mathbb{R}[\xi]$, $a, b$ Hurwitz, in the unknowns $x, y \in \mathbb{R}[\xi]$. These Bezout-type equations have a solution since $a, b$ Hurwitz implies that $a, b^*$ are co-prime. Now use $B = B^*$ to show that $X^*A + A^*Y^* = B$ and conclude that $(X + Y^*)/2$ solves (8).

It is easy to see [3, 9], that if $X$ is a solution, then all the solutions are generated by $X \mapsto X + SA$, where $S \in \mathbb{R}^{q \times q}[\xi]$ ranges over the skew-symmetric elements. Note further that the polynomial part of $(A^*)^{-1}X^* + XA^{-1}$ equals $2I$. The solution $X' = X + SA$, with $S$ such that the polynomial part of $XA^{-1} + S$ equals $I$, yields a solution such that $XA^{-1}$ is bi-proper with polynomial part $I$. The above representation of all solutions also yields the uniqueness of this solution.

To prove the normalization of the polynomial part of $((X^*)^{-1}BX^{-1})_\infty$, note that $(X^*)^{-1}A^* + AX^{-1} = (X^*)^{-1}BX^{-1}$. Since $XA^{-1}$ bi-proper implies $AX^{-1}$ bi-proper, and since their polynomial parts are each other's inverse, it follows that $(X^*)^{-1}BX^{-1}$ is also bi-proper, with polynomial part $2I$. ∎

The following refinement of the above theorem plays also an important role in our analysis of the Newton iteration (3).

**Theorem 3.2** *Assume that $B = B^* \in \mathbb{R}^{q \times q}[\xi]$, that $A \in \mathbb{R}^{q \times q}[\xi]$ is Hurwitz, and that $(A^*)^{-1}BA^{-1}$ is bi-proper, normalized to $((A^*)^{-1}BA^{-1})_\infty = 2I$. Assume further that (2) holds. Let $X$ be the unique solution of (8,9), identified in theorem 3.1. Then $X$ is Hurwitz.*

In the proof we use the following lemmas. The first lemma is proven in [9].

**Lemma 3.1** *Let $B = B^* \in R^{q \times q}[\xi]$. Then there exists $\Phi = \Phi^* \in R^{q \times q}[\zeta, \eta]$ such that $\partial\Phi = B$ and $\Phi > 0$ iff (2) holds.*

**Lemma 3.2** *Let the assumptions of theorem 3.2 be in force, and let $X \in \mathbb{R}^{q \times q}[\xi]$ be the solution to (8,9). Define $\Psi \in \mathbb{R}^{q \times q}[\zeta, \eta]$ by :*

$$\Psi(\zeta, \eta) = \frac{A^T(\zeta)X(\eta) + X^T(\zeta)A(\eta) - \Phi(\zeta, \eta)}{\zeta + \eta} \tag{10}$$

*Let $x = S(\frac{d}{dt})w$ be a minimal state map of the system defined by $A(\frac{d}{dt})w = 0$. Then there exists $K = K^T$ such that $\Psi(\zeta, \eta) = S^T(\zeta)KS(\eta)$.*

**Proof:** Let $\Phi(\zeta, \eta) = M^T(\zeta)M(\eta)$. Use (8) to deduce that $MA^{-1}$ is proper. Therefore $(\zeta + \eta)\Psi(\zeta, \eta)A^{-1}(\eta)$, viewed as a matrix of rational functions in $\eta$ with coefficients in $\mathbb{R}[\zeta]$, is proper. Factor $\Psi(\zeta, \eta)$ as $N^T(\zeta)LN(\eta)$ with $L = L^T \in \mathbb{R}^{\bullet \times \bullet}$. Let $(NA^{-1})_\infty(\eta) = N_k\eta^k + \ldots + N_0$. Then $N^T(\zeta)N_k = 0$. Proceed recursively and obtain $N^T(\zeta)(N_k\eta^k + \ldots + N_0) = 0$. This yields $\Psi(\zeta, \eta) = N^T(\zeta)L\tilde{N}^T(\eta)$ with $\tilde{N}A^{-1}$ strictly proper, and, by a symmetric argument, $\Psi(\zeta, \eta) = \tilde{N}^T(\eta)L\tilde{N}(\eta)$, with $\tilde{N}A^{-1}$ strictly proper. By lemma 2.1 $\tilde{N}(\xi) = FS(\xi)$ for some $F \in \mathbb{R}^{\bullet \times \bullet}$. The result follows. ∎

**Proof of Theorem 3.2:** The proof that $X$ is Hurwitz is based on the Lyapunov theory for high-order differential equations discussed in [9]. Let $\Psi$ be defined by 10, where $\Phi$ is as in lemma 3.1. That $\Psi$ is indeed a matrix of polynomials follows from proposition 1.1 and equation (8). Let $\mathfrak{B}_X \in \mathcal{L}^q$ be the behavior defined by $X(\frac{d}{dt})w = 0$. From the definition of $\Psi$, it follows that $\overset{\bullet}{\Psi}\overset{\mathfrak{B}_X}{=} -\Phi$. Obviously, therefore, $\overset{\bullet}{\Psi}\overset{\mathfrak{B}_X}{<} 0$. Next, we show that $\Psi \overset{\mathfrak{B}_X}{\geq} 0$. Indeed, for all $w \in C^\infty$ that converge to zero together with all its derivatives, there holds

$$Q_\Psi(w)(0) = \int_0^\infty (Q_\Phi(w) + 2 < A(\frac{d}{dt})w, X(\frac{d}{dt})w >)dt.$$

In particular, along solutions of $A(\frac{d}{dt})w = 0$, we have

$$Q_\Psi(w)(0) = \int_0^\infty Q_\Phi(w)dt$$

Hence $\Psi \overset{\mathfrak{B}_A}{\geq} 0$. By lemma 3.2, $\Psi(\zeta, \eta) = S^T(\zeta)KS(\eta)$, with $K = K^T \in R^{\bullet \times \bullet}$, and $S$ a minimal state map for $\mathfrak{B}_A$, and thus, by lemma 2.1, for $\mathfrak{B}_X$. Since the map that takes $w \in \mathfrak{B}_A$ to $X(\frac{d}{dt})w(0)$ is surjective, $\Psi \overset{\mathfrak{B}_A}{\geq} 0$. This implies that $K = K^T \geq 0$. Hence $\Psi \overset{\mathfrak{B}_X}{\geq} 0$ and $\overset{\bullet}{\Psi}\overset{\mathfrak{B}_X}{<} 0$, and therefore by proposition 2.1, $X$ is indeed Hurwitz. ∎

# 4  Convergence analysis

In this section, we will prove theorem 1.2. The result of theorem 3.2 allows to conclude that if $X_0 \in \mathbb{R}^{q \times q}[\xi]$ is Hurwitz, with $((X_0^*)^{-1}BX_0)_\infty = I$, then (3) generates an unique sequence such that $(X_{k+1}X_k^{-1})_\infty = I$. We will now prove that the limit of these $X_k$'s exists and that this limit is a Hurwitz spectral factorization of $B$.

In order to do this, we consider the sequence of two-variable polynomial matrices $\Psi_1, \Psi_2, \cdots$ with $\Psi_k$ defined by:

$$\Psi_{k+1}(\zeta, \eta) = \frac{X_{k+1}^T(\zeta)X_k(\eta) + X_k^T(\zeta)X_{k+1}(\eta) - \Phi(\zeta, \eta) - X_k^T(\zeta)X_k(\eta)}{\zeta + \eta} \tag{11}$$

where $\Phi$ is obtained from $B$ using lemma 3.1.

Let $x = S\left(\frac{d}{dt}\right)w$ be a minimal state map for $X_0\left(\frac{d}{dt}\right)w = 0$. Use lemma 2.1 and the fact that $X_kX_0^{-1}$ is bi-proper to conclude that $S\left(\frac{d}{dt}\right)w$ is hence a minimal state map for all the systems $X_k\left(\frac{d}{dt}\right)w = 0$. Hence, by lemma 3.2, each of the $\Psi_k(\zeta, \eta)$'s is of the form $X_0^T(\zeta)K_kX_0(\eta)$ for suitable $K_k = K_k^T \in \mathbb{R}^{\bullet \times \bullet}$. We will show that $\Psi_1 \geq \Psi_2 \geq \cdots \geq \Psi_k \geq \Psi_{k+1} \geq \cdots \geq 0$. In order to see this, observe that $\Delta_k = \Psi_k - \Psi_{k+1}$ satisfies (in the obvious notation)

$$\overset{\bullet}{\Delta}_k = X_k^*X_{k-1} + X_{k-1}^*X_k - X_{k-1}^*X_{k-1} + X_{k+1}^*X_k + X_k^*X_{k+1} - X_k^*X_k$$

Denote the behavior of $X_k\left(\frac{d}{dt}\right)w = 0$ by $\mathfrak{B}_k$. It follows from the above equation that

$$Q_{\Delta_k}(w)(0) \overset{\mathfrak{B}_k}{=} \int_0^{+\infty} \| X_{k-1}(\frac{d}{dt})w \|^2 \, dt$$

Using the surjectivity of the map $w \in \mathfrak{B}_k \mapsto X(\frac{d}{dt})w(0)$, this yields $\Delta_k \geq 0$. To show that $\Psi_k \geq 0$, use the same reasoning after observing that

$$Q_{\Psi_{k+1}}(w)(0) \overset{\mathfrak{B}_k}{=} \int_0^\infty Q_\Phi(w)dt.$$

The monotone convergence and boundedness from below, imply that the $K_k$'s and hence the $\Psi_k$'s converge.

Note that the $X_k$'s were defined by (3,4), without involving the $\Psi_k$'s. Further, (11) shows how to compute the $\Psi_k$'s using the $X_k$'s. However, it is also possible to deduce the $X_k$'s from the $\Psi_k$'s. Since we know already that the $\Psi_k$'s converge, this will allow us to conclude that the $X_k$'s also converge. In order to obtain this desired relation, write (11) as

$$(\zeta + \eta)S^T(\zeta)K_kS(\eta) =$$
$$X_k^T(\zeta)X_{k-1}(\eta) + X_{k-1}^T(\zeta)X_k(\eta) - M^T(\zeta)M(\eta) - X_{k-1}^T(\zeta)X_{k-1}(\eta)$$

where $M^T(\zeta)M(\eta) = \Phi(\zeta, \eta)$. Now pre-multiply by $(X_0^T(\zeta))^{-1}$ and keep the polynomial part of the left and the right hand side, viewed as rational functions with coefficients in $\mathbb{R}^{\bullet \times \bullet}[\eta]$. Note that, as a consequence of (3), $MX_k^{-1}$ is proper. We obtain $Y_\infty^T K_k S(\eta) = X_k(\eta) - M_\infty^T M(\eta)$, where $Y_\infty = (\xi S(\xi) X_0^{-1}(\xi))_\infty$, and $M_\infty = (MX_0^{-1})_\infty$. This relation and the convergence of the $\Psi_k$'s imply that the $X_k$'s also converge. Let $\tilde{X}$ be this limit. Obviously, $(\tilde{X} X_0^{-1})_\infty = I$. Use this and the fact that each of the $X_k$'s is Hurwitz to conclude that the limit $\tilde{X}$ is also Hurwitz.

That the convergence is quadratic is a general property of convergent Newton iterations.

This ends the proof of theorem 1.2. $\blacksquare$

It is worthwhile to note the more than casual similarity between the above proof and the proof of the convergence of the Newton iteration for computing the solution of the Algebraic Riccati Equation [4].

# 5 Conclusions

In this paper we have given a proof for the convergence of the Newton iteration for spectral factorization studied before in [5, 3, 6]. The proof involves an interesting interplay between one- and two-variable polynomial matrices.The crucial step of the algorithm is the solution of the matrix polynomial Lyapunov equation at each iteration. In [3] a very nice recursive implementation for giving this solution is given, using the Routh array. We are presently investigating whether this equation can be solved using fast algorithms for polynomial equations, similar to FFT algorithms.

# References

[1] P.A. Fuhrmann and J.C. Willems, Factorization indices at infinity for rational matrix functions, *Integral Equations and Operator Theory*, volume 2/3, pages 287-301, 1979.

[2] P.A. Fuhrmann and J.C. Willems, A study of (A,B)-invariant subspaces via polynomial models, *International Journal on Control*, volume 31, pages 467-494, 1980.

[3] J. Ježek and V. Kučera, Efficient algorithm for matrix spectral factorization, *Automatica*, volume 21, pages 663-669, 1985.

[4] D.L. Kleinman, On an iterative technique for Riccati equation computations, *IEEE Transactions on Automatic Control*, volume 13, pages 114-115, 1968.

[5] V. Kučera, *Discrete Linear Control: The Polynomial Equation Approach*, Wiley, 1979.

[6] V. Kučera, *Analysis and Design of Discrete Linear Control Systems*, Prentice Hall, 1991.

[7] J.C. Willems, Paradigms and puzzles in the theory of dynamical systems, *IEEE Transactions on Automatic Control*, volume 36, pages 259-294, 1991.

[8] J.C. Willems and P.A. Fuhrmann, Stability theory of high order equations, *Linear Algebra and its Applications*, volume 167, pages 131-150, 1992.

[9] J.C. Willems and H.L. Trentelman, On quadratic differential forms, *SIAM Journal on Control and Optimization*, submitted.

Jan C. Willems
Mathematics Institute
University of Groningen
P.O. Box 800
9700 AV Groningen
The Netherlands
fax: +31503633976
e-mail: J.C.Willems@math.rug.nl

# Author Index